Statistical Methods for the Reliability of Repairable Systems

WILEY SERIES IN PROBABILITY AND STATISTICS
APPLIED PROBABILITY AND STATISTICS SECTION

Established by WALTER A. SHEWHART and SAMUEL S. WILKS

A complete list of the titles in this series appears at the end of this volume.

Statistical Methods for the Reliability of Repairable Systems

Steven E. Rigdon
Asit P. Basu

A Wiley-Interscience Publication
JOHN WILEY & SONS, INC.
New York / Chichester / Weinheim / Brisbane / Singapore / Toronto

This text is printed on acid-free paper. ♾

Published simultaneously in Canada.

For ordering and customer service, call 1-800-CALL-WILEY.

Library of Congress Cataloging in Publication Data is available.

ISBN 0-471-34941-0

Printed in the United States of America

10 9 8 7 6 5 4 3 2 1

To Teresa (in memoriam) and Duke Dummerth,
and to all of the Dummerth Family

— S.E.R.

In memory of my parents
Hari Pada Basu and Himansu Prabha Basu

— A.P.B.

Contents

Preface

Reliability plays a key role in developing quality products and in enhancing competitiveness. For most products, customers see reliability as one of the most important quality characteristics. In the last several decades, there has been much research on the theory and applications of reliability. Most of this literature, however, is directed toward nonrepairable systems, that is, systems that fail only once and are then discarded. This book covers the reliability of *repairable* systems only, and is intended to give a thorough presentation of

1. Probabilistic models for the reliability of repairable systems
2. Statistical methods, including graphical methods, for analyzing data from repairable systems.

The first part of this book looks much like a book on stochastic processes, although only selected topics from that subject are presented. This part of the book is a reasonably thorough presentation of stochastic point processes. The second part of the book deals with analyzing data from repairable systems, and includes graphical methods, point estimation, interval estimation, hypothesis tests, goodness-of-fit tests, and reliability prediction.

This book is written for reliability engineers, quality engineers, statisticians, quality managers, and anyone who is involved in producing reliable systems. It should be useful as a reference for practioners and researchers in the field. In addition, the book may also serve as a textbook for an upper level or beginning graduate level course on reliability. For this reason we have included numerous exercises, many with real data. Readers with a background consisting of a calculus-based course on probability and statistics should be able to follow much of the book, although some of the proofs will be difficult to follow. Readers with a background that includes random variables (discrete and continuous), marginal and joint probability distributions, expectation, point estimation, confidence intervals, and hypothesis testing should be able to follow nearly all of the book. Some of the more advanced developments, such as the derivation of maximum likelihood estimates, the derivation of Bayesian estimates, and proofs of some of the theorems are optional.

Chapter 1 begins with a discussion of several terms that are often used (and abused) in reliability. The distinction between repairable systems and nonrepairable systems, and the corresponding sets of terms and notation, is also discussed in this chapter. Chapter 2 discusses the Poisson process, including the nonhomogeneous Poisson process, and gives some of its properties. Chapter 3 discusses other probabilistic models that can be applied to the reliability of repairable systems. These models include the renewal process, as well as some other more specialized models. Chapters 4 and 5 discuss the analysis of data from repairable systems. Chapter 4 deals with analyzing data from a single repairable system, and Chapter 5 deals with several systems.

We would like to thank all the students who have read drafts of this book. Michael Gough and Randall Holden deserve special thanks for their careful reading of the manuscript in its latter stages. We also want to thank the staff at John Wiley and Sons, especially Stephen Quigley, Andrew Prince, and Heather Haselkorn, for their hard work and encouragement along the way. Finally, we wish to thank our wives, Pat and Sandra, for their patience and understanding during the years that it took to complete this project.

STEVEN E. RIGDON

Edwardsville, Illinois

ASIT P. BASU

Columbia, Missouri

1

Terminology and Notation for Repairable Systems

1.1 BASIC TERMINOLOGY AND EXAMPLES

A repairable system is a system that, when a failure occurs, can be restored to an operating condition by some repair process other than replacement of the entire system. For example, an automobile is a repairable system because most failures, such as the inability to start because of a bad starter, can be fixed without replacing the entire automobile. Repair needn't involve replacement of any parts. For example, the automobile may fail to start because of bad connectors with the battery. In this case, cleaning the cables and their connection to the battery may solve the problem. A light socket, on the other hand, is not considered a repairable system. The only way to repair a burned-out light is to replace the bulb; that is, replace the entire system.

A nonrepairable system is one which is discarded after failure. A light bulb, for instance, is a nonrepairable system. The heating element in a clothes dryer is also a nonrepairable system. Today, with automated production processes turning out inexpensive products, many products that previously were repaired after failure are now discarded when they fail. Consider for example a small desk-top fan which can be purchased for less than $10 at a discount store. When such a unit fails, we would probably discard it and buy another, because the cost to fix it is greater than the cost to purchase a new fan. Many electrical systems are now nonrepairable, or they are more expensive to repair than to replace. Have you ever had a pocket calculator repaired?

A piece of software may be considered a repairable system. As software is developed and tested, failures (or bugs) will be detected and corrected. After the corrections are made, the software will be put back into use until another failure is observed.

Many real world systems, such as automobiles, airplanes, computers, and air conditioners, are repairable systems. Despite this observation, most books on reliability stress nonrepairable systems, and some cover nonrepairable systems exclusively. This is not to say studying nonrepairable systems is not effective. It *is*, because repairable systems are made up of components that are nonrepairable. Studies that can improve the reliability of the nonrepairable components are sure to improve the reliability of the repairable system comprised of those components.

As we study models for the reliability of repairable systems, we must be clear about what time scale is used to measure the failure times. For a refrigerator that runs constantly, it may be appropriate to measure time as actual elapsed time. For other systems, another measure of time may be appropriate. The number of miles driven is a better measure of "age" of an automobile than the time since it was put into service. For a copier, the number of copies made is appropriate. For other systems, such as jet engines or ship propulsion engines, the age may be expressed in terms of the number of hours of operation.

We reiterate the differences between repairable and nonrepairable systems. For nonrepairable systems, we observe a single failure time for each system. For a repairable system, we observe a number of failures on that one system. We will let $0 < T_1 < T_2 < \cdots$ denote the failure times of the system measured in *global time*, that is, the time since the initial start up of the system. The times between failures, or the *gaps*, will be denoted $X_1, X_2, \ldots$. With this formulation, we have

$$\begin{aligned} X_1 &= T_1 \\ X_2 &= T_2 - T_1 \\ X_3 &= T_3 - T_2 \\ &\vdots \end{aligned} \tag{1.1}$$

This can be inverted to give the T_i's in terms of the X_i's:

$$\begin{aligned} T_1 &= X_1 \\ T_2 &= X_1 + X_2 \\ T_3 &= X_1 + X_2 + X_3 \\ &\vdots \end{aligned} \tag{1.2}$$

We summarize the above discussion in the definitions below.

Failure Number i	Failure Time t_i	TBF $x_i = t_i - t_{i-1}$
1	340	340
2	570	230
3	604	34
4	1358	754
5	1469	111
6	1484	15
7	1619	135
8	1764	145
9	1862	98
10	2155	293

Table 1.1 Failure data from computer system (Cox and Lewis, 1966).

Definition 1 Global time *Failures of a repairable system are measured in* global time *if the failure times are recorded as time since the initial start-up of the system. Failures in global time will be denoted by* $T_1 < T_2 < \cdots$.

Definition 2 Local time *Failure times of a repairable system are measured in* local time *if the failure times are recorded as time since the previous failure. Failures in local time will be denoted by* $X_1, X_2, \ldots$.

Definition 3 Deterioration and improvement *We will say that a repairable system is* deteriorating *if the times between failure tend to get shorter with advancing age. If the times between failure tend to increase, then we will say that the system is* improving.

This simple definition is somewhat vague. It turns out to be rather difficult to give precise meanings to deterioration and improvement. See Ebrahimi (1989) for a thorough treatment of the problem.

Example 1.1 Cox and Lewis (1966, p. 11) gave data for 257 failures of a computer system. The first ten failure times are given in Table 1.1. The entire set of 257 numbers shows no evidence of deterioration or reliability improvement. □

Example 1.2 Jelinski and Moranda (1972) gave the failure times of a software system. The failure times, along with the times between failures are

i	t_i	$x_i = t_i - t_{i-1}$	i	t_i	$x_i = t_i - t_{i-1}$
1	9	9	18	98	3
2	21	12	19	104	6
3	32	11	20	105	1
4	36	4	21	116	11
5	43	7	22	149	33
6	45	2	23	156	7
7	50	5	24	247	91
8	58	8	25	249	2
9	63	5	26	250	1
10	70	7	27	337	87
11	71	1	28	384	47
12	77	6	29	396	12
13	78	1	30	405	9
14	87	9	31	540	135
15	91	4	32	798	258
16	92	1	33	814	16
17	95	3	34	849	35

Table 1.2 Software failure data (Jelinski and Moranda, 1972).

shown in Table 1.2. The software system is clearly improving because the times between failure tend to increase. □

While the two previous examples gave the failure times in terms of hours, this is not always necessary. In many cases the appropriate measure of age is something other than actual clock time.

Example 1.3 Table 1.3 shows the dates when unscheduled maintenance was required for the photocopier in the Department of Mathematics and Statistics at SIUE. For this data set, the age variable t is not measured by calendar time, but by the number of copies made. Since the process of data collection was not started with the brand new system (we inherited the copier from another department) it is reasonable to count the first recorded failure as *time zero.* The fourth column in Table 1.3 gives the number of copies made since the first recorded failure. The fifth column gives the time (measured as the number of copies) between failures. Using the terms defined above, we say that the fourth column gives the failure times in global time, and the fifth column gives the failure times in local time. The data indicate that the copier is neither deteriorating nor improving. □

i	**Date of Failure**	$c_i =$ **Copy Count**	$t_i = c_i - 240497$	$x_i = t_i - t_{i-1}$
1	11-17-92	240497	0	
2	12-3-92	243324	2827	2827
3	2-17-93	253625	13128	10301
4	4-22-93	266229	25732	12604
5	9-20-93	286734	46237	20505
6	12-3-93	298511	58014	11777
7	3-2-94	311662	71165	13151
8	5-3-94	320246	79749	8584
9	7-26-94	331053	90556	10807
10	12-15-94	352816	112319	21763
11	5-25-95	372329	131832	19513
12	8-25-95	383052	142555	10723
13	11-9-95	393774	153277	10722
14	12-20-95	398892	158395	5118
15	1-10-96	400200	159703	1308
16	1-11-96	400217	159720	17
17	5-1-96	417702	177205	17485
18	8-22-96	432158	191661	14456
19	9-25-96	437374	196877	5216

Table 1.3 Failure data from the department's photocopier.

Month of Operation	Number of Operating Hours	Cumulative Operating Hours	Number of Failures
1	541	541	3
2	1171	1712	5
3	1939	3651	4
4	2403	6054	1
5	1718	7772	2
6	2206	9978	2
7	1366	11344	3
8	1529	12873	0
9	1449	14322	2
10	1451	15773	2

Table 1.4 Failure data for ARN-118 system (Balaban, 1978).

In many cases, the exact time of failure is not recorded. Only the number of failures per time period is recorded.

Example 1.4 Balaban (1978) gave failure data for a mechanical system that he labels ARN-118. The number of failures per month for the ARN-118 along with the number of operating hours in that month are shown in Table 1.4. This system shows reliability improvement. □

A simple way to display the failure times of a repairable system is to plot a time line and mark the failure times with a symbol such as a dot (•), cross (×), or diamond (◇). For the last example, it is not possible to mark the exact failure times, because only the number of failures in intervals of operating time were recorded. We may, however, assume that the failure times are equally spaced within each interval, and make a time dot plot. (This was done in Figure 1.4.) Figures 1.1 through 1.4 show time dot plots for the data sets in Examples 1.1 through 1.4.

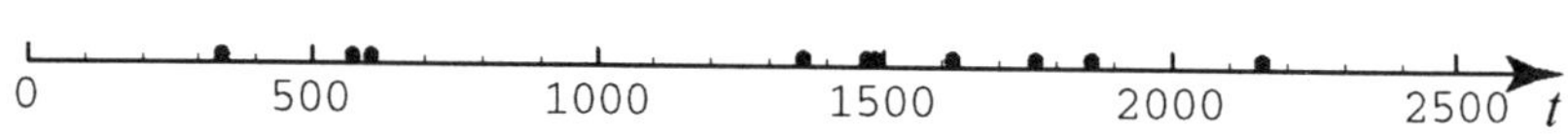

Fig. 1.1 Time dot plot for the first 10 failures from a computer system (Cox and Lewis, 1966).

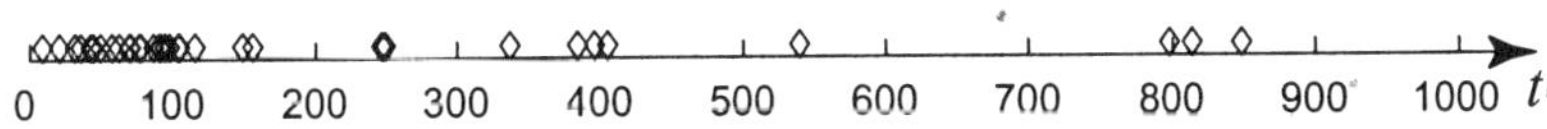

Fig. 1.2 Time dot plot for software failures (Jelinski and Moranda, 1972).

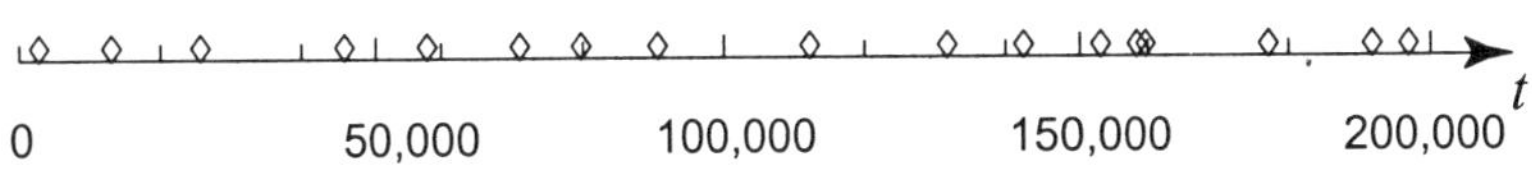

Fig. 1.3 Time dot plot for failure times of copy machine.

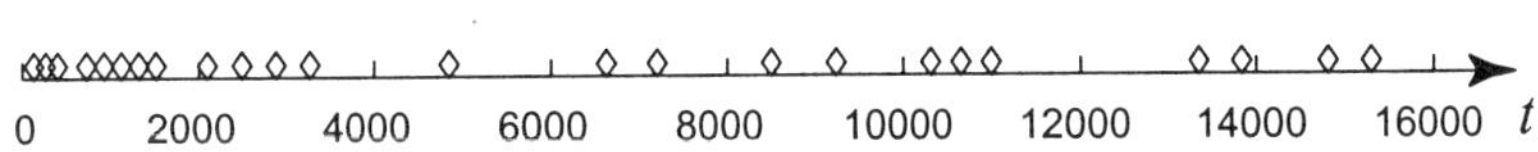

Fig. 1.4 Time dot plot for Balaban's ARN-118 system assuming failures are equally spaced in each interval.

1.2 NONREPAIRABLE SYSTEMS

In this section we discuss the similarities and differences between repairable and nonrepairable systems. There are many terms and concepts that are similar, and the distinctions are subtle.

For a nonrepairable system the lifetime is a random variable. There is no repair, and the system is discarded after its one and only failure. Since the failure of one system does not affect the performance of a similar system located elsewhere, the assumption that different systems have lifetimes that are independent is reasonable. Also, if many copies of the system were produced by the same manufacturing system, then it is reasonable to assume that the system lifetimes have the same distribution. These two assumptions can be combined into the one statement that says the lifetimes are independent and identically distributed (iid) from some distribution having cumulative distribution (cdf) $F(x)$.

Definition 4 Cumulative distribution function *The* cumulative distribution function (cdf) *of a random variable X is defined to be the function*

$$F(x) = P(X \leq x).$$

Since the lifetime must be nonnegative the probability distribution must have positive probability on the positive time axis only. In other words, $F(x) = 0$ for $x < 0$.

Definition 5 Survival function *The* survival function $S(x)$, *also called the* reliability function, *is the probability that a system will carry out its mission through time x.*

The survival function evaluated at x is just the probability that the failure time is beyond time x. Thus the reliability function is related to the cdf in the following way:

$$S(x) = P(X > x) = 1 - P(X \leq x) = 1 - F(x). \tag{1.3}$$

Definition 6 Probability density function *The* probability density function (pdf) *is defined to be the derivative of the cdf, provided the derivative exists. That is,*

$$f(x) = \frac{d}{dx} F(x).$$

From (1.3) we see that

$$f(x) = -\frac{d}{dx} S(x).$$

Another way to express the pdf is through the limit

$$f(x) = \lim_{\Delta x \to 0} \frac{F(x + \Delta x) - F(x)}{\Delta x} = \lim_{\Delta x \to 0} \frac{P(x < X \leq x + \Delta x)}{\Delta x}. \tag{1.4}$$

This, of course, is from the definition of the derivative of $F(x)$.

Another important function related to, but distinct from the pdf, is the hazard function.

Definition 7 Hazard function *The hazard function is*

$$h(x) = \lim_{\Delta x \to 0} \frac{P(x < X \leq x + \Delta x | X > x)}{\Delta x}. \tag{1.5}$$

This is the limit of the probability that a unit fails (for the first and only time) in a small interval given that it survived to the beginning of the interval. Compare the definition of the hazard function $h(x)$ in (1.5) with the result for the pdf given in (1.4). These are nearly the same, except one is a conditional probability and the other is not.

One property of pdf's is that they must integrate to one; that is, since we are dealing with random variables that have all the probability on the nonnegative axis,

$$\int_0^\infty f(x)\, dx = 1.$$

The hazard is defined as the limit of a conditional probability, but it is not a conditional pdf! The hazard function need not integrate to one, and in fact, for most distributions we study, the hazard will *not* integrate to one. For a system whose hazard function is increasing, this means that (in the limit) the probability of failure in a small interval divided by the length of the interval is increasing with time. Thus if we take a small fixed length of time, such as one hour, an increasing hazard would mean that the probability of failing in this one hour, given that the system survived past the start of that hour, increases with the age of the system. In this case we say that the system is *wearing out.* Compare this definition with the definition of *deteriorate* for repairable systems. We say that a repairable system *deteriorates* when the times between failure tend to get smaller, and we say that a nonrepairable system is *wearing out* if the hazard function is increasing. A nonrepairable system with a decreasing hazard function is said to experience *burn-in.* The term "deteriorate" will be reserved for repairable systems and the term "wear out" will be reserved for nonrepairable systems. Similarly, the terms "improvement" and "burn-in" will be reserved for repairable and nonrepairable systems, respectively.

Some books define the hazard function to be the ratio

$$h(x) = \frac{f(x)}{S(x)} \,. \tag{1.6}$$

While this result is true (we will prove it below) it gives the reader no insight into what the hazard function measures. The limit definition of the hazard function *does* give some insight.

Theorem 1 *For a continuous random variable the hazard function is*

$$h(x) = \frac{f(x)}{S(x)} \,.$$

Proof. The proof is direct:

$$\begin{aligned} h(x) &= \lim_{\Delta x \to 0} \frac{P(x < X \le x + \Delta x \,|\, X > x)}{\Delta x} \\ &= \lim_{\Delta x \to 0} \frac{P(x < X \le x + \Delta x \text{ and } X > x)}{\Delta x \, P(X > x)} \\ &= \lim_{\Delta x \to 0} \frac{P(x < X \le x + \Delta x)}{\Delta x \, S(x)} \\ &= \frac{1}{S(x)} \lim_{\Delta x \to 0} \frac{P(x < X \le x + \Delta x)}{\Delta x} \\ &= \frac{f(x)}{S(x)} \,. \end{aligned}$$

■

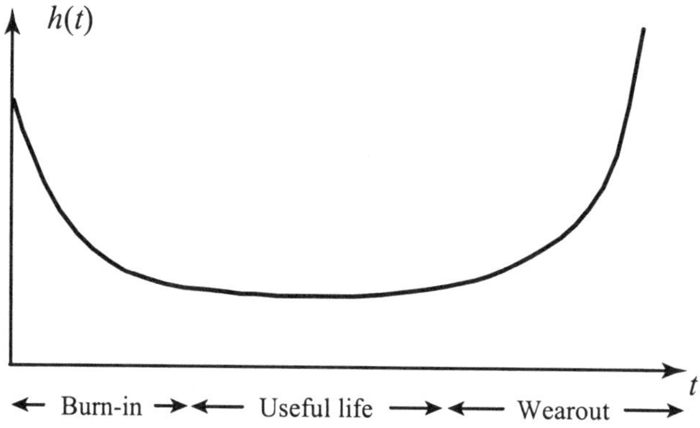

Fig. 1.5 A bathtub hazard function.

Often the choice of which distribution to use to model lifetimes depends on what the hazard function is expected to look like. If units tend to wear out, then the hazard function must be an increasing function. In this case, a distribution with a constant hazard function such as an exponential distribution will not be appropriate. The Weibull and gamma distributions are flexible enough to include distributions that have either increasing or decreasing hazard functions. Some mechanical parts have a high chance of failure in the very early stages, but the hazard function levels off and eventually increases as the system gets old. Such a hazard function, called a bathtub curve, is shown in Figure 1.5. The three stages of the units life are called (1) burn-in, (2) useful life, and (3) wear-out. Neither the Weibull distribution nor the gamma distribution is flexible enough to have a hazard function that is initially decreasing and eventually increasing.

Knowing any one of the pdf $f(x)$, the cdf $F(x)$, the survival function $S(x)$, or the hazard function $h(x)$ is enough to find all of the others. The discussion above shows how to get the hazard from the pdf and the cdf. The next theorem shows how to get the cdf and pdf from the hazard.

Theorem 2 *For a continuous random variable with hazard function $h(x)$ the cdf and the pdf are given by*

$$F(x) = 1 - \exp\left(-\int_0^x h(y)\,dy\right)$$

and

$$f(x) = h(x)\exp\left(-\int_0^x h(y)\,dy\right),$$

respectively.

Proof. Since $f(x) = F'(x) = -S'(x)$, we have

$$h(x) = \frac{f(x)}{S(x)} = -\frac{S'(x)}{S(x)}.$$

This is a first order separable differential equation. To solve it, integrate both sides from 0 to x (and change the variable of integration from x to y). This yields[1]

$$\begin{aligned} -\int_0^x h(y)\,dy &= \int_0^x \frac{S'(y)}{S(y)}\,dy \\ &= \log|S(y)|\Big|_{y=0}^{y=x} \\ &= \log S(x) - \log S(0) \\ &= \log S(x) - \log 1 \\ &= \log S(x). \end{aligned}$$

Exponentiating both sides yields

$$S(x) = \exp\left(-\int_0^x h(y)\,dy\right).$$

Thus

$$F(x) = 1 - \exp\left(-\int_0^x h(y)\,dy\right).$$

The pdf is therefore

$$f(x) = \frac{d}{dx}F(x) = h(x)\exp\left(-\int_0^x h(t)\,dt\right).$$

■

Definition 8 **Cumulative hazard function** *The quantity*

$$H(t) = \int_0^t h(x)\,dx$$

is called the cumulative hazard function.

The next few sections cover some of the commonly used distributions for lifetimes, including the exponential, the Weibull, and the gamma.

[1] Here, and throughout the book, log indicates the natural logarithm; that is, the logarithm to the base e.

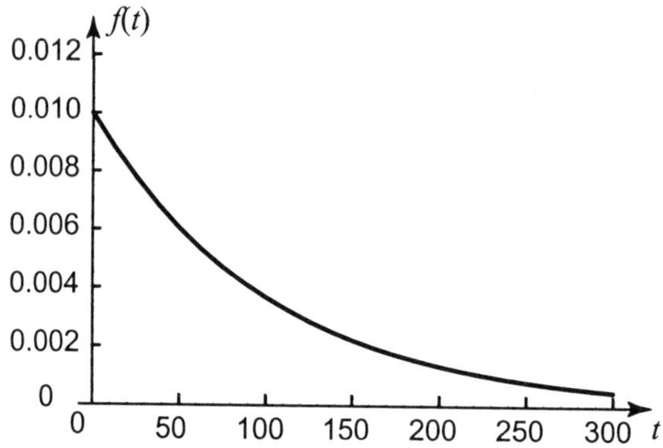

Fig. 1.6 PDF for the exponential distribution with mean $\theta = 100$.

1.2.1 The Exponential Distribution

The simplest model for lifetimes is the exponential distribution.

Definition 9 Exponential distribution *The* exponential distribution *is a continuous distribution having pdf*

$$f(x) = \frac{1}{\theta}\exp\left(-x/\theta\right), \quad x > 0$$

and cdf

$$F(x) = P(X \le x) = \int_0^\infty \frac{1}{\theta}\exp\left(-x/\theta\right)\, dx = 1 - \exp\left(-x/\theta\right), \quad x > 0. \quad (1.7)$$

We write $X \sim EXP(\theta)$ to indicate that the random variable X has an exponential distribution with a cdf given by (1.7).

Figures 1.6 and 1.7 show graphs of the exponential pdf and cdf for the case where $\theta = 100$. The mean and variance of the exponential distribution are given in the next theorem.

Theorem 3 *If $X \sim EXP(\theta)$, then*

$$E(X) = \theta$$

and

$$V(X) = \theta^2$$

Proof. The proof is left as an exercise. ■

The most distinctive feature of the exponential distribution is that it is the only continuous distribution with the *memoryless property*.

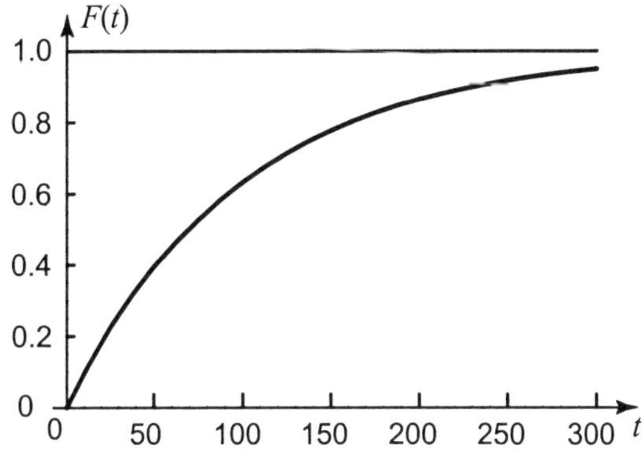

Fig. 1.7 CDF for the exponential distribution with mean $\theta = 100$.

Definition 10 Memoryless property *A distribution has the* memoryless property *if*

$$P(X > t + x \mid X > t) = P(X > x). \tag{1.8}$$

To illustrate this point, let us suppose that $t = 5000$ and $x = 24$ and that the failure times are measured in hours. The conditional probability on the left side of the equal sign is the probability that the unit survives 24 more hours given that it has already survived for 5000 hours. The probability on the right is just the probability that a new system will survive for 24 hours. If these two probabilities are equal, then the distribution has the memoryless property. In other words, if the distribution has the memoryless property, then the probability that an old unit survives one more day will equal the probability that a brand new unit will survive one day. The memoryless property imposes some strong assumptions about the way units age.

Theorem 4 *The exponential distribution has the memoryless property and it is the only continuous distribution with the memoryless property.*

Proof. Assuming that $t > 0$ and $x > 0$, we can write the conditional probability on the left side of (1.8) as

$$P(X > t + x \mid X > t) = \frac{P(X > t + x \text{ and } X > x)}{P(X > t)}.$$

The event in the numerator is just $\{X > x + t\}$ because $X > t + x$ *and* $X > x$ implies that $X > t + x$. The denominator is $S(t) = \exp(-t/\theta)$. Thus if $X \sim$

EXP(θ), then

$$\begin{aligned} P(X > t + x | X > t) &= \frac{P(X > t + x)}{\exp(-t/\theta)} \\ &= \frac{\exp\left[-(t+x)/\theta\right]}{\exp(-t/\theta)} \\ &= \exp\left[-(t+x)/\theta + t/\theta\right] \\ &= \exp\left(-x/\theta\right), \end{aligned}$$

so the exponential distribution has the memoryless property.

To prove that the exponential distribution is the *only* continuous distribution on $(0, \infty)$ that has the memoryless property, we must prove that

$$\frac{P(X > t + x \text{ and } X > x)}{P(X > t)} = 1 - F(x) \tag{1.9}$$

implies $F(x) = 1 - \exp(-x/\theta)$ for some θ. The left side of (1.9) is

$$\frac{P(X > t + x \text{ and } X > x)}{P(X > t)} = \frac{P(X > t + x)}{P(X > t)} = \frac{1 - F(t + x)}{1 - F(t)}.$$

Thus if (1.9) holds, we have

$$\frac{1 - F(t + x)}{1 - F(t)} = 1 - F(x). \tag{1.10}$$

Let $S(x) = 1 - F(x)$. Then (1.10) becomes

$$\frac{S(t + x)}{S(t)} = S(x).$$

This implies that

$$\begin{aligned} S(t + x) &= S(x)S(t) \\ S(t + x) - S(t) &= S(x)S(t) - S(t) \\ \frac{S(t + x) - S(t)}{x} &= \frac{S(t)\left[S(x) - 1\right]}{x} \\ \frac{S(t + x) - S(t)}{x} &= S(t)\frac{S(0 + x) - S(0)}{x}. \end{aligned}$$

Now take the limit as $x \to 0^+$. This yields

$$\begin{aligned} \lim_{x \to 0^+} \frac{S(t + x) - S(t)}{x} &= S(t) \lim_{x \to 0^+} \frac{S(0 + x) - S(0)}{x} \\ S'(t) &= S(t)S'(0). \end{aligned}$$

The expression $S'(0)$ is a negative constant which we will denote as $-\lambda$. Thus we have the differential equation

$$S'(t) = -\lambda S(t)$$

with initial condition $S(0) = 1$. The solution of this initial value problem is

$$S(t) = \exp(-\lambda t).$$

Thus the cdf $F(x)$ must be of the form

$$F(x) = 1 - S(x) = 1 - \exp(-\lambda x)$$

as was to be shown. ■

Another unique feature of the exponential distribution is that it is the only continuous distribution with a constant hazard function.

Theorem 5 *The exponential distribution has a constant hazard function and it is the only distribution with a constant hazard function.*

Proof. The hazard function for the exponential distribution is

$$\begin{aligned} h(x) &= \frac{f(x)}{S(x)} \\ &= \frac{\frac{1}{\theta}\exp(-x/\theta)}{\exp(-x/\theta)} \\ &= \frac{1}{\theta}. \end{aligned}$$

Now we must show that the exponential distribution is the only distribution with this property. Suppose that the hazard function is $h(x) = \lambda$. Then by Theorem 2 the cdf is

$$F(x) = 1 - \exp\left(-\int_0^x \lambda\, dy\right) = 1 - \exp(-\lambda x),$$

which is the cdf of the EXP$(1/\lambda)$ distribution. ■

Occasionally, you will find the pdf and cdf for the exponential distribution written as

$$f(x) = \lambda \exp(-\lambda x), \quad x > 0$$

and

$$F(x) = 1 - \exp(-\lambda x), \quad x > 0.$$

With this representation, the mean and variance are given by

$$E(X) = \frac{1}{\lambda}$$

and

$$V(X) = \frac{1}{\lambda^2}.$$

The hazard function is then

$$h(x) = \lambda.$$

Just remember that for the exponential distribution, the hazard function is constant, and that the mean and hazard are reciprocals of one another.

1.2.2 The Weibull Distribution

We discuss here the Weibull distribution for several reasons. First, it is probably the most widely used distribution for lifetimes. Later, when we study repairable systems, it will be instructive to compare models for repairable systems and nonrepairable systems. Second, the Weibull distribution is related to the power law process, a commonly used model for repairable systems. Third, if repairs bring a system back to a good-as-new state, then the assumption that the times between failures $X_1, X_2, \ldots$ are iid Weibull random variables may be reasonable.

Definition 11 *The Weibull distribution has survival function*

$$S(x) = \exp\left[-\left(\frac{x}{\alpha}\right)^{\eta}\right], \quad x > 0. \tag{1.11}$$

If X is a random variable with this cdf, then we will write $X \sim WEI(\eta, \alpha)$.

The cdf, pdf, and hazard functions are therefore given as follows:

$$F(x) = 1 - S(x) = 1 - \exp\left[-\left(\frac{x}{\alpha}\right)^{\eta}\right], \quad x > 0 \tag{1.12}$$

$$f(x) = F'(x) = \frac{\eta}{\alpha}\left(\frac{x}{\alpha}\right)^{\eta-1} \exp\left[-\left(\frac{x}{\alpha}\right)^{\eta}\right], \quad x > 0 \tag{1.13}$$

$$h(x) = \frac{f(x)}{S(x)} = \frac{\frac{\eta}{\alpha}\left(\frac{x}{\alpha}\right)^{\eta-1}\exp\left[-\left(\frac{x}{\alpha}\right)^{\eta}\right]}{\exp\left[-\left(\frac{x}{\alpha}\right)^{\eta}\right]} = \frac{\eta}{\alpha}\left(\frac{x}{\alpha}\right)^{\eta-1}, \quad x > 0. \tag{1.14}$$

The hazard function h is increasing when $\eta > 1$ and decreasing when $\eta < 1$. When $\eta = 1$, the hazard function is the constant function $h(x) = 1/\alpha$. Thus, the exponential distribution is a special case of the Weibull distribution that occurs when $\eta = 1$.

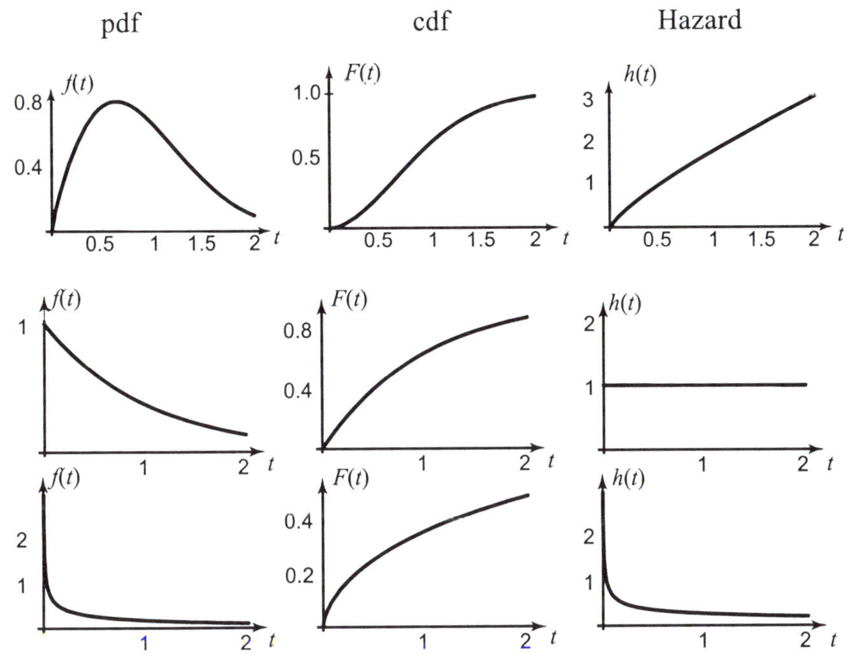

Fig. 1.8 Weibull probability density functions, cumulative distribution functions, and hazard functions for $\beta = 1.8$ (top row), $\beta = 1$ (middle row) and $\beta = 0.6$ (bottom row).

Figure 1.8 shows a number of Weibull hazard functions and the corresponding pdf's and cdf's.

The mean and variance of the Weibull can be expressed in terms of the gamma function which is defined below.

Definition 12 Gamma Function *For $a > 0$ the gamma function is defined to be*

$$\Gamma(a) = \int_0^\infty x^{a-1} e^{-x}\, dx.$$

If $a > 1$, then we can apply integration by parts to this integral. Let $u = x^{a-1}$ and $dv = e^{-x} dx$. Then $du = (a-1)x^{a-2}\, dx$ and $v = -e^{-x}$ and the limits on the integral remain the same. Then

$$\begin{aligned}\Gamma(a) &= \int_0^\infty x^{a-1}e^{-x}\,dx \\ &= \left[-x^{a-1}e^{-x}\right]_0^\infty - \int_0^\infty \left(-e^{-x}\right)(a-1)x^{a-2}dx \\ &= 0 + (a-1)\int_0^\infty x^{a-2}e^{-x}\,dx \\ &= (a-1)\Gamma(a-1).\end{aligned}$$

Since $\Gamma(1) = 1$, we have

$$\begin{aligned}\Gamma(2) &= (2-1)\Gamma(2-1) = 1 \\ \Gamma(3) &= (3-1)\Gamma(3-1) = 2\Gamma(2) = 2 \\ \Gamma(4) &= (4-1)\Gamma(4-1) = 3\Gamma(3) = 3\times 2 = 6 \\ \Gamma(5) &= (5-1)\Gamma(5-1) = 4\Gamma(4) = 4\times 3\times 2 = 24\end{aligned}$$

and in general, $\Gamma(a) = (a-1)!$. The next theorem gives the mean and variance of the Weibull distribution in terms of the gamma function.

Theorem 6 *If $X \sim WEI(\eta, \alpha)$, then*

$$E(X) = \alpha\Gamma\left(1 + \frac{1}{\eta}\right) \tag{1.15}$$

and

$$V(X) = \alpha^2\left[\Gamma\left(1+\frac{2}{\eta}\right) - \left(\Gamma\left(1+\frac{1}{\eta}\right)\right)^2\right]. \tag{1.16}$$

Proof. The expectation is

$$\begin{aligned}E(X) &= \int_0^\infty x\,f(x)\,dx \\ &= \int_0^\infty x\,\frac{\eta}{\alpha}\left(\frac{x}{\alpha}\right)^{\eta-1}\exp\left[-\left(\frac{x}{\alpha}\right)^\eta\right]\,dx.\end{aligned}$$

We make the following change of variable in the integrand. Let $y = (x/\alpha)^\eta$; then $x = \alpha y^{1/\eta}$ and $dy = (\eta/\alpha)(x/\alpha)^{\eta-1}dx$, and the limits of integration remain the same. We then have

$$\begin{aligned}E(X) &= \int_0^\infty \alpha y^{1/\eta}e^{-y}\,dy \\ &= \alpha\int_0^\infty y^{(1+1/\eta)-1}e^{-y}\,dy \\ &= \alpha\Gamma\left(1+\frac{1}{\eta}\right).\end{aligned}$$

The derivation of the variance for the Weibull distribution is similar, and is left as an exercise. ■

1.2.3 Gamma Distribution

The gamma distribution is useful as a model for the lifetimes of nonrepairable systems, and it has a direct connection with the Poisson process.

Definition 13 *The pdf for the gamma distribution can be written as*

$$f(x) = \frac{x^{\kappa-1}}{\theta^{\kappa}\Gamma(\kappa)} \exp\left(-x/\theta\right), \quad x > 0.$$

We will write $X \sim GAM(\kappa, \theta)$ if the random variable X has this pdf. Another useful form for the gamma pdf is obtained by substituting $1/\lambda$ for θ; this gives

$$f(x) = \frac{\lambda^{\kappa} x^{\kappa-1}}{\Gamma(\kappa)} \exp\left(-\lambda x\right), \quad x > 0. \tag{1.17}$$

The cdf and the survival function, and hence also the hazard function, cannot be written in closed form. We can write the cdf as

$$F(x) = \int_0^x \frac{\lambda^{\kappa} w^{\kappa-1}}{\Gamma(\kappa)} \exp\left(-\lambda w\right)\, dw.$$

If we make the transformation $y = \lambda w$, then this becomes

$$\begin{aligned} F(x) &= \int_0^{\lambda x} \frac{\lambda^{\kappa}(y/\lambda)^{\kappa-1}}{\Gamma(\kappa)} e^{-y} \frac{1}{\lambda}\, dy \\ &= \frac{\lambda^{\kappa}}{\lambda^{\kappa-1}\Gamma(\kappa)} \frac{1}{\lambda} \int_0^{\lambda x} y^{\kappa-1} e^{-y}\, dy \\ &= \frac{1}{\Gamma(\kappa)} \int_0^{\lambda x} y^{\kappa-1} e^{-y}\, dy. \end{aligned}$$

This last expression can be written in terms of the incomplete gamma function G, which is defined by

$$G(a, x) = \frac{1}{\Gamma(a)} \int_0^x y^{a-1} e^{-y}\, dy.$$

(See Abramowitz and Stegun (1965, §6.5).) Thus the cdf for the gamma distribution is

$$F(x) = G(\kappa, \lambda x).$$

The hazard function is therefore

$$\begin{aligned} h(x) &= \frac{f(x)}{1 - F(x)} \\ &= \frac{\dfrac{\lambda^{\kappa} x^{\kappa-1}}{\Gamma(\kappa)} \exp\left(-\lambda x\right)}{1 - \dfrac{1}{\Gamma(\kappa)} \displaystyle\int_0^{\lambda x} y^{\kappa-1} e^{-y}\, dy}. \end{aligned}$$

This hazard function is increasing when $\kappa > 1$, decreasing when $\kappa < 1$, and constant when $\kappa = 1$. If $\kappa = 1$, then the pdf from (1.17) is

$$f(x) = \frac{\lambda x^{1-1}}{\Gamma(1)} \exp(-\lambda x) = \lambda \exp(-\lambda x), \quad x > 0,$$

which is the pdf of the exponential distribution. When $\kappa \neq 1$, the hazard function is not a constant, but it converges to the constant λ as $x \to \infty$. To see this, we apply L'Hopital's rule to evaluate the limit. This gives

$$\begin{aligned}
\lim_{x\to\infty} h(x) &= \lim_{x\to\infty} \frac{x^{\kappa-1}\exp(-\lambda x)}{\dfrac{\Gamma(\kappa)}{\lambda^\kappa} - \displaystyle\int_0^x w^{\kappa-1}\exp(-\lambda w)\,dw} \\
&= \lim_{x\to\infty} \frac{\dfrac{d}{dx} x^{\kappa-1}\exp(-\lambda x)}{\dfrac{d}{dx}\left[\dfrac{\Gamma(\kappa)}{\lambda^\kappa} - \displaystyle\int_0^x w^{\kappa-1}\exp(-\lambda w)\,dw\right]} \\
&= \lim_{x\to\infty} \frac{(\kappa-1)x^{\kappa-2}e^{-\lambda x} - x^{\kappa-1}\lambda e^{-\lambda x}}{-x^{\kappa-1}\exp(-\lambda x)} \\
&= \lim_{x\to\infty} \frac{(\kappa-1)-\lambda x}{-x} \\
&= \lambda.
\end{aligned}$$

Figure 1.9 shows some hazard functions, pdf's and cdf's for some gamma random variables.

The next theorem gives a connection between the exponential distribution and the gamma distribution. This result is important because, as we will see in the next chapter, the homogeneous Poisson process implies that the times between failures are iid exponential random variables. Thus, for the homogeneous Poisson process, the time of the nth failure has a gamma distribution.

Theorem 7 *If $X_1, X_2, \ldots, X_n$ are iid EXP$(1/\lambda)$ random variables, then the random variable $T_n = X_1 + X_2 + \cdots + X_n$ has the GAM$(n, 1/\lambda)$ distribution, which has pdf*

$$g_n(t_n) = \frac{\lambda^n t_n^{n-1}}{(n-1)!}\exp(-\lambda t_n), \qquad t_n > 0. \tag{1.18}$$

Proof. There are a number of ways to prove this result, including one method that uses moment generating functions. We shall prove the result by induction. If $n = 1$ then $T_1 = X_1$, which has an exponential distribution with mean $1/\lambda$. The pdf is therefore

$$g_1(t_1) = \lambda \exp(-\lambda t_1), \qquad t_n > 0,$$

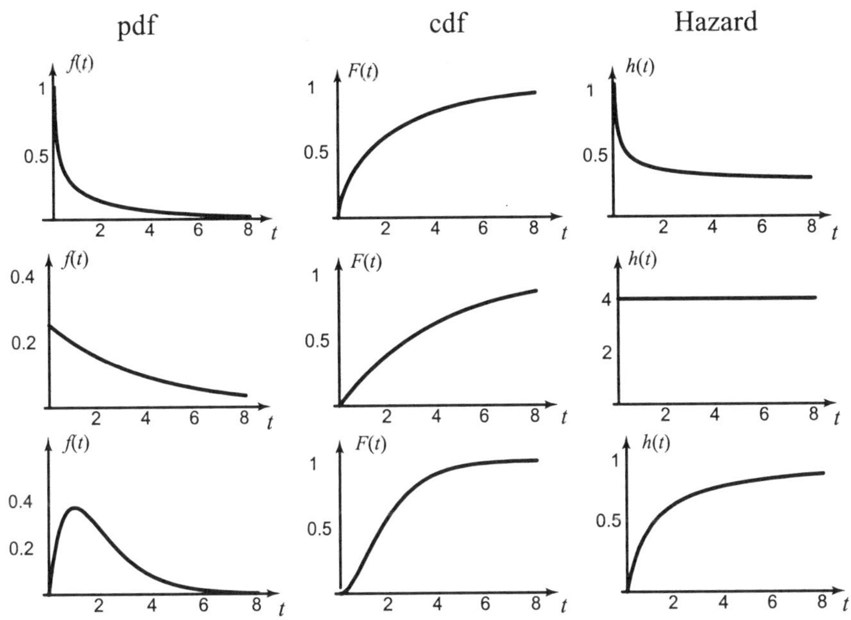

Fig. 1.9 Gamma probability density functions, cumulative distribution functions, and hazard functions for $\kappa = 0.6$ (top row), $\kappa = 1$ (middle row) and $\kappa = 2$ (bottom row).

which is (1.18) with $n = 1$.

Suppose now that T_k has the GAM$(k, 1/\lambda)$ distribution; that is, the pdf of T_k is

$$g_k(t_k) = \frac{\lambda^k t_k^{k-1}}{(k-1)!} \exp(-\lambda t_k), \qquad t_k > 0. \tag{1.19}$$

The pdf of X_{k+1} is

$$p(x_{k+1}) = \lambda \exp(-\lambda x_{k+1}), \qquad x_{k+1} > 0$$

and the pdf of T_k is given by (1.19). Furthermore, T_k and X_{k+1} are independent random variables so their joint pdf is

$$\begin{aligned} f(t_k, x_{k+1}) &= g_k(t_k) p(x_{k+1}) \\ &= \frac{\lambda^k t_k^{k-1}}{(k-1)!} \exp(-\lambda t_k) \lambda \exp(-\lambda x_{k+1}) \\ &= \frac{\lambda^{k+1} t_k^{k-1}}{(k-1)!} \exp[-\lambda(t_k + x_{k+1})]. \end{aligned}$$

Now make the transformation of variables

$$\begin{aligned} U &= T_k \\ V &= T_{k+1} = T_k + X_{k+1}. \end{aligned}$$

Then

$$\begin{aligned} T_k &= U \\ X_{k+1} &= V - U. \end{aligned}$$

The Jacobian for this transformation is

$$\begin{aligned} J &= \begin{vmatrix} \dfrac{\partial t_k}{\partial u} & \dfrac{\partial t_k}{\partial v} \\ \dfrac{\partial x_{k+1}}{\partial u} & \dfrac{\partial x_{k+1}}{\partial v} \end{vmatrix} \\ &= \begin{vmatrix} 1 & 0 \\ -1 & 1 \end{vmatrix} \\ &= 1. \end{aligned}$$

The joint pdf for U and V is therefore

$$h(u, v) = \frac{\lambda^{k+1} u^{k-1}}{(k-1)!} \exp(-\lambda v), \qquad 0 < u < v.$$

The marginal distribution of $V = T_k + X_{k+1} = T_{k+1}$ is

$$\begin{aligned} g_{k+1}(v) &= \int_0^v \frac{\lambda^{k+1} u^{k-1}}{(k-1)!} \exp(-\lambda v)\, du \\ &= \frac{\lambda^{k+1}}{(k-1)!} \exp(-\lambda v) \int_0^v u^{k-1}\, du \\ &= \frac{\lambda^{k+1}}{(k-1)!} \exp(-\lambda v) \frac{v^k}{k} \\ &= \frac{\lambda^{k+1} v^k}{k!} \exp(-\lambda v). \end{aligned}$$

Thus T_{k+1} has a GAM$(k+1, 1/\lambda)$ distribution. ∎

There is also a connection between the gamma distribution and the Poisson distribution. If $X \sim$GAM$(n, 1/\lambda)$, where n is a positive integer, then the cdf can be expressed as the sum of Poisson probabilities. The proof of this theorem will be deferred to the next chapter.

Theorem 8 *If $X \sim GAM(n, 1/\lambda)$, where n is a positive integer, then the cdf of X is*

$$F(x) = 1 - \sum_{i=0}^{n-1} \frac{(\lambda x)^i}{i!} \exp(-\lambda x), \quad x > 0.$$

1.3 BASIC THEORY OF POINT PROCESSES

A point process is a stochastic model that describes the occurrence of events in time. These occurrences are thought of as points on the time axis. In general, the times between occurrences are neither independent nor identically distributed. For our purposes, these "occurrences in time" will be the failure times of a repairable system, although the models discussed in this book may be applicable in a number of other situations.

Definition 14 Counting random variable *Let $N(t)$ be the random variable that denotes the number of failures in the interval $[0, t]$. When N has as its argument an interval, such as $N(a, b]$, the result is the number of failures in that interval. N is called the* counting random variable.

We can write the number of failures in the interval $(a, b]$ as

$$N(a, b] = N(b) - N(a).$$

To specify the stochastic model for a point process, we must give the joint distribution of the random variables $N(t_1), N(t_2), \ldots, N(t_n)$ for any n and for any $t_1, t_2, \ldots, t_n$.

Definition 15 Mean function of a point process *The* mean function of a point process *is defined to be the expectation*

$$\Lambda(t) = E\left(N(t)\right).$$

Thus, $\Lambda(t)$ is the expected number of failures through time t, and so it must be a nondecreasing function. Because $N(t)$ is a nondecreasing step function (with probability one) it is clear that the mean function must be nondecreasing. If simultaneous failures are possible, then Λ will not be continuous, but Λ will be continuous from the right.

Theorem 9 *The mean function $\Lambda(t)$ of a point process $N(t)$ is right continuous.*

Proof.

$$\begin{aligned}\lim_{t\to a^+} \Lambda(t) &= \lim_{t\to a^+} E(N(t)) \\ &= E\left(\lim_{t\to a^+} N(t)\right).\end{aligned}$$

The limit inside the expectation is

$$\lim_{t\to a^+} N(t) = N(a)$$

because N is right continuous. Thus

$$\begin{aligned}\lim_{t \to a^+} \Lambda(t) &= E\left(N(a)\right) \\ &= \Lambda(a),\end{aligned}$$

so we conclude that Λ is right continuous. ∎

Definition 16 Rate of Occurrence of Failures (ROCOF) *When Λ is differentiable we define the* rate of occurrence of failures (ROCOF) *as*

$$\mu(t) = \frac{d}{dt}\Lambda(t).$$

The ROCOF can be interpreted as the instantaneous rate of change in the expected number of failures.

To see what happens when the mean function Λ is discontinuous, suppose that a point process has mean function

$$\Lambda(t) = \begin{cases} 2t, & 0 \le t < 3 \\ 8 + (t-3)^2, & t \ge 3 \end{cases}.$$

Then

$$\begin{aligned} E\left(N(3-\varepsilon, 3]\right) &= E\left(N(3) - N(3-\varepsilon)\right) \\ &= E\left(N(3)\right) - E\left(N(3-\varepsilon)\right) \\ &= \Lambda(3) - \Lambda(3-\varepsilon) \\ &= 8 - 2(3-\varepsilon) \\ &= 2 + 2\varepsilon. \end{aligned}$$

Thus, no matter how small ε is, the expected number of failures in the interval $(3-\varepsilon, 3]$ is greater than two. Therefore, the probability of having two or more failures at time $t = 3$ is strictly positive.

Alternative methods of specifying the point process model are to give the joint pdf of the failure times $0 < T_1 < T_2 < \cdots < T_n$ or the gap times

$$\begin{aligned} X_1 &= T_1 \\ X_2 &= T_2 - T_1 \\ X_3 &= T_3 - T_2 \\ &\vdots \end{aligned}$$

To see why these alternative specifications are equivalent, we will make use of the results in the following theorem.

Theorem 10 *For any stochastic point process*

(1) $T_i > v \Leftrightarrow N(v) < i$
(2) $T_i \leq w \Leftrightarrow N(w) \geq i$
(3) $v < T_i \leq w \Leftrightarrow N(v) < i \leq N(w)$.

Proof. Figure 1.10 helps to illustrate the proof of the theorem. If $T_i > v$, then the ith failure occurred beyond time v so fewer than i failures occurred through time v. Thus, $N(v) < i$.

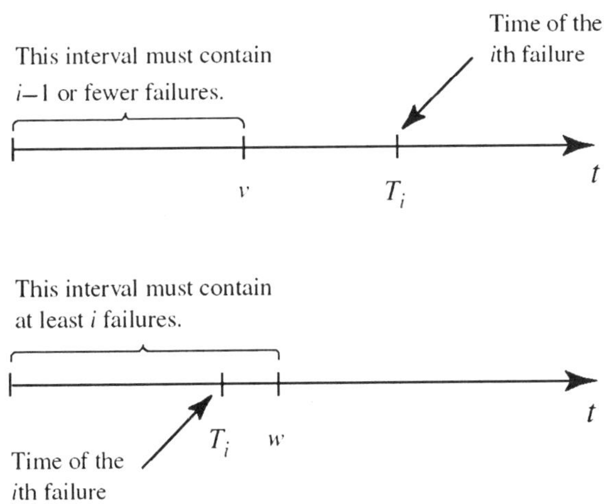

Fig. 1.10 Partition of failures as needed in the proof of Theorem 10.

Conversely, suppose $N(w) < i$. In this case, the number of failures in the interval $[0, s]$ is $0, 1, 2, \ldots$, or $i - 1$. If T_i were less than or equal to w then the number of failures in $[0, w]$ would be at least i, which is a contradiction. The ith failure must have therefore occurred past time w, that is, $T_i > w$.

The proof of the second result follows a similar argument and is left as an exercise. The third result follows directly from the first and second. ∎

Using the results from this lemma, we can prove that a point process may be specified in any one of three ways.

Theorem 11 *The joint pdf of any of the following sets of random variables determines the joint pdf of the others.*

1. $N(u_1), N(u_2), \ldots, N(u_n)$ *for any* n *and for any* $u_1, u_2, \ldots, u_n$,
2. $T_1, T_2, \ldots, T_n$ *for any* n,
3. $X_1, X_2, \ldots, X_n$ *for any* n.

Proof. We will prove (1) $\Leftrightarrow$ (2) and (2) $\Leftrightarrow$ (3).

To prove (1) $\Rightarrow$ (2), we let $0 < r_1 < s_1 < r_2 < s_2 < \cdots < r_m < s_m$. If condition (1) holds, then the joint distribution of $N(r_1)$, $N(s_1)$, ..., $N(r_m)$, $N(s_m)$ is determined. The probability of the events

$$N(r_i) < i \le N(s_i), \quad i = 1, 2, \ldots, m$$

is thus determined. These events are, by Theorem 10, equivalent to the events

$$r_i < T_i \le s_i, \quad i = 1, 2, \ldots, m.$$

Since m, r_i and s_i are arbitrary, the probabilities of these events will completely determine the joint distribution of $T_1, T_2, \ldots, T_m$ for any m.

Now we consider the converse (2) $\Rightarrow$ (1). If the joint distribution of $T_1, T_2, \ldots, T_n$ is specified, then we can compute probabilities involving the random variables $N(u_1), N(u_2), \ldots, N(u_n)$ as follows:

$$\begin{aligned} &P\left(N(u_1) = k_1, N(u_2) = k_2, \ldots, N(u_n) = k_n\right) \\ &= P\left(T_{k_1} \le u_1 < T_{k_1+1}, T_{k_2} \le u_2 < T_{k_2+1}, \ldots, T_{k_n} \le u_n < T_{k_n+1}\right). \end{aligned}$$

Finally, we consider (2) $\Leftrightarrow$ (3). Given the joint pdf of $T_1, T_2, \ldots, T_n$, we can use the Jacobian method to get the joint pdf of $X_1, X_2, \ldots, X_n$ for any n, and conversely given the joint pdf of $X_1, X_2, \ldots, X_n$ for any n we can get the pdf of $T_1, T_2, \ldots, T_n$. ■

When studying repairable systems, we will often ask whether the pattern of failures is stable or stationary over time, or whether the times between failures tend to increase (for an improving system) or decrease (for a deteriorating system). We will need the following definition to clarify this concept.

Definition 17 Stationary Increments *A point process has* stationary increments *if for all k*

$$P\left(N(t, t+s] = k\right)$$

is independent of t.

We shall also make use of the following definitions.

Definition 18 Independent Increments *A point process has* independent increments *if for all n and for all $r_1 < s_1 \le r_2 < s_2 \le \cdots < r_n \le s_n$, the random variables $N(r_1, s_1]$, $N(r_2, s_2]$, ..., $N(r_n, s_n]$ are independent. In other words,*

$$P\left(N(r_1, s_1] = k_1, \ldots, N(r_n, s_n] = k_n\right) = \prod_{i=1}^{n} P\left(N(r_i, s_i] = k_i\right).$$

Joint Probability

Definition 19 Intensity Function *The* intensity function *of a point process is*

$$\lambda(t) = \lim_{\Delta t \to 0} \frac{P\left(N(t, t + \Delta t] \geq 1\right)}{\Delta t}.$$

Roughly speaking, the intensity function is the probability of failure in a small interval divided by the length of the interval. Thus, there will be many failures over intervals on which $\lambda(t)$ is large, and fewer failures over intervals on which $\lambda(t)$ is small. It is instructive to compare the definitions of the hazard function (from Definition 7) and the intensity function (from Definition 19). The hazard function is the limit of a conditional probability; the intensity function is not. The hazard function is the conditional probability that the one and only one failure will occur in a small interval, divided by the length of the interval. This probability is conditioned on survival to the beginning of the interval. The intensity function is the unconditional probability of a failure (not necessarily the first) in a small interval divided by the length of the interval.

Clearly, the intensity function and the ROCOF are measures of the reliability of a repairable system. It turns out that these two functions are equal, provided that simultaneous failures cannot occur. (Recall that simultaneous failures can occur only at points at which the mean function $\Lambda(t)$ is discontinuous.) The proof that the intensity function equals the ROCOF function requires the following lemma.

Lemma 12 *Let $g_k(t)$ denote the density of T_k, the time of the kth failure. Then the ROCOF function is*

$$\mu(t) = \sum_{k=1}^{\infty} g_k(t).$$

Proof. Let G_k denote the cdf of T_k. Then

$$P\left(N(t) \geq k\right) = P\left(T_k \leq t\right) = G_k(t)$$

and

$$P\left(N(t) \geq k + 1\right) = P\left(T_{k+1} \leq t\right) = G_{k+1}(t).$$

Subtraction yields

$$\begin{aligned} P\left(N(t) = k\right) &= P\left(N(t) \geq k\right) - P\left(N(t) \geq k + 1\right) \\ &= G_k(t) - G_{k+1}(t). \end{aligned}$$

The mean function Λ is then

$$\begin{aligned}
\Lambda(t) &= E\left(N(t)\right) \\
&= \sum_{k=0}^{\infty} kP\left(N(t)=k\right) \\
&= \sum_{k=0}^{\infty} k\left[G_k(t) - G_{k+1}(t)\right] \\
&= 1\left[G_1(t) - G_2(t)\right] + 2\left[G_2(t) - G_3(t)\right] + 3\left[G_3(t) - G_4(t)\right] + \cdots \\
&= \sum_{k=1}^{\infty} G_k(t).
\end{aligned}$$

Differentiating both sides with respect to t gives

$$\mu(t) = \Lambda'(t) = \frac{d}{dt}\sum_{k=1}^{\infty} G_k(t) = \sum_{k=1}^{\infty}\frac{d}{dt}G_k(t) = \sum_{k=1}^{\infty} g_k(t)$$

as was to be shown. ■

We use this previous result to show that when simultaneous failures cannot occur the intensity function and the ROCOF are equal.

Theorem 13 *If the probability of simultaneous failures is zero , then* $\lambda(t) = \mu(t)$.

Proof. The key to the proof is that if simultaneous failures are impossible, then the event $N(t, t+\Delta t] \geq 1$ is equivalent to the event that the *first* failure occurs in $(t, t+\Delta t]$, or the *second* failure occurs in $(t, t+\Delta t]$, or the *third* failure occurs in $(t, t+\Delta t]$, or ....

The intensity function is therefore

$$\begin{aligned}
\lambda(t) &= \lim_{\Delta t \to 0} \frac{P\left(N(t, t+\Delta t] \geq 1\right)}{\Delta t} \\
&= \lim_{\Delta t \to 0} \frac{P(t < T_1 \leq t+\Delta t \text{ or } t < T_2 \leq t+\Delta t \text{ or } t < T_3 \leq t+\Delta t \text{ or } \ldots)}{\Delta t} \\
&= \lim_{\Delta t \to 0} \sum_{k=1}^{\infty} \frac{P(t < T_k \leq t+\Delta t)}{\Delta t} \\
&= \sum_{k=1}^{\infty} \lim_{\Delta t \to 0} \frac{P(t < T_k \leq t+\Delta t)}{\Delta t} \\
&= \sum_{k=1}^{\infty} g_k(t).
\end{aligned}$$

By the previous lemma, this last expression is equal to the ROCOF $\mu(t)$. ■

The intensity function measures the unconditional probability that a unit will fail in a small interval. For some models, it will be more appropriate to consider the conditional probability given the failure history of the process. Let $\mathcal{H}_t$ denote the entire history of the failure process through time t. This history can be represented by the set of failure times $\{t_i : i = 1, 2, \ldots, N(t)\}$.

Definition 20 Complete Intensity Function *The* complete intensity function *is*

$$\lambda(t) = \lim_{\Delta t \to 0} \frac{P(N(t, t + \Delta t] \geq 1 \,|\, \mathcal{H}_t)}{\Delta t}.$$

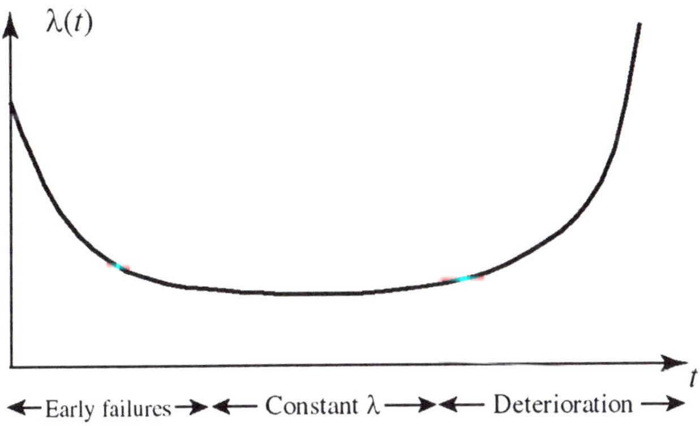

Fig. 1.11 Bathtub-shaped intensity function.

Many repairable systems have a ROCOF or intensity function that has the bathtub shape as shown in Figure 1.5. For small values of t, that is, when the system is young, the ROCOF is high and failures are frequent. After the bugs are removed, or after some of the weakest components fail, the ROCOF will be smaller, and it will remain at this level throughout its useful life. Then as the system ages, the ROCOF begins to increase. At this stage, the system is deteriorating.

Figures 1.5 and 1.11 may look identical, but their interpretations are different. The bathtub hazard function indicates that there is a high chance that the system will fail (for the first and only time) early in its life. This may be because a few of the systems have serious defects that will cause this *infant* mortality. Remember, the hazard function is the limit of a conditional probability. For a system that is wearing out, the probability of failure in $(x_0, x_0 + \Delta x_0]$ conditioned on survival past time x_0 will be smaller than the probability of failure in $(x_1, x_1 + \Delta x_1]$ conditioned on survival past time x_1.

The bathtub hazard function expresses conditional probabilities of the one and only failure of the system. The bathtub ROCOF or intensity indicates that the system will experience many failures early in its life, which will be followed by a time when the ROCOF is constant. Finally, as the system ages, the failures will become more frequent.

1.4 OVERVIEW OF MODELS

A probabilistic or statistical model for a repairable system must describe the occurrence of events in time. The assumption of independent and identically distributed times between failures is usually invalid for repairable systems. Thus we must consider models for which the assumption of independence or the assumption of an identical distribution (or both) is invalid. The intensity function, or more broadly the complete intensity function, plays an important role in selecting a model for a repairable system because this function contains the information about the likelihood of a failure at or around any time t. The complete intensity function also contains information about how a system ages over time. Typically, the assumptions we make concerning the way a system ages and how it is affected by a failure and repair will drive the choice of a repairable system model.

Definition 21 Minimal Repair Minimal repair *means that the repair done on a system leaves the system in exactly the same condition as it was just before the failure.*

The assumption of minimal repair leads to the nonhomogeneous Poisson process (NHPP). The NHPP, which is discussed in the next chapter, is often a good model for repairable systems because it can model systems that are deteriorating or improving.

Definition 22 Renewal (or Perfect) Repair *A* renewal *(or* perfect*) repair means that the system is brought to a like new state after the repair.*

If every repair is a renewal repair, then the times between failure *are* independent and identically distributed. The renewal process, discussed in Chapter 3, is an appropriate model for the system.

A number of the models that we discuss can be thought of as a compromise between the minimal repair model (which leads to the NHPP) and the renewal repair model (which leads to the renewal process).

1.5 EXERCISES

1. Make a time dot plot for the failure times given in the following table. Is the system deteriorating or improving (or neither)?

i	T_i
1	341
2	502
3	510
4	620
5	678
6	705
7	722

2. Make a time dot plot for the failure times given in the following table. Is the system deteriorating or improving (or neither)?

i	T_i
1	34
2	50
3	178
4	388
5	401
6	621
7	722

3. Make a time dot plot for the failure times given in the following table. Is the system deteriorating or improving (or neither)?

i	T_i
1	101
2	199
3	304
4	411
5	523
6	605
7	722

4. Show that the mean and variance of the exponential distribution are, respectively, θ and θ^2.

5. Give an example of data from a repairable system that is initially improving and eventually deteriorating

6. What discrete distribution(s) has (have) the memoryless property ?

7. Derive the formula for the variance of the Weibull distribution.

8. Derive the mean and variance for the GAM$(\kappa, 1/\lambda)$ distribution.

9. Thompson (1988) defines the term *without aftereffect* as follows. A point process is without aftereffect if for all n and all $t_1 < t_2 < \cdots < t_n$ we have

$$\begin{aligned} &P\left(N(t_{n-1}, t_n] = k \,|\, N(t_1) = k_1, N(t_2) = k_2, \ldots, N(t_{n-1}) = k_{n-1}\right) \\ &= P\left(N(t_{n-1}, t_n] = k\right). \end{aligned}$$

   (a) Discuss the property "without aftereffect" as it applies to repairable systems.

   (b) Show that this property is equivalent to the independent increments property.

10. Prove Palm's formula:

$$P\left(N(a, b] \geq 1\right) = \int_a^b P\left(N(t, b] = 0 \,|\, N\{t\} \geq 1\right) \lambda(t)\, dt.$$

Hints: Recall that $N\{t\}$ is the number of failures that occur *at* time t. Partition the interval $[a, b]$ and write $\{N(a, b] \geq 1\}$ as the union of mutually exclusive events.

2

Probabilistic Models: The Poisson Process

This chapter will develop some of the elementary properties of the Poisson process. We consider both the nonhomogeneous Poisson process (NHPP) and the homogeneous Poisson process (HPP).

2.1 THE POISSON PROCESS

The Poisson distribution, a useful model for a number of applications, is a discrete distribution on the nonnegative integers. It plays a central role in the study of Poisson processes, which form a class of models for repairable systems.

Definition 23 *A random variable X has a* Poisson distribution *if it is a discrete random variable that has probability mass function (pmf)*

$$p(x) = P(X = x) = \frac{\phi^x \exp(-\phi)}{x!}, \qquad x = 0, 1, 2, \ldots .$$

If X is a discrete random variable having this pmf, we will write $X \sim POI(\phi)$.

The next theorem gives the mean and variance of the Poisson distribution.

Theorem 14 *If $X \sim POI(\phi)$ then $E(X) = \phi$ and $V(X) = \phi$.*

Proof. The proof is direct:

$$
\begin{aligned}
E(X) &= \sum_{x=0}^{\infty} x\, p(x) \\
&= \sum_{x=0}^{\infty} x \, \frac{\phi^x \exp(-\phi)}{x!} \\
&= \exp(-\phi) \sum_{x=1}^{\infty} \frac{x\,\phi^x}{x!} \\
&= \phi \exp(-\phi) \sum_{x=1}^{\infty} \frac{\phi^{x-1}}{(x-1)!} \\
&= \phi \exp(-\phi) \sum_{x=0}^{\infty} \frac{\phi^x}{x!}.
\end{aligned}
$$

The series in this last expression is just the power series for $\exp(\phi)$. Thus

$$
E(X) = \phi \exp(-\phi) \exp(\phi) = \phi.
$$

To find the variance of X, we first find $E(X^2)$:

$$
\begin{aligned}
E(X^2) &= \sum_{x=0}^{\infty} x^2\, p(x) \\
&= \sum_{x=0}^{\infty} x^2 \, \frac{\phi^x \exp(-\phi)}{x!} \\
&= \sum_{x=1}^{\infty} x \, \frac{\phi^x \exp(-\phi)}{(x-1)!} \\
&= \exp(-\phi) \sum_{x=0}^{\infty} \frac{(x+1)\,\phi^{x+1}}{x!} \\
&= \phi \exp(-\phi) \sum_{x=0}^{\infty} \left(\frac{x\,\phi^x}{x!} + \frac{\phi^x}{x!} \right) \\
&= \phi \exp(-\phi) \sum_{x=0}^{\infty} \left(\frac{x\,\phi^x}{x!} + \frac{\phi^x}{x!} \right) \\
&= \phi \exp(-\phi) \left(\sum_{x=0}^{\infty} \frac{x\,\phi^x}{x!} + \sum_{x=0}^{\infty} \frac{\phi^x}{x!} \right) \\
&= \phi \exp(-\phi) \left[\phi \exp(\phi) + \exp(\phi) \right] \\
&= \phi^2 + \phi.
\end{aligned}
$$

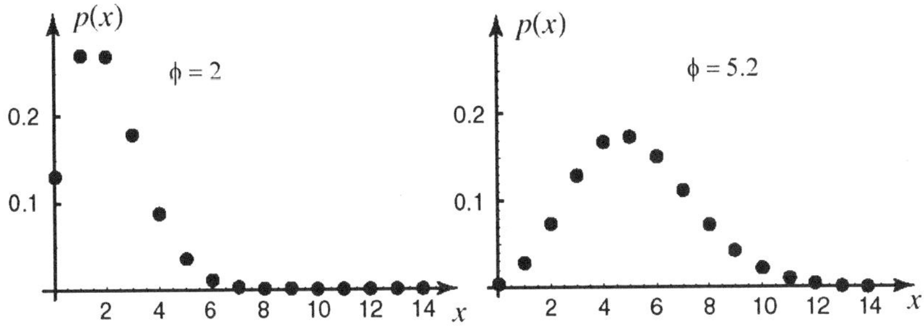

Fig. 2.1 Probability mass functions for two Poisson distributions.

Thus

$$\begin{aligned} V(X) &= E(X^2) - [E(X)]^2 \\ &= \phi^2 + \phi - \phi^2 \\ &= \phi. \end{aligned}$$

■

Figure 2.1 shows the probability mass functions for the Poisson distributions with means $\phi = 2$ and $\phi = 5.2$. Zero is the smallest possible value for the random X and there is no upper bound for the outcome X.

We first define the NHPP and then specialize it to the case in which the intensity function is a constant. As in the first chapter, we let $N(t)$ denote the number of failures through time t, and we let $N(a, b]$ denote the number of failures in the interval $(a, b]$.

Definition 24 Poisson Process *A counting process $N(t)$ is said to be a* Poisson process *if*

1. $N(0) = 0$

2. *For any $a < b \leq c < d$ the random variables $N(a, b]$ and $N(c, d]$ are independent. This is called the* independent increments property.

3. *There is a function λ such that*

$$\lambda(t) = \lim_{\Delta t \to 0} \frac{P\left(N(t, t + \Delta t] = 1\right)}{\Delta t}.$$

The function λ is called the intensity function *of the Poisson process.*

4.

$$\lim_{\Delta t \to 0} \frac{P\left(N(t, t + \Delta t] \geq 2\right)}{\Delta t} = 0.$$

The last property precludes the possibility of simultaneous failures.

As minimal as these assumptions are, it is possible to deduce the property that the number of failures in an interval is a random variable having a Poisson distribution. This property is shown in the next two theorems.

Theorem 15 *Properties (1) through (4) of the Poisson process imply that*

$$P(N(t)=n)=\frac{1}{n!}\left(\int_0^t \lambda(x)\,dx\right)^n \exp\left(-\int_0^t \lambda(x)\,dx\right), \qquad n=0,\,1,\,2,\,\ldots.$$

Proof. Define the function P_n to be

$$P_n(t)=P(N(t)=n).$$

We consider first the function $P_0(t)$. For any $\Delta t>0$ we have

$$\begin{aligned}P_0(t+\Delta t)&=P(N(t+\Delta t)=0)\\&=P(N(t)=0 \text{ and } N(t,t+\Delta t]=0).\end{aligned}$$

Since the intervals $[0,t]$ and $(t,t+\Delta t]$ are nonoverlapping, the random variables $N(t)$ and $N(t,t+\Delta t]$ are independent. Thus

$$\begin{aligned}P_0(t+\Delta t)&=P(N(t)=0)\,P(N(t,t+\Delta t]=0)\\&=P(N(t)=0)\,[1-P(N(t,t+\Delta t]=1)-P(N(t,t+\Delta t]\ge 2)]\\&=P(N(t)=0)-P(N(t)=0)\,P(N(t,t+\Delta t]=1)\\&\quad -P(N(t)=0)\,P(N(t,t+\Delta t]\ge 2)\\&=P_0(t)-P_0(t)P(N(t,t+\Delta t]=1)-P_0(t)P(N(t,t+\Delta t]\ge 2).\end{aligned}$$

Therefore,

$$P_0(t+\Delta t)-P_0(t)=-P_0(t)P(N(t,t+\Delta t]=1)-P_0(t)P(N(t,t+\Delta t]\ge 2).$$

Now divide both sides by Δt and take the limit as Δt goes to zero. This yields

$$\begin{aligned}\lim_{\Delta t\to 0}\frac{P_0(t+\Delta t)-P_0(t)}{\Delta t}&=-P_0(t)\lim_{\Delta t\to 0}\frac{P(N(t,t+\Delta t]=1)}{\Delta t}\\&\quad -P_0(t)\lim_{\Delta t\to 0}\frac{P(N(t,t+\Delta t]\ge 2)}{\Delta t}.\end{aligned}$$

The limit on the left is $P_0'(t)$. The limits on the right are $\lambda(t)$ and 0, respectively. This leads to the differential equation

$$P_0'(t)=-P_0(t)\lambda(t)$$

with initial condition

$$P_0(0)=P(N(0)=0)=1.$$

Therefore, we have

$$\frac{P_0'(t)}{P_0(t)} = -\lambda(t).$$

Switching the variable to x and integrating from 0 to t gives

$$\begin{aligned}\int_0^t \frac{P_0'(x)}{P_0(x)}\,dx &= -\int_0^t \lambda(x)\,dx \\ \log P_0(t) &= -\int_0^t \lambda(x)\,dx.\end{aligned}$$

Thus

$$P_0(t) = \exp\left(-\int_0^t \lambda(x)\,dx\right),$$

which can be written in the form

$$P_0(t) = \frac{1}{0!}\left(\int_0^t \lambda(x)\,dx\right)^0 \exp\left(-\int_0^t \lambda(x)\,dx\right).$$

This is clearly the probability that a Poisson random variable with mean $\int_0^t \lambda(x)\,dx$ is equal to zero.

Before doing the inductive part, we work out the case for $n = 1$ because it illustrates the key features of the inductive proof. For any $\Delta t > 0$ we have

$$\begin{aligned}P_1(t+\Delta t) &= P\left(N(t+\Delta t) = 1\right) \\ &= P\left(N(t) = 0 \text{ and } N(t, t+\Delta t] = 1\right) \\ &\qquad + P\left(N(t) = 1 \text{ and } N(t, t+\Delta t] = 0\right) \\ &= P\left(N(t) = 0\right) P\left(N(t, t+\Delta t] = 1\right) \\ &\qquad + P\left(N(t) = 1\right) P\left(N(t, t+\Delta t] = 0\right) \\ &= P_0(t) P\left(N(t, t+\Delta t] = 1\right) \\ &\qquad + P_1(t)\left[1 - P\left(N(t, t+\Delta t] = 1\right) - P\left(N(t, t+\Delta t] \geq 2\right)\right].\end{aligned}$$

Subtracting $P_1(t)$ from both sides yields

$$\begin{aligned}P_1(t+\Delta t) - P_1(t) = P_0(t) P\left(N(t, t+\Delta t] = 1\right) \\ - P_1(t) P\left(N(t, t+\Delta t] = 1\right) \\ - P_1(t) P\left(N(t, t+\Delta t] \geq 2\right).\end{aligned}$$

Now divide both sides by Δt and take the limit as Δt goes to zero, to obtain

$$\begin{aligned}\lim_{\Delta t\to 0}\frac{P_1(t+\Delta t)-P_1(t)}{\Delta t} &= P_0(t)\lim_{\Delta t\to 0}\frac{P\left(N(t,t+\Delta t]=1\right)}{\Delta t}\\ &\quad - P_1(t)\lim_{\Delta t\to 0}\frac{P\left(N(t,t+\Delta t]=1\right)}{\Delta t}\\ &\quad - P_1(t)\lim_{\Delta t\to 0}\frac{P\left(N(t,t+\Delta t]\geq 2\right)}{\Delta t}.\end{aligned}$$

This gives

$$P_1'(t) = P_0(t)\lambda(t) - P_1(t)\lambda(t).$$

Substituting the expression we just obtained for $P_0(t)$ gives, after some rearranging,

$$P_1'(t) + \lambda(t)P_1(t) = \lambda(t)\exp\left(-\int_0^t \lambda(x)\,dx\right).$$

The integrating factor for this differential equation is

$$\exp\left(\int_0^t \lambda(x)\,dx\right).$$

Multiplying both sides by this integrating factor yields

$$\exp\left(\int_0^t \lambda(x)\,dx\right)P_1'(t) + \exp\left(\int_0^t \lambda(x)\,dx\right)\lambda(t)P_1(t) = \lambda(t).$$

The left side is now the derivative of a product. Thus

$$\frac{d}{dt}\left[\exp\left(\int_0^t \lambda(x)\,dx\right)P_1(t)\right] = \lambda(t).$$

Integrating both sides from 0 to y gives

$$\int_0^y \frac{d}{dt}\left[\exp\left(\int_0^t \lambda(x)\,dx\right)P_1(t)\right]dt = \int_0^y \lambda(t)\,dt$$

$$\exp\left(\int_0^y \lambda(x)\,dx\right)P_1(y) - \exp\left(\int_0^0 \lambda(x)\,dx\right)P_1(0) = \int_0^y \lambda(t)\,dt.$$

Since the initial condition is now $P_1(0) = P\left(N(0)=1\right) = 0$, we have

$$\exp\left(\int_0^y \lambda(x)\,dx\right)P_1(y) = \int_0^y \lambda(t)\,dt$$

and so

$$P_1(y) = \exp\left(-\int_0^y \lambda(x)\,dx\right)\int_0^y \lambda(x)\,dx.$$

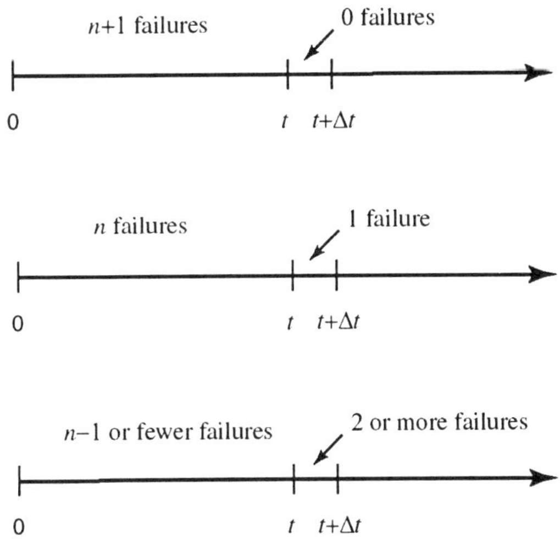

Fig. 2.2 Illustration of partitioning the number of failures in the nonoverlapping intervals $[0, t)$ and $[t, t + \Delta t)$.

After switching the variable from y to t, and after doing some manipulation, we get

$$P_1(t) = \frac{1}{1!}\left(\int_0^t \lambda(x)\,dx\right)^1 \exp\left(-\int_0^t \lambda(x)\,dx\right),$$

which is equal to the probability that a $\text{POI}\left(\int_0^t \lambda(x)\,dx\right)$ random variable is equal to 1.

Now for the inductive step of the proof. Suppose that

$$P_n(t) = \frac{1}{n!}\left(\int_0^t \lambda(x)\,dx\right)^n \exp\left(-\int_0^t \lambda(x)\,dx\right)$$

for some $n \geq 1$. We must then prove that

$$P_{n+1}(t) = \frac{1}{(n+1)!}\left(\int_0^t \lambda(x)\,dx\right)^{n+1} \exp\left(-\int_0^t \lambda(x)\,dx\right).$$

As before, we consider the probability $P_{n+1}(t+\Delta t) = P(N(0, t+\Delta t] = n+1)$. This probability can be decomposed into the three mutually exclusive events (see Figure 2.2):

$$\{N(0,t] = n+1 \text{ and } N(t,t+\Delta t] = 0\}$$
$$\{N(0,t] = n \text{ and } N(t,t+\Delta t] = 1\}$$
$$\{N(0,t] \leq n-1 \text{ and } N(t,t+\Delta t] \geq 2\}.$$

Each of these probabilities involves the number of failures in nonoverlapping intervals. Thus

$$\begin{aligned}
P_{n+1}(t+\Delta t) &= P\left(N(0,t] = n+1\right) P\left(N(t,t+\Delta t] = 0\right) \\
&\quad + P\left(N(0,t] = n\right) P\left(N(t,t+\Delta t] = 1\right) \\
&\quad + P\left(N(0,t] \leq n-1\right) P\left(N(t,t+\Delta t] \geq 2\right) \\
&= P_{n+1}(t)\left[1 - P\left(N(t,t+\Delta t] = 1\right) - P\left(N(t,t+\Delta t] \geq 2\right)\right] \\
&\quad + P_n(t) P\left(N(t,t+\Delta t] = 1\right) + P\left(N(0,t] \leq n-1\right) P\left(N(t,t+\Delta t] \geq 2\right) \\
&= P_{n+1}(t) - P_{n+1}(t) P\left(N(t,t+\Delta t] = 1\right) \\
&\quad + P_n(t) P\left(N(t,t+\Delta t] = 1\right) \\
&\quad + \left[P\left(N(0,t] \leq n-1\right) - P_{n+1}(t)\right] P\left(N(t,t+\Delta t] \geq 2\right).
\end{aligned}$$

Now, subtract $P_{n+1}(t)$ from both sides, divide both sides by Δt, and take the limit as Δt goes to zero. This yields

$$\begin{aligned}
\lim_{\Delta t \to 0} & \frac{P_{n+1}(t+\Delta t) - P_{n+1}(t)}{\Delta t} \\
&= -P_{n+1}(t) \lim_{\Delta t \to 0} \frac{P\left(N(t,t+\Delta t] = 1\right)}{\Delta t} \\
&\quad + P_n(t) \lim_{\Delta t \to 0} \frac{P\left(N(t,t+\Delta t] = 1\right)}{\Delta t} \\
&\quad + \left[P\left(N(0,t] \leq n-1\right) - P_{n+1}(t)\right] \lim_{\Delta t \to 0} \frac{P\left(N(t,t+\Delta t] \geq 2\right)}{\Delta t}.
\end{aligned}$$

Evaluating the limits gives us the differential equation

$$P'_{n+1}(t) = -P_{n+1}(t)\lambda(t) + P_n(t)\lambda(t) - 0.$$

Rearranging the terms, and applying the inductive hypothesis gives

$$P'_{n+1}(t) + P_{n+1}(t)\lambda(t) = \frac{\lambda(t)}{n!}\left(\int_0^t \lambda(x)\,dx\right)^n \exp\left(-\int_0^t \lambda(x)\,dx\right).$$

This is a first-order linear differential equation with initial condition $P_{n+1}(0) = 0$. Multiplying both sides by the integrating factor $\exp\left(\int_0^t \lambda(x)\,dx\right)$ yields

$$P'_{n+1}(t)\exp\left(\int_0^t \lambda(x)\,dx\right) + P_{n+1}(t)\lambda(t)\exp\left(\int_0^t \lambda(x)\,dx\right) = \frac{\lambda(t)}{n!}\left(\int_0^t \lambda(x)\,dx\right)^n.$$

The left side is the derivative of a product. Thus

$$\frac{d}{dt}\left[P_{n+1}(t)\exp\left(\int_0^t \lambda(x)\,dx\right)\right] = \frac{\lambda(t)}{n!}\left(\int_0^t \lambda(x)\,dx\right)^n.$$

Now, integrate both sides from 0 to y. This gives

$$\int_0^y \frac{d}{dt}\left[P_{n+1}(t)\exp\left(\int_0^t \lambda(x)\,dx\right)\right] dt = \int_0^y \frac{\lambda(t)}{n!}\left(\int_0^t \lambda(x)\,dx\right)^n dt.$$

Hence

$$P_{n+1}(y)\exp\left(\int_0^y \lambda(x)\,dx\right) = \frac{1}{n!}\frac{\left(\int_0^y \lambda(x)\,dx\right)^{n+1}}{n+1}$$

and so

$$P_{n+1}(t) = \frac{1}{(n+1)!}\left(\int_0^t \lambda(x)\,dx\right)^{n+1} \exp\left(-\int_0^t \lambda(x)\,dx\right),$$

as was to be shown. ■

The result that $N(a,b]$ has a Poisson distribution for any $a < b$ is a slight generalization of this theorem.

Corollary 16 *For a Poisson process, the random variable $N(a,b]$ has a Poisson distribution with mean*

$$\int_a^b \lambda(x)\,dx.$$

Proof. If X is a Poisson distribution with mean ϕ, then its moment generating function is

$$M(s) = E\left(e^{sX}\right) = \exp\left[\phi(e^s - 1)\right].$$

Let $a < b$. Then, by the previous theorem,

$$N(a) \sim \text{POI}\left(\int_0^a \lambda(x)\,dx\right)$$

and

$$N(b) \sim \mathrm{POI}\left(\int_0^b \lambda(x)\,dx\right).$$

The moment generating functions of $N(a)$ and $N(b)$ are therefore

$$M_{N(a)}(s) = \exp\left[\left(\int_0^a \lambda(x)\,dx\right)(e^s - 1)\right]$$

and

$$M_{N(b)}(s) = \exp\left[\left(\int_0^b \lambda(x)\,dx\right)(e^s - 1)\right].$$

By the independent increments property, the random variables $N(a)$ and $N(a,b]$ are independent. Using the result that the moment generating function of the sum of independent random variables is the product of their moment generating functions (Bain and Engelhardt, 1992), we have

$$\begin{aligned} M_{N(b)}(s) &= M_{N(a)+N(a,b]}(s) \\ &= M_{N(a)}(s) \times M_{N(a,b]}(s). \end{aligned}$$

Thus

$$\exp\left[\left(\int_0^b \lambda(x)\,dx\right)(e^s - 1)\right] = \exp\left[\left(\int_0^a \lambda(x)\,dx\right)(e^s - 1)\right] M_{N(a,b]}(s).$$

Solving for $M_{N(a,b]}(s)$ yields

$$\begin{aligned} M_{N(a,b]}(s) &= \frac{\exp\left[\left(\int_0^b \lambda(x)\,dx\right)(e^s - 1)\right]}{\exp\left[\left(\int_0^a \lambda(x)\,dx\right)(e^s - 1)\right]} \\ &= \exp\left[\left(\int_0^b \lambda(x)\,dx - \int_0^a \lambda(x)\,dx\right)(e^s - 1)\right] \\ &= \exp\left[\left(\int_a^b \lambda(x)\,dx\right)(e^s - 1)\right]. \end{aligned}$$

This is the moment generating function for a Poisson random variable with mean

$$\int_a^b \lambda(x)\,dx.$$

By the uniqueness property of moment generating functions, we conclude that

$$N(a,b] \sim \text{POI}\left(\int_a^b \lambda(x)\,dx\right).$$

■

Example 2.1 Suppose that a Poisson process has intensity function

$$\lambda(t) = 0.002t^{1.25}.$$

The mean function is

$$\Lambda(t) = \int_0^t \lambda(x)\,dx = \int_0^t 0.002x^{1.25}\,dx = 0.00089\,t^{2.25}.$$

The number of failures N in the interval $[50, 100]$ has a Poisson distribution with mean

$$\Lambda(100) - \Lambda(50) = \int_{50}^{100} 0.002x^{1.25}\,dx = 22.2.$$

Thus the probability of having 30 or more failures in the interval $[50, 100]$ is

$$\sum_{n=30}^{\infty} P(N = n) = \sum_{n=30}^{\infty} \frac{\exp(-22.2)\,22.2^n}{n!} \approx 0.0658.$$

□

The next theorem gives another way to characterize a Poisson process.

Theorem 17 *A counting process $N(t)$ is a Poisson process if and only if*
1) $N(0) = 0$,
2) The process has the independent increments property, and
3) For any $a < b$, $N(a,b] \sim POI\left(\int_a^b \lambda(x)\,dx\right)$.

Proof. The previous theorem stated that under the four conditions of Definition 24, the random variable $N(a,b]$ has a Poisson distribution with mean $\int_a^b \lambda(x)\,dx$. We must prove the converse. Properties 1 and 2 of the Poisson property are identical to the first two conditions of the theorem, so there is nothing to prove there. We must therefore prove that if, for $a < b$, the random variable $N(a,b]$ has a Poisson distribution, then

$$\lim_{\Delta t \to 0} \frac{P(N(t, t+\Delta t] = 1)}{\Delta t} = \lambda(t)$$

and

$$\lim_{\Delta t \to 0} \frac{P(N(t, t+\Delta t] \geq 2)}{\Delta t} = 0.$$

Consider the limit of $P(N(t,t+\Delta t] \geq 1)/\Delta t$:

$$\lim_{\Delta t \to 0} \frac{P(N(t,t+\Delta t] \geq 1)}{\Delta t} = \lim_{\Delta t \to 0} \frac{1}{\Delta t}\left[1 - P(N(t,t+\Delta t] = 0)\right]$$

$$= \lim_{\Delta t \to 0} \frac{1 - \exp\left(-\int_t^{t+\Delta t} \lambda(x)\,dx\right)}{\Delta t}.$$

Substitution of $\Delta t = 0$ yields an indeterminate form. Application of L'Hopital's Rule along with the Fundamental Theorem of Calculus gives

$$\lim_{\Delta t \to 0} \frac{P(N(t,t+\Delta t] \geq 1)}{\Delta t} = \lim_{\Delta t \to 0} \frac{0 + \lambda(t+\Delta t)\exp\left(-\int_t^{t+\Delta t} \lambda(x)\,dx\right)}{1}$$

$$= \lambda(t).$$

Using this result, we find that the second limit is

$$\lim_{\Delta t \to 0} \frac{P(N(t,t+\Delta t] \geq 2)}{\Delta t}$$

$$= \lim_{\Delta t \to 0} \frac{1 - P(N(t,t+\Delta t] = 0) - P(N(t,t+\Delta t] = 1)}{\Delta t}$$

$$= \lim_{\Delta t \to 0} \frac{1 - \exp\left(-\int_t^{t+\Delta t}\lambda(x)\,dx\right) - \left(\int_t^{t+\Delta t}\lambda(x)\,dx\right)\exp\left(-\int_t^{t+\Delta t}\lambda(x)\,dx\right)}{\Delta t}$$

$$= \lim_{\Delta t \to 0} \frac{1 - \exp\left(-\int_t^{t+\Delta t} \lambda(x)\,dx\right)}{\Delta t}$$

$$- \lim_{\Delta t \to 0} \frac{\left(\int_t^{t+\Delta t} \lambda(x)\,dx\right)}{\Delta t} \times \lim_{\Delta t \to 0} \exp\left(-\int_t^{t+\Delta t} \lambda(x)\,dx\right)$$

$$= \lambda(t) - \lambda(t)\exp(0)$$

$$= 0.$$

Thus

$$\lim_{\Delta t \to 0} \frac{P(N(t,t+\Delta t] = 1)}{\Delta t}$$

$$= \lim_{\Delta t \to 0} \left[\frac{P(N(t,t+\Delta t] \geq 1)}{\Delta t} - \frac{P(N(t,t+\Delta t] \geq 2)}{\Delta t}\right]$$

$$= \lambda(t) - 0$$

$$= \lambda(t).$$

■

2.2 THE HOMOGENEOUS POISSON PROCESS

Definition 25 Homogeneous Poisson Process *The* homogeneous Poisson process (HPP) *is a Poisson process with constant intensity function.*

This model is the simplest one possible for repairable systems, but it should be applied with caution. Since the intensity function is constant, the HPP cannot be used to model systems that deteriorate or improve. For these situations another model, such as a Poisson process with nonconstant intensity function, must be applied.

The HPP is closely related to the exponential distribution as illustrated by the following theorem.

Theorem 18 *A process is an HPP with intensity λ, if and only if the times between failure are iid exponential random variables with mean $1/\lambda$.*

Proof. We first assume that the process is an HPP and prove that the times between failures are iid exponential random variables with mean $1/\lambda$. Let $T_1 < T_2 < \cdots < T_n$ denote the failure times, and let $X_i = T_i - T_{i-1}$ (where $T_0 = 0$) be the times between failure. We will first derive the joint pdf of the random variables $T_1, T_2, \ldots, T_n$ by using

$$f(t_1, t_2, \ldots, t_n) = f_1(t_1) f_2(t_2|t_1) f_3(t_3|t_1, t_2) \cdots f_n(t_n|t_1, t_2, \ldots, t_{n-1}),$$

which is valid for $0 < t_1 < t_2 < \cdots < t_n$. The survival function for T_1 is

$$\begin{aligned} S_1(t_1) &= P(T_1 > t_1) \\ &= P(N(0, t_1] = 0) \\ &= \exp\left(-\int_0^{t_1} \lambda \, dx\right) \\ &= \exp(-\lambda t_1), \quad t_1 > 0. \end{aligned}$$

The pdf is thus

$$f_1(t_1) = -S_1'(t_1) = \lambda \exp(-\lambda t_1), \quad t_1 > 0.$$

The conditional survival function of T_2 given $T_1 = t_1$ is

$$\begin{aligned} S_2(t_2|t_1) &= P(T_2 > t_2|t_1) \\ &= P(N(t_1, t_2] = 0) \\ &= \exp\left(-\int_{t_1}^{t_2} \lambda \, dx\right) \\ &= \exp[-\lambda(t_2 - t_1)], \quad t_2 > t_1 > 0. \end{aligned}$$

The conditional pdf of T_2 given $T_1 = t_1$ is thus

$$\begin{aligned} f_2(t_2 \mid t_1) &= -\frac{d}{dt_2} S_2(t_2 \mid t_1) \\ &= \lambda \exp\left[-\lambda(t_2 - t_1)\right], \quad t_2 > t_1 > 0. \end{aligned}$$

Before we give the result for the general case, let's look at one more special case. Because of the independent increments property, the conditional pdf of T_3 given $T_1 = t_1$ and $T_2 = t_2$ is independent of t_1. The conditional survival function for T_3 is therefore

$$\begin{aligned} S_3(t_3 \mid t_1, t_2) &= P\left(T_3 > t_3 \mid T_1 = t_1, T_2 = t_2\right) \\ &= P(T_3 > t_3 \mid T_2 = t_2) \\ &= P\left(N(t_2, t_3] = 0\right) \\ &= \exp\left(-\int_{t_2}^{t_3} \lambda \, dx\right) \\ &= \exp\left[-\lambda(t_3 - t_2)\right], \quad t_3 > t_2 > 0. \end{aligned}$$

In general, we have

$$f_n(t_n \mid t_1, t_2, \ldots, t_{n-1}) = \lambda \exp\left[-\lambda(t_n - t_{n-1})\right], \quad t_n > t_{n-1} > 0.$$

The joint pdf of $T_1, T_2, \ldots, T_n$ is thus

$$\begin{aligned} f(t_1, t_2, \ldots, t_n) &= \left[\lambda \exp(-\lambda t_1)\right] \left\{\lambda \exp\left[-\lambda(t_2 - t_1)\right]\right\} \left\{\lambda \exp\left[-\lambda(t_3 - t_2)\right]\right\} \\ &\quad \times \cdots \times \left\{\lambda \exp\left[-\lambda(t_n - t_{n-1})\right]\right\} \\ &= \lambda^n \exp\left(-\lambda t_n\right), \qquad t_n > t_{n-1} > \cdots > t_2 > t_1 > 0. \end{aligned}$$

Solving for the T_i's in terms of the X_i's yields

$$\begin{aligned} T_1 &= X_1 \\ T_2 &= X_1 + X_2 \\ &\vdots \\ T_n &= X_1 + X_2 + \cdots + X_n. \end{aligned}$$

The joint pdf of $X_1, X_2, \ldots, X_n$ is therefore

$$g(x_1, x_2, \ldots, x_n) = f(x_1, x_1 + x_2, \ldots, x_1 + x_2 + \cdots + x_n) \left|J\right|,$$

where the Jacobian J is equal to the determinant

$$
\begin{aligned}
J &= \begin{vmatrix} \dfrac{\partial t_1}{\partial x_1} & \dfrac{\partial t_1}{\partial x_2} & \cdots & \dfrac{\partial t_1}{\partial x_n} \\ \dfrac{\partial t_2}{\partial x_1} & \dfrac{\partial t_2}{\partial x_2} & \cdots & \dfrac{\partial t_2}{\partial x_n} \\ \vdots & \vdots & & \vdots \\ \dfrac{\partial t_n}{\partial x_1} & \dfrac{\partial t_n}{\partial x_2} & \cdots & \dfrac{\partial t_n}{\partial x_n} \end{vmatrix} \\
&= \begin{vmatrix} 1 & 0 & \cdots & 0 \\ 1 & 1 & \cdots & 0 \\ \vdots & \vdots & & \vdots \\ 1 & 1 & \cdots & 1 \end{vmatrix} \\
&= 1.
\end{aligned}
$$

Hence,

$$
\begin{aligned}
g(x_1, x_2, \ldots, x_n) &= \lambda^n \exp\left[-\lambda(x_1 + x_2 + \cdots + x_n)\right] |1| \\
&= \prod_{i=1}^{n} \left[\lambda \exp\left(-\lambda x_i\right)\right], \quad x_i > 0.
\end{aligned}
$$

This result implies that the times between failures are iid exponential random variables with mean $1/\lambda$.

Now we consider the converse. Suppose that the times between failures are iid exponential with mean $1/\lambda$. We must prove that the process is an HPP. We actually prove the slightly weaker result that for any $0 < a < b$, the random variables $N(0, a]$ and $N(a, b]$ are independent Poisson random variables with means λa and $\lambda(b-a)$, respectively. The proof can be extended in a straightforward fashion to prove the general case. If the times between failure $X_1, X_2, \ldots, X_n$ are iid exponential random variables then the joint pdf of $T_1, T_2, \ldots, T_n$ is

$$
f(t_1, t_2, \ldots, t_n) = \lambda^n \exp\left(-\lambda t_n\right), \qquad 0 < t_1 < t_2 < \cdots < t_{n-1} < t_n.
$$

This result can be proved using a transformation argument similar to that given in the first part of this proof. The joint pdf for the first $r+s+1$ failure times is thus

$$
\begin{aligned}
f(t_1, t_2, \ldots, t_{r+s+1}) &= \lambda^{r+s+1} \exp\left(-\lambda t_{r+s+1}\right), \\
&\qquad 0 < t_1 < t_2 < \cdots < t_{r+s} < t_{r+s+1}.
\end{aligned}
$$

The probability that $N(0, a] = r$ and $N(a, b] = s$ is then the integral over that region of $(r+s+1)$-space in which the t_i's satisfy this condition. We then have

$$
\begin{aligned}
&P\left(N(0,a]=r,\ N(a,b]=s\right)\\
&= P\left(0<T_1<T_2<\cdots T_r<a<T_{r+1}<\cdots<T_{r+s}<b<T_{r+s+1}\right)\\
&= \underset{\{0<t_1<\cdots<t_r<a\}}{\int\cdots\int}\ \underset{\{a<t_{r+1}<\cdots<t_{r+s}<b\}}{\int\cdots\int}\ \int_b^\infty \lambda^{r+s+1}\, e^{-\lambda t_{r+s+1}}\\
&\qquad\qquad dt_{r+s+1}\, dt_{r+s}\cdots dt_{r+1}\, dt_r\cdots dt_1\\
&= \underset{\{0<t_1<\cdots<t_r<a\}}{\int\cdots\int}\ \underset{\{a<t_{r+1}<\cdots<t_{r+s}<b\}}{\int\cdots\int} \lambda^{r+s}\, e^{-\lambda b}\, dt_{r+s}\cdots dt_{r+1}\, dt_r\cdots dt_1\\
&= \lambda^{r+s}e^{-\lambda b} \underset{\{0<t_1<\cdots<t_r<a\}}{\int\cdots\int} \left(\underset{\{a<t_{r+1}<\cdots<t_{r+s}<b\}}{\int\cdots\int} 1\, dt_{r+s}\cdots dt_{r+1}\right) dt_r\cdots dt_1\\
&= \lambda^{r+s}e^{-\lambda b} \left(\underset{\{a<t_{r+1}<\cdots<t_{r+s}<b\}}{\int\cdots\int} 1\, dt_{r+s}\cdots dt_{r+1}\right)\\
&\qquad \times \left(\underset{\{0<t_1<\cdots<t_r<a\}}{\int\cdots\int} 1\, dt_r\cdots dt_1\right).
\end{aligned}
$$

Each of these multiple integrals is over one of the $n!$ simplex regions in an n-dimensional hypercube. The first hypercube has length $b-a$ on each side, and the second has length a on each side. To see what the volume of such a region is, suppose we were to partition the hypercube defined by $c < x_i < d$, $i = 1, 2, \ldots, n$, into the $n!$ partitions defined by $c < x_{i_1} < x_{i_2} < \cdots < x_{i_n} < d$. Since these are disjoint sets whose union (together with the boundaries) is the whole hypercube, the sum of the volumes of the $n!$ regions must equal the volume of the hypercube, that is $(d-c)^n$. Since each region has the same volume, we conclude that each region must have volume $(d-c)^n/n!$. Thus these two multiple integrals are $(b-a)^s/s!$ and $a^r/r!$, respectively. The above probability is therefore

$$
\begin{aligned}
P\left(N(0,a]=r,\ N(a,b]=s\right) &= \lambda^{r+s}\exp\left(-\lambda b\right)\frac{(b-a)^s}{s!}\frac{a^r}{r!}\\
&= \frac{(\lambda a)^r\exp\left(-\lambda a\right)}{r!}\ \frac{[\lambda(b-a)]^s\exp\left[-\lambda(b-a)\right]}{s!}.
\end{aligned}
$$

Thus the random variables $N(0,a]$ and $N(a,b]$ are independent Poisson random variables with means λa and $\lambda(b-a)$, respectively.

The method described here can be extended to prove that the random variables $N(0, a_1]$, $N(a_1, a_2]$, ..., $N(a_{n-1}, a_n]$ are independent Poisson random variables with respective means λa_1, $\lambda(a_2 - a_1)$, ..., $\lambda(a_n - a_{n-1})$. This would then be sufficient to conclude using Theorem 17 that the process is an HPP. ■

In the proof of the last theorem we derived the joint distribution of the first n failure times. This result is worth stating as a corollary.

Corollary 19 *If $0 < T_1 < T_2 < \cdots < T_n$ are the failure times from an HPP, then the joint pdf of $T_1, T_2, \ldots, T_n$ is*

$$f(t_1, t_2, \ldots, t_n) = \lambda^n \exp(-\lambda t_n), \quad 0 < t_1 < t_2 < \cdots < t_n.$$

Theorem 18 implies that for an HPP, the time of the nth failure

$$T_n = X_1 + X_2 + \cdots + X_n$$

is the sum of independent exponential random variables. Theorem 7 from Chapter 1 states that the sum of n EXP$(1/\lambda)$ random variables has a GAM$(n, 1/\lambda)$ distribution. This leads to the following corollary.

Corollary 20 *The time to the* nth *failure from a system modeled by an HPP has a* GAM$(n, 1/\lambda)$ *distribution.*

Example 2.2 Suppose that the failure times of a repairable system occur according to an HPP with $\lambda = 0.0125$ failures per hour.

(a) The number of failures N_1 in the first $t = 100$ hours has a Poisson distribution with mean $\lambda t = 0.0125 \times 100 = 1.25$. The probability of observing no failures in the first 100 hours is

$$P(N_1 = 0) = \frac{e^{-1.25} 1.25^0}{0!} \approx 0.287$$

and the probability of observing 4 or more failures in the first 100 hours is

$$\begin{aligned} P(N_1 \geq 4) &= 1 - P(N_1 \leq 3) \\ &= 1 - \sum_{n=0}^{3} \frac{e^{-1.25} 1.25^n}{n!} \\ &\approx 0.0383. \end{aligned}$$

(b) The number of failures N_2 between time 900 and 1000 also has a Poisson distribution with mean $\lambda(1000 - 900) = 1.25$. This is the same distribution as the number of failures in the *first* 100 hours. In fact, for any interval of length 100, the number of failures will have a Poisson distribution with mean 1.25. The probabilities of observing $N_2 = 0$ and $N_2 \geq 4$ are therefore 0.287 and 0.0383, respectively. The random variables N_1 and N_2 are independent by the independent increments property.

(c) By Theorem 18, the times between failures, $X_1, X_2, \ldots$ are iid exponential random variables with mean $1/\lambda = 1/0.0125 = 80$ hours.

(d) Suppose a maintenance contract calls for repairs for the first ten failures. The maintenance contract will then expire at the time of the tenth failure, that is, at time $T_{10} = X_1 + X_2 + \cdots + X_{10}$. Using Theorem 20 we know that T_{10} has a GAM(10, 80) distribution. Thus, the probability that the tenth failure will occur before time 500 is

$$\int_0^{500} \frac{t^{10-1}}{80^{10}\Gamma(10)} \exp\left(-t/80\right) \, dt \approx 0.102.$$

The expected value of the time of the tenth failure is $10 \times 80 = 800$ and the standard deviation is $\sqrt{10 \times 80^2} = \sqrt{64000} \approx 253.0$. □

When we study statistical inference for repairable systems, we will need the conditional distributions given in the next two theorems.

Theorem 21 *If the failure times of a homogeneous Poisson process are $T_1 < T_2 < \cdots < T_n$, then, conditioned on $T_n = t_n$, the random variables $T_1 < T_2 < \cdots < T_{n-1}$ are distributed as $n-1$ order statistics from a uniform distribution on the interval $(0, t_n)$.*

Proof. The conditional pdf of $T_1, T_2, \ldots, T_{n-1}$ given $T_n = t_n$ is given by

$$p(t_1, t_2, \ldots, t_{n-1} | t_n) = \frac{f(t_1, t_2, \ldots, t_{n-1}, t_n)}{p(t_n)}.$$

By Corollary 19 the numerator is

$$f(t_1, t_2, \ldots, t_{n-1}, t_n) = \lambda^n \exp(-\lambda t_n), \quad 0 < t_1 < t_2 < \cdots < t_n$$

and by Corollary 20 the denominator is

$$p(t_n) = \frac{\lambda^n t_n^{n-1}}{\Gamma(n)} \exp\left(-\lambda t_n\right), \quad t_n > 0.$$

Thus

$$\begin{aligned} p(t_1, t_2, \ldots, t_{n-1} | t_n) &= \frac{\lambda^n \exp(-\lambda t_n)}{\dfrac{\lambda^n t_n^{n-1}}{\Gamma(n)} \exp\left(-\lambda t_n\right)} \\ &= \frac{(n-1)!}{t_n^{n-1}}, \quad 0 < t_1 < t_2 < \cdots < t_{n-1} < t_n. \end{aligned}$$

This is the distribution of $(n-1)$ order statistics from a uniform distribution on the interval $[0, t_n]$. ■

A similar result holds for the failure times conditioned on the number of failures $N(t)$ through time t.

Theorem 22 *For an HPP, conditioned on $N(t) = n$, the failure times $T_1 < T_2 < \ldots < T_n$ are distributed as order statistics from the UNIF$(0,t)$ distribution.*

Proof. In this proof we write

$$\lim_{\Delta t_i \to 0}$$

to mean the limit in n-space

$$\lim_{\Delta t_1 \to 0} \lim_{\Delta t_2 \to 0} \cdots \lim_{\Delta t_n \to 0} .$$

The joint pdf of $T_1, T_2, \ldots, T_n$ given $N(t) = n$ is

$$f(t_1, t_2, \ldots, t_n \mid n)$$

$$= \lim_{\Delta t_i \to 0} \frac{P(t_1 < T_1 \le t_1 + \Delta t_1, \ldots, t_n < T_n \le t_n + \Delta t_n \mid N(t) = n)}{\prod_{i=1}^{n} \Delta t_i}$$

$$= \lim_{\Delta t_i \to 0} \frac{P(t_1 < T_1 \le t_1 + \Delta t_1, \ldots, t_n < T_n \le t_n + \Delta t_n \, , \, N(t) = n)}{P\left(N(t) = n\right) \prod_{i=1}^{n} \Delta t_i}$$

$$= \lim_{\Delta t_i \to 0} P\left\{N(0, t_1] = 0, N(t_1, t_1 + \Delta t_1] = 1, \ldots, \right.$$
$$\left. N(t_n, t_n + \Delta t_n] = 1, N(t_n + \Delta t_n, t] = 0\right\} \Big/ \left(P\left(N(t) = n\right) \prod_{i=1}^{n} \Delta t_i \right)$$

$$= \lim_{\Delta t_i \to 0} P\left(N(0, t_1] = 0\right) P\left(N(t_1, t_1 + \Delta t_1] = 1\right) \times \cdots \times$$
$$P\left(N(t_n, t_n + \Delta t_n] = 1\right) P\left(N(t_n + \Delta t_n, t] = 0\right) \Big/ \left(P\left(N(t) = n\right) \prod_{i=1}^{n} \Delta t_i \right)$$

$$= \lim_{\Delta t_i \to 0} \exp\left(-\lambda t_1\right) \left[(\lambda \Delta t_1) \exp\left(-\lambda \Delta t_1\right)\right] \times \cdots \times$$
$$\left[(\lambda \Delta t_n) \exp\left(-\lambda \Delta t_n\right)\right] \exp\left[-\lambda(t - t_n - \Delta t_n)\right] \Big/ \left(P\left(N(t) = n\right) \prod_{i=1}^{n} \Delta t_i \right)$$

$$= \lim_{\Delta t_i \to 0} \frac{\exp(-\lambda t)\, \lambda^n \Delta t_1 \Delta t_2 \cdots \Delta t_n}{\frac{1}{n!} \exp(-\lambda t)\, (\lambda t)^n \prod_{i=1}^{n} \Delta t_i}$$

$$= \frac{n!}{t^n}, \qquad 0 < t_1 < t_2 < \cdots < t_n < t.$$

This is the pdf of order statistics from a sample of size n from the UNIF$(0, t)$ distribution. ■

2.2.1 Gap Lengths for the HPP

Suppose we fix on a time t, and look forward (from time t) to the next failure or backward to the most recent failure. These concepts lead to the forward and backward recurrence times.

Definition 26 Forward Recurrence Time *The* forward recurrence time *is* $F_t = T_{N(t)+1}$ *(that is, the waiting time to the next failure if we begin observing the system at time t).*

Definition 27 Backward Recurrence Time *The* backward recurrence *time is* $B_t = T_{N(t)}$ *(that is, the time since the most recent failure).*

The time between the most recent failure and the next failure is $X_{N(t)+1} = B_t + F_t$. The survival function for B_t can be found as follows:

$$\begin{aligned} P(B_t > s) &= P(N(t-s, t] = 0) \\ &= \exp\left(-\int_{t-s}^{t} \lambda\, dx\right) \\ &= \exp(-\lambda s), \qquad s > 0. \end{aligned}$$

Thus $B_t \sim$ EXP$(1/\lambda)$. Similarly

$$\begin{aligned} P(F_t > s) &= P\left(N(t, t+s] = 0\right) \\ &= \exp\left(-\int_{t}^{t+s} \lambda\, ds\right) \\ &= \exp(-\lambda s), \qquad s > 0. \end{aligned}$$

Thus $F_t \sim$ EXP$(1/\lambda)$. Since the current gap length $X_{N(t)+1}$ is equal to $B_t + F_t$ we conclude that $X_{N(t)}$ has the GAM$(2, 1/\lambda)$ distribution. The expected value of $X_{N(t)+1}$ is 2λ. But wait! Theorem 18 says that for an HPP the time between failures are iid EXP$(1/\lambda)$ random variables and $X_{N(t)+1}$ is just one of these times between failure. Therefore, $X_{N(t)+1} \sim$ EXP$(1/\lambda)$.

So which is it? Is $X_{N(t)+1} \sim \text{GAM}(2, 1/\lambda)$ or is $X_{N(t)+1} \sim \text{EXP}(1/\lambda)$? This apparent contradiction is called Feller's paradox (Feller, 1971, pp. 11–13), named after the probabilist William Feller. Its resolution is as follows. The correct distribution is $\text{GAM}(2, 1/\lambda)$. The gap length $X_{N(t)+1}$ is not just *one of* the X_i's; it is the gap that covers the point t. If t is chosen before the process begins, it is more likely that the point t will be covered by a large gap than by a small gap. (For those readers who have studied order statistics, the following analogy may be helpful. Let $U_1, U_2, \ldots, U_n$ be iid random variables having a uniform distribution over the interval $(0, 1)$ and let $V = \max(U_1, U_2, \ldots, U_n)$. Even though V is equal to one of the U_i's, it does not have a uniform distribution. The random variable V is not just one of the U_i's.)

2.3 THE NONHOMOGENEOUS POISSON PROCESS

Definition 28 Nonhomogeneous Poisson Process *The* nonhomogeneous Poisson process (NHPP) *is a Poisson process whose intensity function is non-constant.*

Example 2.3 Suppose that a repairable system is modeled by an NHPP with intensity function

$$\lambda(t) = 0.02t^{0.8}.$$

The mean function is

$$\Lambda(t) = \int_0^t 0.02x^{0.8}\,dx = \frac{1}{90}t^{1.8}.$$

Figure 2.3 shows the intensity and mean functions for this NHPP.

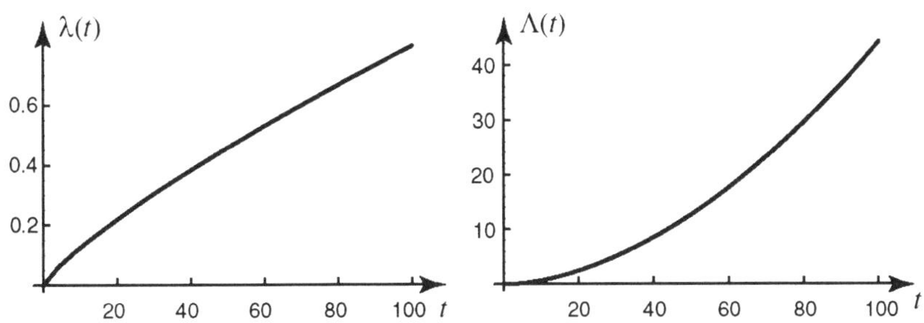

Fig. 2.3 Intensity function and mean function for the NHPP with $\lambda(t) = 0.02\,t^{0.8}$.

The distribution of the number of failures in the interval $[a, b]$ has a Poisson distribution with mean

$$\int_a^b 0.02t^{0.8}\,dt = \frac{b^{1.8} - a^{1.8}}{90}.$$

The expected numbers of failures in the intervals $[0, 20]$, $[20, 40]$, $[40, 60]$, and $[60, 80]$ are

$$E\left(N[0,20]\right) = \frac{20^{1.8} - 0^{1.8}}{90} \approx 2.44$$

$$E\left(N[20,40]\right) = \frac{40^{1.8} - 20^{1.8}}{90} \approx 6.06$$

$$E\left(N[40,60]\right) = \frac{60^{1.8} - 40^{1.8}}{90} \approx 9.14$$

$$E\left(N[60,80]\right) = \frac{80^{1.8} - 60^{1.8}}{90} \approx 11.96.$$

Since the expected number of failures in these 20-hour intervals increases as the unit ages, this system is deteriorating. □

POWER LAW PROCESS

When the intensity function has the form $\lambda(t) = (\beta/\theta)(t/\theta)^{\beta-1}$, where $\beta > 0$ and $\theta > 0$, the process is called the **power law process**, a model that we discuss in Chapters 4 and 5.

2.3.1 Likelihood Functions

In subsequent chapters the likelihood function given a set of failure times will be needed. In this section, we derive the likelihood function, that is, the joint density function of the observed random variables, for the situations where testing stops after a predetermined number of failures, or at a predetermined time.

2.3.1.1 Failure Truncated Case

Definition 29 Failure Truncation *When testing stops after a predetermined number of failures, the data are said to be* failure truncated.

For failure truncated data, the likelihood is the joint pdf of the failure times $T_1, T_2, \ldots, T_n$. This result is given in the next theorem.

Theorem 23 *The joint pdf of the failure times $T_1, T_2, \ldots, T_n$ from an NHPP with intensity function $\lambda(t)$ is*

$$f(t_1, t_2, \ldots, t_n) = \left(\prod_{i=1}^{n} \lambda(t_i)\right) \exp\left(-\int_0^{t_n} \lambda(x)\,dx\right).$$

Proof. We will derive this result using the relationship

$$f(t_1, t_2, \ldots, t_n) = f_1(t_1) f_2(t_2 \,|\, t_1) f_3(t_3 \,|\, t_1, t_2) \cdots f_n(t_n | t_1, t_2, \ldots, t_{n-1}),$$

where $0 < t_1 < t_2 < \cdots < t_n$. The survival function for T_1 is

$$\begin{aligned} S_1(t_1) &= P\,(T_1 > t_1) \\ &= P\,(N(0, t_1] = 0) \\ &= \exp\left(-\int_0^{t_1} \lambda(x)\,dx\right), \qquad t_1 > 0. \end{aligned}$$

The pdf of T_1 is thus

$$\begin{aligned} f_1(t_1) &= -S_1'(t_1) \\ &= -\frac{d}{dt_1} \exp\left(-\int_0^{t_1} \lambda(x)\,dx\right) \\ &= \lambda(t_1) \exp\left(-\int_0^{t_1} \lambda(x)\,dx\right). \end{aligned}$$

The conditional survival function of T_2 given $T_1 = t_1$ is

$$\begin{aligned} S_2(t_2 | t_1) &= P(T_2 > t_2 \,|\, t_1) \\ &= P\,(N(t_1, t_2] = 0) \\ &= \exp\left(-\int_{t_1}^{t_2} \lambda(x)\,dx\right), \qquad t_2 > t_1 > 0. \end{aligned}$$

The conditional pdf of T_2 given $T_1 = t_1$ is thus

$$\begin{aligned} f_2(t_2 \,|\, t_1) &= -\frac{d}{dt_2} S_2(t_2 \,|\, t_1) \\ &= \lambda(t_2) \exp\left[-\int_{t_1}^{t_2} \lambda(x)\,dx\right], \qquad t_2 > t_1 > 0. \end{aligned}$$

By an argument similar to that used in the proof of Theorem 18, we conclude that the conditional survival function of T_k given $T_1 = t_1$, $T_2 = t_2$, $\ldots, T_{k-2} = t_{k-2}$, $T_{k-1} = t_{k-1}$ is independent of t_1, t_2, $\ldots, t_{k-2}$. Thus in general, we have

$$\begin{aligned} S_k(t_k \,|\, t_1, t_2, \ldots, t_{k-1}) &= S(t_k \,|\, t_{k-1}) \\ &= P\,(T_k > t_k \,|\, T_{k-1} = t_{k-1}) \\ &= P\,(N(t_{k-1}, t_k] = 0) \\ &= \exp\left(-\int_{t_{k-1}}^{t_k} \lambda(x)\,dx\right), \qquad t_k > t_{k-1}. \end{aligned}$$

Thus

$$f(t_k \mid t_{k-1}) = \lambda(t_k) \exp\left(- \int_{t_{k-1}}^{t_k} \lambda(x)\, dx \right), \qquad t_k > t_{k-1}.$$

The joint pdf of $T_1, T_2, \ldots, T_n$ is therefore

$$\begin{aligned} f(t_1, t_2, \ldots, t_n) &= \left[\lambda(t_1) \exp\left(- \int_0^{t_1} \lambda(x)\, dx \right)\right] \left[\lambda(t_2) \exp\left(- \int_{t_1}^{t_2} \lambda(x)\, dx \right)\right] \\ &\quad \times \cdots \times \left[\lambda(t_n) \exp\left(- \int_{t_{n-1}}^{t_n} \lambda(x)\, dx \right)\right] \\ &= \left(\prod_{i=1}^{n} \lambda(t_i)\right) \exp\left(- \int_0^{t_n} \lambda(x)\, dx \right), \\ &\qquad t_n > t_{n-1} > \cdots > t_2 > t_1 > 0. \end{aligned}$$

This is the desired result. ■

When we derive goodness-of-fit tests in Chapter 4, we will need the conditional distribution of the first $n-1$ failure times given the time of the nth failure time. This result is proved in the next theorem. The proof of this result, however, requires the following lemma.

Lemma 24 *The time of the nth occurrence of an NHPP with intensity function $\lambda(t)$ has pdf*

$$f_n(t) = \frac{\lambda(t) \left[\Lambda(t)\right]^{n-1}}{\Gamma(n)} \exp\left[-\Lambda(t)\right], \qquad t > 0.$$

Proof. The proof hinges on Theorem 10 from Chapter 1. This theorem asserts that $T_n > t$ if and only if $N(t) < n$. Thus

$$\begin{aligned} f_n(t) &= -\frac{d}{dt} S_{T_n}(t) \\ &= -\frac{d}{dt} P(T_n > t) \\ &= -\frac{d}{dt} P\left(N(t) < n\right) \\ &= -\frac{d}{dt} \sum_{i=0}^{n-1} P\left(N(t) = i\right) \\ &= -\frac{d}{dt} \sum_{i=0}^{n-1} \frac{\left[\Lambda(t)\right]^i \exp\left[-\Lambda(t)\right]}{i!} \end{aligned}$$

$$= -\sum_{i=0}^{n-1} \frac{d}{dt} \frac{[\Lambda(t)]^i \exp[-\Lambda(t)]}{i!}$$

$$= -\sum_{i=0}^{n-1} \frac{1}{i!} \left\{ \exp[-\Lambda(t)]\, i\, [\Lambda(t)]^{i-1} \Lambda'(t) - [\Lambda(t)]^i \exp[-\Lambda(t)]\, \Lambda'(t) \right\}$$

$$= \lambda(t) \exp[-\Lambda(t)] \sum_{i=0}^{n-1} \frac{1}{i!} \left\{ [\Lambda(t)]^i - i\, [\Lambda(t)]^{i-1} \right\}.$$

The first term in this series is $(1-0)$. The rest of the terms are telescoping, so we have

$$f_n(t) = \lambda(t) \exp[-\Lambda(t)] \left((1-0) + \sum_{i=1}^{n-1} \left\{ \frac{1}{i!} [\Lambda(t)]^i - \frac{1}{(i-1)!} [\Lambda(t)]^{i-1} \right\} \right)$$

$$= \lambda(t) \exp[-\Lambda(t)] \left\{ [1-0] + [\Lambda(t) - 1] + \left[\tfrac{1}{2} (\Lambda(t))^2 - \Lambda(t) \right] + \cdots \right.$$

$$+ \left[\tfrac{1}{(n-2)!} (\Lambda(t))^{n-2} - \tfrac{1}{(n-3)!} (\Lambda(t))^{n-3} \right]$$

$$\left. + \left[\tfrac{1}{(n-1)!} (\Lambda(t))^{n-1} - \tfrac{1}{(n-2)!} (\Lambda(t))^{n-2} \right] \right\}$$

$$= \frac{\lambda(t) [\Lambda(t)]^{n-1}}{\Gamma(n)} \exp[-\Lambda(t)].$$

■

When the intensity function has the form

$$\lambda(t) = \frac{\beta}{\theta} \left(\frac{t}{\theta} \right)^{\beta - 1}$$

the above pdf is

$$f(t) = \frac{\frac{\beta}{\theta} \left(\frac{t}{\theta} \right)^{\beta-1} \left[\left(\frac{t}{\theta} \right)^{\beta} \right]^{n-1}}{\Gamma(n)} \exp\left[-\left(\frac{t}{\theta} \right)^{\beta} \right], \quad t > 0$$

$$= \frac{1}{\Gamma(n)} \frac{\beta}{\theta} \left(\frac{t}{\theta} \right)^{n\beta - 1} \exp\left[-\left(\frac{t}{\theta} \right)^{\beta} \right], \quad t > 0, \tag{2.1}$$

which is the pdf of the **generalized gamma distribution**. This distribution, which we abbreviate GGAM(n, β, θ), is a generalization of both the Weibull distribution and the gamma distribution. When $n = 1$, the GGAM(n, β, θ)

becomes the WEI(β, θ) distribution, and when $\beta = 1$ the GGAM(n, β, θ) becomes the GAM(n, θ) distribution.

The result of the next theorem is needed to derive goodness-of-fit tests in Chapter 4.

Theorem 25 *If the failure times of a nonhomogeneous Poisson process are $T_1 < T_2 < \cdots < T_n$ then conditioned on $T_n = t_n$, the random variables $T_1 < T_2 < \cdots < T_{n-1}$ are distributed as $n-1$ order statistics from the distribution with cumulative distribution function*

$$G(y) = \begin{cases} 0, & y \le 0 \\ \Lambda(y)/\Lambda(t_n), & 0 < y \le t_n \\ 1, & y > t_n \end{cases}.$$

Proof. The joint pdf of order statistics from a distribution with cdf $G(y)$ is

$$\begin{aligned} g(y_1, y_2, \ldots, y_n) &= (n-1)! \prod_{i=1}^{n-1} G'(y_i) \\ &= (n-1)! \prod_{i=1}^{n-1} \frac{\Lambda'(y_i)}{\Lambda(t_n)}, \quad 0 < y_1 < y_2 < \cdots < y_{n-1} < y_n. \end{aligned}$$

The remainder of the proof involves applying Theorem 23 and Lemma 24. Proceeding directly, we find that the pdf of $T_1 < T_2 < \cdots < T_{n-1}$ given $T_n = t_n$ is

$$\begin{aligned} f(t_1, t_2, \ldots, t_{n-1} \,|\, t_n) &= \frac{f(t_1, t_2, \ldots, t_{n-1}, t_n)}{f(t_n)} \\ &= \frac{\left(\prod_{i=1}^{n} \lambda(t_i)\right) \exp\left[-\Lambda(t_n)\right]}{\frac{\lambda(t)\left[\Lambda(t_n)\right]^{n-1}}{\Gamma(n)} \exp\left[-\Lambda(t_n)\right]} \\ &= \frac{(n-1)! \left(\prod_{i=1}^{n-1} \lambda(t_i)\right)}{\left[\Lambda(t_n)\right]^{n-1}} \\ &= (n-1)! \prod_{i=1}^{n-1} \frac{\lambda(t_i)}{\Lambda(t_n)}, \quad 0 < y_1 < y_2 < \cdots < y_{n-1} < y_n. \end{aligned}$$

This last expression is the joint pdf of a sample of $n-1$ order statistics from the distribution with cdf $G(y)$. ■

2.3.2 Time Truncated Case

Definition 30 *Data are said to be* time truncated *when testing stops at a predetermined time t.*

When we observe a system until a predetermined time t, the number of failures $N(t)$ is a random variable. This randomness must be taken into account when we determine the likelihood function.

The following theorem will be needed to find the likelihood function for an NHPP. This result is also needed to derive goodness-of-fit tests for the NHPP.

Theorem 26 *If a NHPP with intensity function $\lambda(t)$ is observed until time t, and if the failure times are $T_1 < T_2 < \cdots < T_{N(t)}$ where $N(t)$ is the random number of failures in the interval $(0,t]$, then conditioned on $N(t) = n$, the random variables $T_1 < T_2 < \cdots < T_n$ are distributed as n order statistics from the distribution with cdf*

$$G(y) = \begin{cases} 0, & y \le 0 \\ \Lambda(y)/\Lambda(t), & 0 < y \le t \\ 1, & y > t \end{cases} .$$

Proof. The proof follows the same line of argument as the proof of Theorem 22. (See Figure 2.4.) The joint pdf of $T_1 < T_2 < \cdots < T_n$ given $N(t) = n$ is

$$\begin{aligned}
f(t_1, \ldots, t_n \,|\, n) &= \lim_{\Delta t_i \to 0} \frac{P(t_1 < T_1 \le t_1 + \Delta t_1, \ldots, t_n < T_n < t_n + \Delta t_n \,|\, n)}{\displaystyle\prod_{i=1}^{n} \Delta t_i} \\
&= \lim_{\Delta t_i \to 0} \frac{P(t_1 < T_1 \le t_1 + \Delta t_1, \ldots, t_n < T_n < t_n + \Delta t_n \,,\, n)}{P\left(N(0,t] = n\right) \displaystyle\prod_{i=1}^{n} \Delta t_i} \\
&= \lim_{\Delta t_i \to 0} P\left(N(0,t_1] = 0\right) \; P\left(N(t_1, t_1 + \Delta t_1] = 1\right) P\left(N(t_1 + \Delta t_1, t_2] = 0\right) \\
&\quad \times \cdots \times P\left(N(t_n, t_n + \Delta t_n] = 1\right) P\left(N(t_n + \Delta t_n, t] = 0\right) \\
&\quad \Bigg/ \left(P\left(N(0,t] = n\right) \prod_{i=1}^{n} \Delta t_i \right) \\
&= \lim_{\Delta t_i \to 0} \exp\left[-\Lambda(t_1)\right] \left(\exp\left[-\Lambda(t_1 + \Delta t_1) + \Lambda(t_1)\right] \left[\Lambda(t_1 + \Delta t_1) - \Lambda(t_1)\right]^1 \right) \\
&\quad \times \exp\left[-\Lambda(t_2) + \Lambda(t_1 + \Delta t_1)\right] \\
&\quad \times \cdots \times \left(\exp\left[-\Lambda(t_n + \Delta t_n) + \Lambda(t_n)\right] \left[\Lambda(t_n + \Delta t_n) - \Lambda(t_n)\right]^1 \right) \\
&\quad \Bigg/ \left(\frac{\exp\left[-\Lambda(t)\right] \left[\Lambda(t)\right]^n}{n!} \prod_{i=1}^{n} \Delta t_i \right) .
\end{aligned}$$

After combining the terms in the exponents, and collecting the expressions of the form $\Lambda(t_i + \Delta t_i) - \Lambda(t_i)$ we find that the above conditional density is equal to

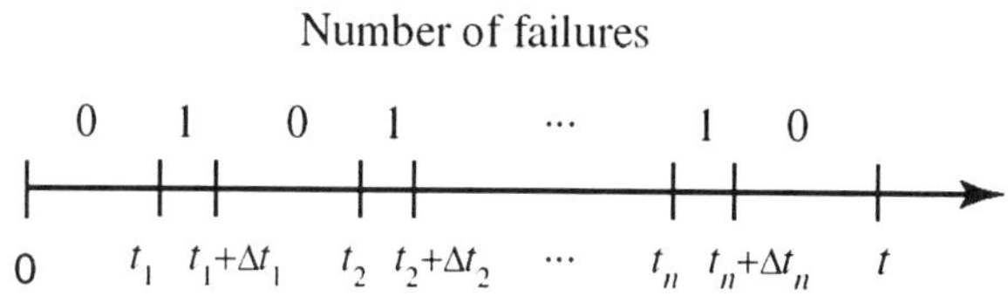

Fig. 2.4 Distribution of failures for proof of Theorem 26.

$$\lim_{\Delta t_i \to 0} \Big(\exp\left[-\Lambda(t_1) - \Lambda(t_1+\Delta t_1) + \Lambda(t_1) - \Lambda(t_2) + \Lambda(t_1+\Delta t_1)\right.$$
$$\left.\cdots - \Lambda(t_n + t_{n+1}) + \Lambda(t_n) - \Lambda(t) + \Lambda(t_n + \Delta t_n)\right]$$
$$\times \frac{\Lambda(t_1+\Delta t_1) - \Lambda(t_1)}{\Delta t_1} \frac{\Lambda(t_2+\Delta t_2) - \Lambda(t_2)}{\Delta t_2} \times \cdots \times \frac{\Lambda(t_n+\Delta t_n) - \Lambda(t_n)}{\Delta t_n}\Big)$$
$$\Big/ \left(\frac{\exp\left[-\Lambda(t)\right]\left[\Lambda(t)\right]^n}{n!}\right)$$

$$= \exp\left[-\Lambda(t)\right] \left(\lim_{\Delta t_1 \to 0} \frac{\Lambda(t_1+\Delta t_1) - \Lambda(t_1)}{\Delta t_1}\right) \left(\lim_{\Delta t_2 \to 0} \frac{\Lambda(t_2+\Delta t_2) - \Lambda(t_2)}{\Delta t_2}\right)$$
$$\times \cdots \times \left(\lim_{\Delta t_n \to 0} \frac{\Lambda(t_n+\Delta t_n) - \Lambda(t_n)}{\Delta t_n}\right) \Big/ \left(\frac{\exp\left[-\Lambda(t)\right]\left[\Lambda(t)\right]^n}{n!}\right).$$

Each of these limits is the derivative of Λ; hence,

$$f(t_1, t_2, \ldots, t_n \mid n) = \frac{n! \prod_{i=1}^{n} \Lambda'(t_i)}{\left[\Lambda(t)\right]^n} = n! \prod_{i=1}^{n} \frac{\lambda(t_i)}{\Lambda(t)}.$$

This last expression is the joint pdf of n order statistics from a sample of size n from the distribution with cdf $G(y)$. ■

2.4 EXERCISES

1. Make a table giving the expected numbers of failures in the intervals $[0, 20]$, $[20, 40]$, $[40, 60]$, and $[60, 80]$ for the nonhomogeneous Poisson processes with intensities:

   (a) $\lambda(t) = \dfrac{3}{16}\sqrt{t+1}$

   (b) $\lambda(t) = \dfrac{91}{80}$

(c) $\lambda(t) = \frac{91}{3200}t$

(d) $\lambda(t) = \frac{91}{8\sqrt{5t}}$.

Which of the systems are improving? Deteriorating?

2. For each system in Problem 1, find the probability of observing 2 or fewer failures in

   (a) $[0, 5]$
   (b) $[75, 80]$

3. Consider a homogeneous Poisson process with intensity $\lambda = \frac{1}{100}$. Find

   (a) $\Lambda(t)$
   (b) $E(N(200))$
   (c) $P(N(200) \leq 5)$
   (d) an approximation to $P(N(0, 50] \leq N(50, 100])$.

4. Repeat parts (a)-(d) of Problem 3 for the nonhomogeneous Poisson process with intensity $\lambda(t) = \frac{2+t}{200}$.

5. Failures of a repairable system are modeled by a Poisson process with $\lambda(t) = \frac{1}{50}$, $0 \leq t \leq 365$, where time is measured in days. The company that owns the machine has three options for a maintenance plan for one year. Which among the following three options will have the smallest expected cost?

   Maintenance Plan #1: \$12,000 with no charge for each repair.

   Maintenance Plan #2: \$4,000 plus \$1000 per repair

   Maintenance Plan #3: \$1,500 per repair.

6. Repeat Problem 5 for the case where the failures are modeled by an NHPP with $\lambda(t) = \frac{1}{50} + \frac{1}{1000}t$.

7. Consider the maintenance plans in Problem 5 for a repairable system modeled by the NHPP with $\lambda = \frac{1}{50} + kt$. What values of k will make Maintenance Plan #1 optimal?

8. Suppose that failures of a repairable system are modeled by a HPP with intensity λ. Find the probability distribution for the number of failures in the interval $[0, s]$ given that n failures occurred in $[0, t]$ if

(a) $s < t$
(b) $s > t$

9. Suppose that failures of a repairable system are modeled by a HPP with intensity λ. Find the probability density functions for the random variables:

   (a) T_1
   (b) T_1/t
   (c) T_k
   (d) T_k/t

10. **Superimposed Homogeneous Poisson Processes** Suppose that the failures of one repairable system follow an HPP with intensity λ_1 and the failures of another repairable system follow an HPP with intensity λ_2. Show that the **superposition** of the two processes (that is, the process for which an "event" occurs whenever either of the two systems fail) is an HPP with intensity $\lambda_1 + \lambda_2$. Generalize to n systems with intensities $\lambda_1, \lambda_2, \ldots, \lambda_n$.

11. Consider two repairable systems whose failures follow separate HPPs with respective intensities λ_1 and λ_2. Suppose it is known that n failures have occurred in the interval $[0, t]$. Let X be the number of times that Machine 1 failed in $[0, t]$. Show that X has a binomial distribution with parameters n and $\lambda_1 / (\lambda_1 + \lambda_2)$.

12. Suppose that a repair person is in charge of ten identical and independent systems, each having a failure intensity of $\frac{1}{24}$ failures per hour. Each worker works 12-hour shifts and repair is instantaneous.

   (a) Find the distribution for the time between the beginning of a workers shift and the first failure.
   (b) Find the distribution for the time between the last failure on the previous shift and the end of that shift.
   (c) What is the distribution for thc time between the last failure on one shift and the first failure on the next? Explain why this distribution is not exponential even though the superimposed process is an HPP.

13. Repeat Problem 8 for an NHPP with intensity $\lambda(t)$.

14. Repeat Problem 9 for an NHPP with intensity $\lambda(t)$.

15. **Another Derivation of the Homogeneous Poisson Process** Partition the interval $[0, a] : 0 = t_0 < t_1 < \cdots < t_n = a$. Let X_i be the number of events in the interval $(t_{i-1}, t_i]$. Suppose that the X's are independent with $P(X_i = 1) = \lambda \Delta t = \lambda a/n$ and $P(X_i = 0) = 1 - P(X_i = 1)$. Let $n \to \infty$ and derive the distribution for the number of events in $[0, a]$.

16. A repairable system is observed over the interval $[0, 200]$. Failures occurred at times $t = 2, 5,$ and 7. Do you believe that the HPP is a reasonable model for the failure process? Explain your reasoning.

17. Consider the NHPP with intensity function $\lambda_1(t) = \dfrac{\beta}{\theta}\left(\dfrac{t}{\theta}\right)^{\beta-1}$ where $0 < \beta < 1,$.

    (a) Find $\Lambda_1(t)$ and $\Lambda_2(t)$.
    (b) Find $E(N_1(t))$ and $E(N_2(t))$.
    (c) Find $E(N_1(\infty))$ and $E(N_2(\infty))$.

18. Repeat all three parts of Problem 17 for the NHPP with intensity function $\lambda_2(t) = \exp(\gamma + \delta t)$, where $\delta < 0$.

19. A company sets aside \$1200 for repairs to its copy machine. Each repair costs \$200 and the failure process is an HPP with $\lambda = \dfrac{1}{20}$ failures per week. Find the pdf for the time until the repair budget is spent.

20. Consider an HPP with $\lambda = \dfrac{1}{5}$. Let T_n be the time of the nth failure. Find the following:

    (a) $P(T_1 < 8)$
    (b) $P(T_2 < 16)$
    (c) $P(T_3 < 24)$
    (d) $P(T_4 < 32)$
    (e) $P(T_{10} < 58)$
    (f) $P(T_{100} < 508)$
    (g) $P(T_{1000} < 5008)$

21. Derive the mean and variance of the generalized gamma distribution.

22. Consider an NHPP with intensity function $\lambda(t) = 0.2t^{-1/2}$. Let T_n be the time of the nth failure. Find the following:

    (a) the pdf of T_1 and $E(T_1)$
    (b) the pdf of T_2 and $E(T_2)$
    (c) the pdf of T_3 and $E(T_3)$
    (d) the pdf of T_n and $E(T_n)$.

23. Repeat Problem 22 for the NHPP with intensity function $\lambda(t) = 0.1t^{1/2}$.

minimal repair
- bringing a failed unit back to the same condition it was in just before the failure. leads to ⟹ NHPP model

perfect repair
- repairing a unit to a condition that is good as new.

3

Probabilistic Models: Renewal and Other Processes

In Chapter 1 we distinguished between minimal repair and perfect (or renewal) repair. To summarize, minimal repair means bringing a failed unit back to the same condition it was in just before the failure, and perfect repair means repairing a unit to a condition that is good as new. The minimal repair model led to the NHPP model, which we studied in Chapter 2. In this chapter we will study the renewal process and some of its generalizations.

3.1 RENEWAL PROCESS

When a perfect repair is effected after each failure, the times between failure must be identically distributed. If we add the assumption of independence we have a renewal process.

Definition 31 Renewal Process *If the times between failure $X_1, X_2, \ldots$ are iid, then the failure process is said to be a* renewal process.

We denote the failure times by

$$\begin{aligned} T_1 &= X_1 \\ T_2 &= X_1 + X_2 \\ T_3 &= X_1 + X_2 + X_3 \\ &\vdots \end{aligned}$$

The complete intensity function, $\lambda(t) = \lim_{\Delta t \to 0} P(N(t, t+\Delta t] \geq 1 \,|\, \mathcal{H}_t)/\Delta t$, depends on the history $\mathcal{H}_t$ of the process only through x, the time since the most recent failure. Figure 3.1 illustrates the complete intensity function, although care must be taken to interpret it correctly. The complete intensity function shown in Figure 3.1 is conditioned on failures occurring at times $t = 2$, 4.5, 8.5, 10.5, 13, and 15. Notice that after each failure, the intensity returns to what it was at time 0, and moreover, its behavior mimics that of the intensity function near $t = 0$; this is true for the first failure as well as the sixth. Thus a renewal process cannot be used to model a system that is deteriorating, because each failure returns the system to a like-new condition.

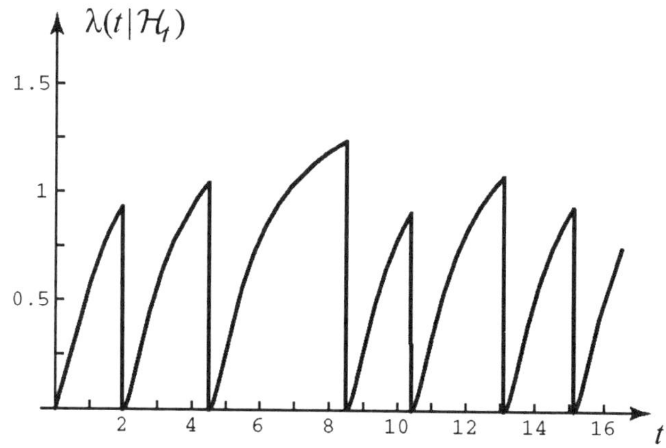

Fig. 3.1 Complete intensity function, conditioned on failures at times $t = 2$, 4.5, 8.5, 10.5, 13, and 15.

The discrete random variable $N(t)$ is defined as the number of failures in the interval $[0, t]$. It satisfies

$$P\left(N(t) \geq k\right) = P\left(T_k \leq t\right) = P\left(\sum_{i=1}^{k} X_k \leq t\right). \tag{3.1}$$

Thus to find the distribution of $N(t)$, we need to be able to find the distribution of a sum of independent random variables. The concept of convolution is useful in determining the distribution of such a sum.

Definition 32 Convolution *Let f and g be pdf's satisfying $f(x) = g(x) = 0$ for $x \leq 0$. The* convolution *of f and g is defined to be*

$$(f * g)(x) = \int_0^x f(y)\, g(x-y)\, dy.$$

The following lemma shows how that the pdf of the sum of continuous (positive) independent random variables is the convolution.

Lemma 27 *If X and Y are independent continuous random variables with $P(X \leq 0) = P(Y \leq 0) = 0$, then the pdf of the sum $S = X + Y$ is given by*

$$f_S(s) = (f_X * f_Y)(s).$$

Proof. Let $R = X$ and $S = X + Y$. Then $X = R$ and $Y = S - R$. The Jacobian for this transformation is

$$J = \begin{vmatrix} 1 & 0 \\ -1 & 1 \end{vmatrix} = 1.$$

Thus the joint pdf of (R, S) is

$$f_{R,S}(r, s) = f_X(r)\, f_Y(s - r)\, |1|\,.$$

The marginal pdf of S is then obtained by integrating over r:

$$f_S(s) = \int_0^s f_X(r)\, f_Y(s - r)\, dr = (f_X * f_Y)(s).$$

This completes the proof. ∎

For a few special cases, the result of Lemma 27 can be used to find the distribution of T_n, the time of the nth failure. For example, if the times between failure are iid GAM$(2, \theta)$, then the distribution of T_n is GAM$(2n, \theta)$; you are asked to prove this in Exercise 3.

From equation (3.1) we see that the random variable $N(t)$ satisfies

$$P(N(t) \geq k) = P\left(\sum_{i=1}^{k} X_k \leq t\right) = F^{(k)}(t),$$

where $F^{(k)}(t)$ is the cdf of the k-fold convolution of the density f. The mean function for a renewal process is thus

$$\begin{aligned}\Lambda(t) = E(N(t)) &= \sum_{k=0}^{\infty} k\, P(N(t) = k) \\ &= \sum_{k=1}^{\infty} k\left[P(N(t) \geq k) - P(N(t) \geq k + 1)\right] \\ &= \sum_{k=1}^{\infty} P(N(t) \geq k) \\ &= \sum_{k=1}^{\infty} F^{(k)}(t)\,.\end{aligned}$$

The ROCOF for a renewal process is therefore

$$\mu(t) = \frac{d}{dt}\Lambda(t) = \frac{d}{dt}\sum_{k=1}^{\infty} F^{(k)}(t) = \sum_{k=1}^{\infty} f^{(k)}(t).$$

Only in some special cases can the density of the sum be written as a simple expression. The next example shows one such case.

Example 3.1 Suppose that the times between failure are iid GAM$(3, \theta)$. In exercise 3 you are asked to show that T_n, the time of the nth failure, has a GAM$(3n, \theta)$ distribution. The ROCOF is therefore

$$\begin{aligned}\mu(t) &= \sum_{k=1}^{\infty} f^{(k)}(t) \\ &= \sum_{k=1}^{\infty} \frac{t^{3k-1}}{\theta^{3k}\Gamma(3k)} e^{-t/\theta}.\end{aligned}$$

Figure 3.2 shows the pdf for the first five failure times assuming $\theta = 1$. For the case of the GAM$(3, 1)$ distribution, the mean and standard deviation of the ith failure time are $3\theta i = 3i$ and $\sqrt{3i}\theta = \sqrt{3}\,i$. The ROCOF $\mu(t)$ is shown in Figure 3.3. Notice that the ROCOF converges very quickly to the constant $1/\eta = 1/3$. □

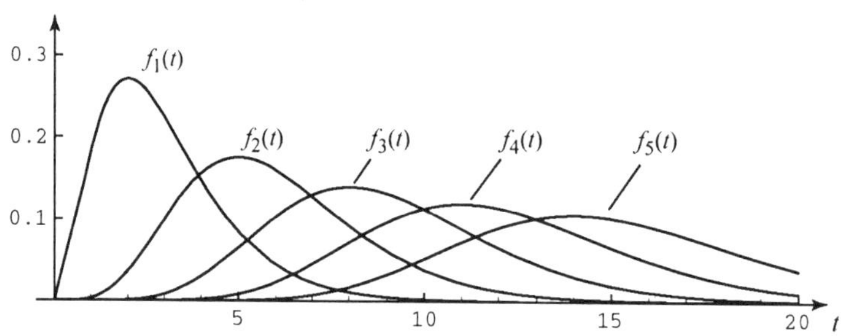

Fig. 3.2 Probability density functions for first five failures from a GAM$(3, 1)$ renewal process.

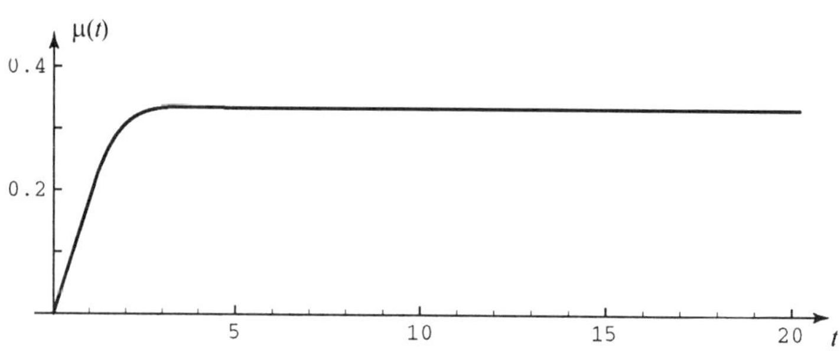

Fig. 3.3 ROCOF function for a GAM$(3, 1)$ renewal process.

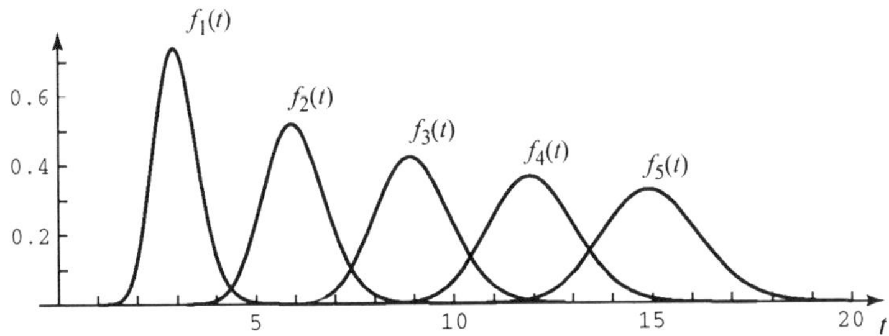

Fig. 3.4 Probability density functions for first five failures from a GAM(30, 0.1) renewal process.

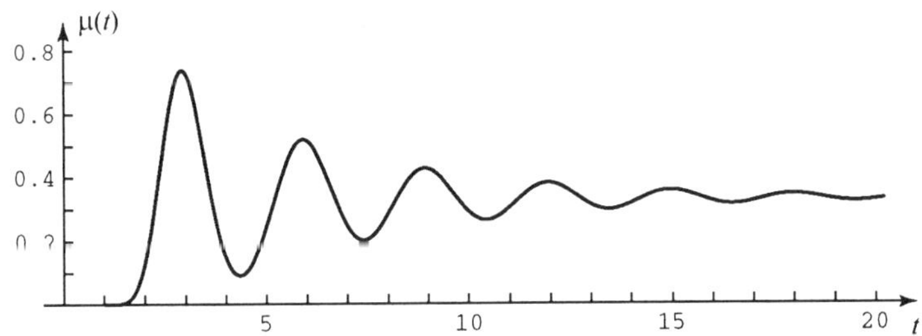

Fig. 3.5 ROCOF function for a GAM(30, 0.1) renewal process.

Figures 3.4 and 3.5 show the situation where the coefficient of variation σ/η for the distribution of times between failures is much smaller. In this case, $\kappa = 30$ and $\theta = 0.1$; this leads to the same mean time between failures, namely, $\eta = 3$, but the standard deviation is smaller. Figures 3.3 and 3.5 show the ROCOF for the two situations, GAM(3,1) and GAM(30,0.1). These figures suggest that the ROCOF, also called the renewal density when the process is a renewal process, converges to the value $1/\eta$. This is confirmed in the next theorem.

Theorem 28 *For a renewal process $X_1, X_2, \ldots,$ with $\eta = E(X_i)$ and $\sigma^2 = V(X_i)$,*

(1) $\displaystyle\lim_{t\to\infty} \frac{\Lambda(t)}{t} = \frac{1}{\eta}$

(2) $\displaystyle\lim_{t\to\infty} \mu(t) = \frac{1}{\eta}.$

Proof. The proof uses Laplace transforms. If f is a function defined on $[0,\infty)$, the **Laplace transform** of f is defined to be

$$\mathcal{L}\left\{f(t)\right\}(s) = f^*(s) = \int_0^\infty e^{-st} f(t)\,dt.$$

Exercise 2 asks you to prove the following properties of the Laplace transform:

1. $\mathcal{L}$ is a linear operator.
2. The Laplace transform of a convolution is the product of the Laplace transforms.
3. If f is a pdf and F is the corresponding cdf, and if $f^*(s) = \mathcal{L}\left\{f(t)\right\}(s)$, then $\mathcal{L}\left\{F(t)\right\}(s) = f^*(s)/s$.
4. $f^*(s) = 1 - s\eta + s^2\left(\mu^2+\sigma^2\right)/2 + o(s^2)$.

Using these facts, we can write

$$\begin{aligned}
\Lambda^*(s) &= \mathcal{L}\left\{\Lambda(t)\right\}(s) \\
&= \mathcal{L}\left\{\sum_{k=1}^{\infty} F^{(k)}(t)\right\}(s) \\
&= \sum_{k=1}^{\infty} \mathcal{L}\left\{F^{(k)}(t)\right\}(s) \\
&= \sum_{k=1}^{\infty} \frac{\mathcal{L}\left\{f^{(k)}(t)\right\}(s)}{s} \\
&= \frac{1}{s}\sum_{k=1}^{\infty} \left[f^*(s)\right]^k \\
&= \frac{1}{s}\frac{f^*(s)}{1-f^*(s)}.
\end{aligned} \tag{3.2}$$

Now, using the fourth result from above, we have

$$\Lambda^*(s) = \frac{1}{s}\frac{1 - s\eta + s^2(\mu^2+\sigma^2) + o(s^2)}{s\eta - s^2(\mu^2+\sigma^2) + o(s^2)}.$$

An exercise in long division (see Figure 3.6) yields

$$\begin{aligned}
\Lambda^*(s) &= \frac{1}{s}\left[\frac{1}{s\eta} + \frac{\sigma^2-\eta^2}{2\eta^2} + s\frac{\left(\sigma^2+\eta^2\right)^2}{4\eta^3} + o(s^2)\right] \\
&= \frac{1}{s^2\eta} + \frac{\sigma^2-\eta^2}{s\left(2\eta^2\right)} + o(s^{-1}).
\end{aligned}$$

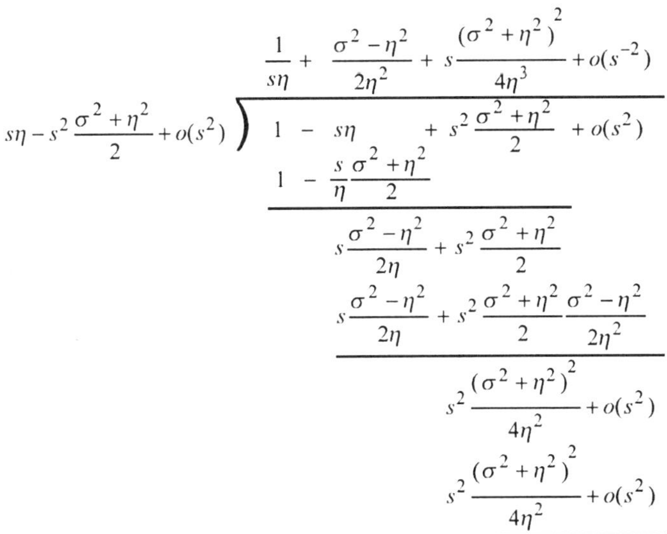

Fig. 3.6 Long division showing quotient.

Behavior of $f(t)$ for large t depends on the behavior of $f^*(s)$ for small s. Thus, we are led to study $f^*(s)$ for small values of s. If we formally invert $\Lambda^*(s)$ we obtain

$$\begin{aligned}\Lambda(t) &= \mathcal{L}^{-1}\left\{\Lambda^*(s)\right\}(t)\\ &= \frac{1}{\eta}\mathcal{L}^{-1}\left\{s^{-2}\right\}(t) + \frac{\sigma^2-\eta^2}{2\eta^2}\mathcal{L}^{-1}\left\{s^{-1}\right\}(t) + \mathcal{L}^{-1}\left\{o\left(s^{-1}\right)\right\}\\ &= \frac{t}{\eta} + \frac{\sigma^2-\eta^2}{2\eta^2} + o(1).\end{aligned}$$

A rigorous proof of this last result is beyond the scope of this book; see Cox and Miller (1965) for an explanation of what is required for the proof. Assuming this result, we conclude that

$$\begin{aligned}\lim_{t\to\infty}\frac{\Lambda(t)}{t} &= \lim_{t\to\infty}\left[\frac{\dfrac{t}{\eta} + \dfrac{\sigma^2-\eta^2}{2\eta^2} + o(1)}{t}\right]\\ &= \lim_{t\to\infty}\left[\frac{1}{\eta} + \frac{\sigma^2-\eta^2}{2\eta^2 t} + \frac{o(1)}{t}\right] = \frac{1}{\eta},\end{aligned}$$

proving the first part of the theorem.

To prove the second part, notice that

$$\begin{aligned}\mu^*(s) &= \mathcal{L}\{\mu(t)\}(s) \\ &= \mathcal{L}\left\{\frac{d}{dt}\Lambda(t)\right\}(s) \\ &= s\mathcal{L}\{\Lambda(t)\}(s) \\ &= \frac{f^*(s)}{1-f^*(s)} \qquad (3.3) \\ &= \frac{1}{s\eta} + \frac{\sigma^2-\eta^2}{2\eta^2} + s\frac{\left(\sigma^2+\eta^2\right)^2}{4\eta^3} + o(s^{-2}).\end{aligned}$$

Inverting yields

$$\mu(t) = \frac{1}{\eta} + o(1).$$

Thus

$$\lim_{t\to\infty} \mu(t) = \frac{1}{\eta},$$

proving the second part of the theorem. ■

The Laplace transforms in (3.2) and (3.3) can be inverted to give interesting relationships between $\Lambda(t)$ and $F(t)$ and between $\mu(t)$ and $f(t)$. Equation (3.2) can be written as

$$(1 - f^*(s))\,\Lambda^*(s) = \frac{f^*(s)}{s}$$

or

$$\Lambda^*(s) = \frac{f^*(s)}{s} + \Lambda^*(s) f^*(s).$$

After inverting, we find that

$$\Lambda(t) = F(t) + \int_0^t \Lambda(t-y)\, f(t)\, dy. \qquad (3.4)$$

This is an integral equation involving the unknown function $\Lambda(t)$. An integral equation of this type is a special case of a class of equations called **Volterra equations**. Similarly, the equation in (3.2) can be written as

$$\mu^*(s) = \frac{f^*(s)}{1-f^*(s)}$$

or

$$\mu^*(s) = f^*(s) + \mu^*(s) f^*(s),$$

which, upon inversion, yields the **renewal equation**

$$\mu(t) = f(t) + \int_0^t \mu(t-y)\, f(y)\, dy. \tag{3.5}$$

The integral equations in (3.4) and (3.5) can also be derived through a probabilistic argument, avoiding the introduction of Laplace transforms. See Exercise 4.

By a direct application of the Central Limit Theorem, we know that the random variable

$$\frac{T_n - n\eta}{\sigma\sqrt{n}}$$

is asymptotically standard normal. This result can be used to obtain an asymptotic result for the random variable $N(t)$ for large t.

Theorem 29 *For a renewal process where times between events have mean η and variance σ^2,*

$$\lim_{t\to\infty} P\left(N(t) < a(t)\right) = \Phi(y),$$

where $a(t) = t/\eta + y\sigma\sqrt{t/\eta^3}$.

Proof. The proof uses the asymptotic result for T_n:

$$\begin{aligned}
P\left(N(t) < a(t)\right) &= P\left(T_{a(t)} > t\right) \\
&= P\left(\frac{T_{a(t)} - \eta a(t)}{\sigma\sqrt{a(t)}} > \frac{t - \eta a(t)}{\sigma\sqrt{a(t)}}\right) \\
&= P\left(\frac{T_{a(t)} - \eta a(t)}{\sigma\sqrt{a(t)}} > \frac{t - \eta\left[t/\eta + y\sigma\sqrt{t/\eta^3}\right]}{\sigma\sqrt{t/\eta + y\sigma\sqrt{t/\eta^3}}}\right) \\
&= P\left(\frac{T_{a(t)} - \eta a(t)}{\sigma\sqrt{a(t)}} > \frac{-\eta y\sigma\sqrt{t/\eta^3}}{\sigma\sqrt{t/\eta + y\sigma\sqrt{t/\eta^3}}}\right) \\
&= P\left(\frac{T_{a(t)} - \eta a(t)}{\sigma\sqrt{a(t)}} > -\frac{y}{\sqrt{1 + y\sigma/\sqrt{\eta t}}}\right) \\
&= P\left(\sqrt{1 + y\sigma/\sqrt{\eta t}}\,\frac{T_{a(t)} - \eta a(t)}{\sigma\sqrt{a(t)}} > -y\right).
\end{aligned}$$

Now let $t \to \infty$ in such a way that $a(t)$ is an integer. The constant $\sqrt{1 + y\sigma/\sqrt{\eta t}}$ converges to 1, and the random variable $\left(T_{a(t)} - \eta a(t)\right)/\sigma\sqrt{a(t)}$ converges to the standard normal distribution. Thus

$$\lim_{t\to\infty} P\left(N(t) < a(t)\right) = \lim_{t\to\infty} P\left(\sqrt{1+y\sigma/\sqrt{\eta t}}\frac{T_{a(t)} - \eta a(t)}{\sigma\sqrt{a(t)}} > -y\right)$$
$$= \Phi(y).$$

Here, Φ is the cdf of the standard normal distribution. ■

Example 3.2 A copy shop uses ordinary copy machines to make copies for customers. Some days have a heavier demand than others, so the lifetime X of a toner cartridge varies from one to the next. The distribution of the toner life, measured in days, is WEI(6, 16). What is the probability that an order of 25 toner cartridges will last a full year?

The required probability, with a continuity correction, is

$$P(N(365) \le 25) = P(N(365) < 25.5).$$

The expected time to failure and variance of the time to failure are, respectively,

$$\eta = \theta\Gamma(1+1/\beta) = 16\Gamma(7/6) \approx 14.84$$

$$\sigma^2 = \theta^2\left[\Gamma(1+2/\beta) - (\Gamma(1+1/\beta))^2\right] = 16^2[\Gamma(4/3) - (\Gamma(7/6))^2] \approx 8.273.$$

The value of $a(365)$ satisfies

$$25.5 = a(365) = \frac{365}{\eta} + y\sigma\sqrt{\frac{365}{\eta^3}}.$$

Solving for y gives

$$y = \frac{25.5 - 365/\eta}{\sigma\sqrt{365/\eta^3}} \approx \frac{25.5 - 365/14.84}{\sqrt{8.273}\sqrt{365/14.84^3}} \approx 0.947.$$

Applying Theorem 3.2 gives

$$P(N(365) \le 25.5) \approx \Phi(y) \approx \Phi(0.947) \approx 0.83.$$

The probability is therefore approximately 0.83 that the order of 25 cartridges will last a year. □

3.2 THE PIECEWISE EXPONENTIAL MODEL

Independent exponentially distributed times between failure does not imply the Poisson process. What is missing is the assumption that the times between failures have *identical* independent exponential distributions. If we relax the "identical" assumption and allow the times between failure to have different distributions (although possibly in the same family of distributions), we can obtain a useful model for repairable systems. One such model was proposed and described by Sen and Bhattarcharyya (1993) and Sen (1998). Their model, called the piecewise exponential (PEXP) model, assumes that the times between failures $X_1, X_2, \ldots, X_n$ are independent exponential random variables with

$$E(X_i) = \frac{\delta}{\mu} i^{\delta-1}, \qquad i = 1, 2, \ldots,$$

where $\delta > 0$ and $\mu > 0$. When $\delta = 1$, we have the iid exponential case because the expectation of X_i is equal to μ, which is independent of i. When $\delta > 1$, then $E(X_i)$ is an increasing function of i; this corresponds to reliability improvement. When $\delta < 1$, then $E(X_i)$ is a decreasing function of i; this corresponds to reliability deterioration.

The PEXP model differs from the NHPP model in the following important way. The PEXP assumes that the system reliability remains unchanged between failures, and that it takes a jump at the time of a failure. On the other hand, the NHPP assumes that the system reliability changes *continuously* as the system ages, and that it remains unchanged at the time of a failure. In the reliability growth case, where improvements are made each time a prototype system fails, the PEXP model may be more appropriate than the NHPP. This situation is often called "test, analyze, and fix" (TAAF).

3.3 MODULATED PROCESSES

Cox (1972) proposed a model that he called the modulated renewal processes (MRP). Let t denote global time, that is, time since the initial start-up of the system, and let B_t denote the backward recurrence time. The complete intensity function for the MRP is

$$\lambda(t \mid \mathcal{H}_t) = h(B_t) \exp\left[\beta_1 z_1(t) + \cdots + \beta_p z_p(t)\right].$$

Note that if all of the β's are equal to 0, then $\lambda(t \mid \mathcal{H}_t) = h(B_t)$. In this case, the probability of failure in a small interval depends only on the time since the last failure; this means that the process is a renewal process. On the other hand, if h is a constant, then the intensity function depends only on the global time t. When h is constant, Cox calls the model a modulated Poisson process. If in addition to a constant h we assume that the $z_j(t)$ functions are known ahead of time, then the process is an NHPP. Thus Cox's MRP is a generalization of both the renewal process and the NHPP.

Cox suggested a number of possibilities for the functions $z_j(t)$.

1. **Trend** If $h(t) = \lambda$, $p = 1$, and $z_1(t) = t$, then

$$\begin{aligned}\lambda(t \mid \mathcal{H}_t) &= \lambda \exp\left[\beta_1 t\right] \\ &= \exp\left[\beta_0 + \beta_1 t\right].\end{aligned}$$

   This model, an NHPP, can be used to model a trend; if $\beta_1 < 0$, then the failures will tend to get further apart, and if $\beta_1 > 0$, the failures will get closer together.

2. **Periodic Phenomenon** The choice of $z_1(t) = \cos \omega t$ and $z_2(t) = \sin \omega t$ lead to a model that can account for periodic fluctuations in the intensity. Often, the value of ω will be known from the system's environment (e.g., yearly), although we could treat ω as an unknown parameter.

3. **Load Factors** The variables $z_j(t)$ could indicate an environmental load or stress to which the system is exposed. Heavier loads would presumably lead to a higher failure intensity.

The inhomogeneous gamma processes, developed by Berman (1981), supposes that shocks occur according to a NHPP with intensity function $\lambda(t)$ and mean function $\Lambda(t)$, but that failures occur, not at every shock, but at every κth shock, where κ is a positive integer. He showed that the pdf of the first n failure times,

$$f_n(t_1, t_2, \ldots, t_n) = \left\{ \prod_{i=1}^{n} \lambda(t_i) \left[\Lambda(t_i) - \Lambda(t_{i-1})\right]^{\kappa - 1} \right\} \exp\left[-\Lambda(t_n)\right] \, / \, \left[\Gamma(\kappa)\right]^n$$

is a pdf for any positive value of κ, so there is no need to restrict attention to positive integers. If $\kappa = 1$, then a failure occurs at every shock, and the model is an NHPP. On the other hand, if $\lambda(t)$ is a constant, then the process is a gamma renewal process, that is, a renewal process for which the times between failure have a gamma distribution.

Berman then generalized this model to what he called a modulated gamma process. A modulated gamma process is an inhomogeneous gamma process for which the shocks have intensity

$$\lambda(t) = \rho \exp\left[\beta_1 z_1(t) + \cdots + \beta_p z_p(t)\right].$$

If $\kappa = 1$, then the modulated gamma process reduces to Cox's modulated Poisson process.

The modulated power law process (MPLP) is a generalization of the power law process and a generalization of the gamma renewal process. It is a special case of the modulated gamma process proposed by Berman (1981). The MPLP was proposed as a model for repairable systems by Lakey and Rigdon (1992). Statistical inference for the parameters of the MPLP were discussed in Black and Rigdon (1994). The MPLP can be viewed as a compromise between the nonhomogeneous Poisson process, for which a repaired system is "bad-as-old," and the renewal process, for which a repaired system is "good-as-new."

The MPLP can be derived in two ways. Shocks occur according to a power law process, but a failure occurs only at every κth failure. If $\kappa = 1$ then we have the usual power law process, so a repaired unit would be in exactly the same condition after a repair as it was just before the failure. If $\kappa > 1$, say $\kappa = 3$, then a repaired system will in better condition than it was just before the failure, since it must then accumulate three shocks in order to fail. If $\beta = 1$ and $\kappa > 1$ then the shocks occur according to the homogeneous Poisson process. Since the times between shocks would have independent and identical exponential distributions, and since the sum of independent exponentials has a gamma distribution, the process of system failures would follow a gamma renewal process. If we have both $\kappa = 1$ and $\beta = 1$ then the process is a homogeneous Poisson process.

A second way of deriving the MPLP i s as follows. Let $X_1, X_2, \ldots$ be iid random variables that have a gamma distribution with pdf

$$f(x|\theta, \kappa) = \frac{x^{\theta-1}}{\theta^\kappa \Gamma(\kappa)} \exp\{-x/\theta\}, \quad x > 0.$$

Then the times

$$\begin{aligned} Y_1 &= X_1 \\ Y_2 &= X_1 + X_2 \\ &\vdots \\ Y_n &= X_1 + X_2 + \cdots + X_n \end{aligned}$$

represent a gamma renewal process. If $\kappa > 1$ then the gamma distribution has an increasing hazard function, so a repaired unit is in better condition than it was just before the failure. Thus the larger κ is, the more evenly spaced the Y_is will be.

Figure 3.7 shows the complete intensity function for a number of different cases. Care must be taken to interpret correctly the graph in this figure; the graphs are conditioned on failures occurring at times $t = 4.5$, 8.5, 9.8, 10.4, 13.1, 13.6, and 15.1. Notice the behavior of the complete intensity function after a failure. When $\kappa = 6$, the complete intensity stays very low just after a failure, then it increases rapidly. When $\beta = 1.6$ (the larger value of β among the plots shown in Figure 3.7) the system's behavior after the later failures is different from the behavior after the earlier failures; for the later failures, the complete intensity function rises much more rapidly. This means that an older system will deteriorate more rapidly after a failure and repair than a younger system.

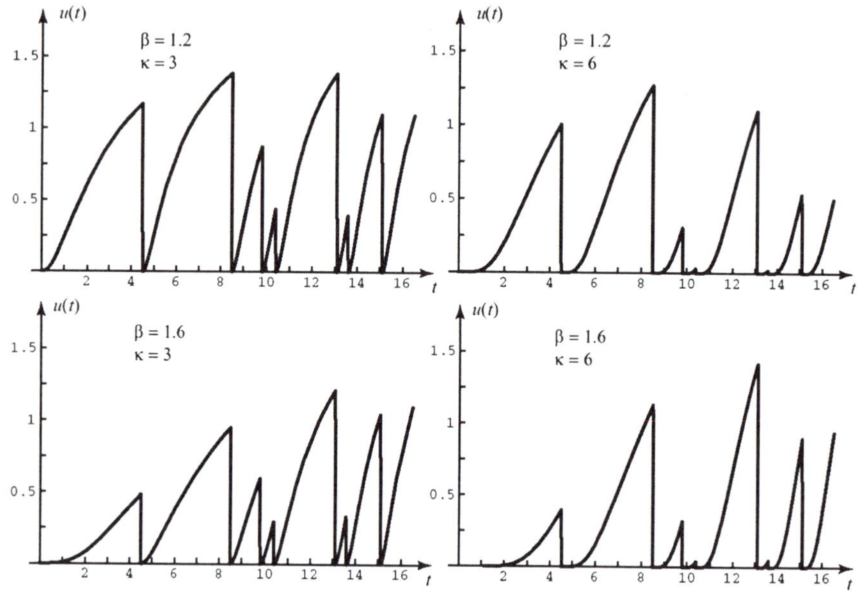

Fig. 3.7 Complete intensity functions for the MPLP for various combinations of β and κ. The complete intensity function is conditioned on failures occurring at times $t = 4.5$, 8.5, 9.8, 10.4, 13.1, 13.6, and 15.1.

3.4 THE BRANCHING POISSON PROCESS

In some cases, failures tend to be bunched together. This may indicate that the system is not correctly repaired, causing a number of subsequent failures for much the same reason as the original failure. A model that can account for this phenomenon is the branching Poisson process (BPP).

The BPP was originally proposed by Bartlett (1963) as a model for traffic flow where a slow moving vehicle may be followed immediately by a number of other vehicles. This is analogous to a single failure causing a number of subsequent failures. Lewis (1964) applied the BPP to failure times of computers. To be more precise, suppose that **primary failures** are generated according to a Poisson process with rate λ. After each failure, there is probability $1 - r$ that the repair will be done correctly; in this case the next failure will occur when the next primary failure occurs. With probability r, the repair is not done correctly; in this case, the primary failure will spawn a finite renewal process of **subsidiary failures**. The number of subsidiary failures that follow from a primary failure is a discrete random variable. Lewis (1964) and Cox and Lewis (1966, Section 7.4) describe the cases where this random variable has a geometric, negative binomial, and Poisson distribution. It is assumed that both kinds of failure (primary and subsidiary) are *indistinguishable*.

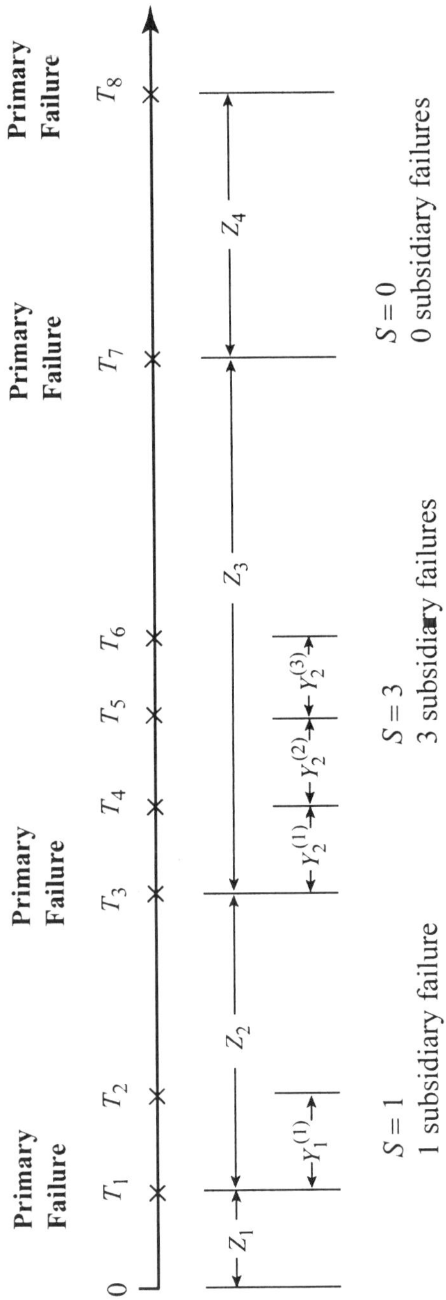

Fig. 3.8 Illustration of a branching Poisson process (BPP).

Define G to be the number of subsidiary failures corresponding to a single primary failure conditioned on there being at least one such subsidiary failure, and let S be the (unconditional) number of subsidiary failures. Note that $S = 0$ when the repair is done correctly. Let $Z_1, Z_2, \ldots$ denote the times between the primary failures, and let $Y_i^{(1)}, Y_i^{(2)}, \ldots, Y_i^{(s)}$ denote the times between the subsidiary failures that are triggered by the ith primary failure. Finally, let $T_1 < T_2 < \cdots$ denote the failure times regardless of the type. In practice, we observe only the T_i's and not the type of failure. The notation is illustrated in Figure 3.8.

The ROCOF function for the BPP can be obtained as follows. Let $H(t)$ denote the expected number of subsidiary failures from a single primary failure in an interval of length t, and let $F_k(t)$ and $f_k(t)$ denote the cdf and pdf, respectively, of the random variable $Z_1 + Z_2 + \cdots + Z_k$ (i.e., the *time* of the kth primary failure).

Given that $Z_1 = z$, the expected number of events in the interval $[0, t]$ from the first subsidiary process is then $H(t - z)$. If we denote by $N^{(1)}(t)$ the number of failures in $[0, t]$ due to the first subsidiary process, then

$$\begin{aligned} E\left[N^{(1)}(t)\right] &= E\left\{E\left[N^{(1)}(t) \mid Z_1\right]\right\} \\ &= \int_0^t E\left[N^{(1)}(t) \mid Z_1 = z\right] f_1(z)\, dz \\ &= \int_0^t H(t-z)\, f_1(z)\, dz. \end{aligned}$$

A similar analysis could be applied to the time of the kth failure. The expected number of failures $N^{(k)}(t)$ in $[0, t]$ due to the kth subsidiary process is

$$E\left[N^{(k)}(t)\right] = \int_0^t H(t-z)\, f_k(z)\, dz.$$

The expected number of failures of any type in $[0, t]$ is thus

$$\begin{aligned} \Lambda(t) &= E\left[N(t)\right] \\ &= E\,(\text{number of primary failures in } [0,t]) \\ &+ \sum_{k=1}^{\infty} E\,(\text{number of subsidiary failures in } [0,t] \text{ from } k\text{th primary failure}) \\ &= \Lambda_Z(t) + \sum_{k=1}^{\infty} \int_0^t H(t-z)\, f_k(z)\, dz \\ &= \Lambda_Z(t) + \int_0^t H(t-z) \left[\sum_{k=1}^{\infty} f_k(z)\right] dz \\ &= \Lambda_Z(t) + \int_0^t H(t-z)\, \Lambda_Z'(z)\, dz\,. \end{aligned} \tag{3.6}$$

The ROCOF for the branching Poisson process is therefore

$$\begin{aligned}\mu(t) &= \Lambda'(t) \\ &= \Lambda'_Z(t) + \frac{d}{dt}\int_0^t H(t-z)\,\Lambda'_Z(z)\,dz\,.\end{aligned}$$

Kreider, Kuller, Ostberg, and Perkins (1965, p. 670) give a method for evaluating the derivative of the above integral. This yields

$$\begin{aligned}\mu(t) &= \mu_Z(t) + \int_0^t \frac{\partial}{\partial t} H(t-z)\,\mu_Z(z)\,dz + H(t-t)\left(\frac{d}{dt}t\right) \\ &= \mu_Z(t) + \int_0^t H(t-z)\,\mu_Z(z)\,dz + 0 \times 1 \\ &= \mu_Z(t) + \int_0^t H(t-z)\,\mu_Z(z)\,dz\,. \qquad (3.7)\end{aligned}$$

Here, we have written $\mu_Z(z)$ for $\Lambda'_Z(z)$.

If the primary process is a HPP with rate λ, then $\Lambda_Z(t) = \lambda t$, and $\mu_Z(t) = \lambda$. Equation (3.6) becomes

$$\Lambda(t) = \lambda t + \int_0^t H(t-z)\,\lambda\,dz - \lambda\left[t + \int_0^t H(t-z)\,dz\right].$$

The ROCOF is then obtained—either from (3.7) or by differentiating (3.6)—and found to be

$$\begin{aligned}\mu(t) &= \lambda + \lambda\int_0^t \frac{\partial}{\partial t} H(t-z)\,dz \\ &= \lambda\left\{1 + \Big[-H(t-z)\Big]_{z=0}^{z=t} + H(0)\right\} \\ &= \lambda\,[1 + H(t)]\,.\end{aligned}$$

Notice that $H(t)$ is equal to the expected value of the number of failures in a finite renewal process (the subsidiary process). Conditioned on $S = s$ (that is, knowing that there are to be exactly s failures in the subsidiary process) the expectation $E(H(t) \mid s)$ should tend to s since eventually, all s failures are bound to occur. Thus

$$\lim_{t\to\infty} E\left[N^{(1)}(t) \mid s\right] = s.$$

Therefore

$$\begin{aligned}H(t) &= E\left[N^{(1)}(t)\right] \\ &= E\left\{E\left[N^{(1)}(t) \mid S\right]\right\} \\ &= \sum_{s=0}^{\infty} E\left[N^{(1)}(t) \mid s\right]\,P(S = s).\end{aligned}$$

Taking the limit as t tends to ∞ gives

$$\begin{aligned}\lim_{t\to\infty} H(t) &= \lim_{t\to\infty} \sum_{s=0}^{\infty} E\left[N^{(1)}(t)\,|\,s\right] P(S=s)\\ &= \sum_{s=0}^{\infty} \lim_{t\to\infty} E\left[N^{(1)}(t)\,|\,s\right] P(S=s)\\ &= \sum_{s=0}^{\infty} s\,P(S=s)\\ &= E(S).\end{aligned}$$

The ROCOF for the BPP therefore satisfies

$$\lim_{t\to\infty} \mu(t) = \lim_{t\to\infty} \lambda\,[1+H(t)] = \lambda\,[1+E(S)]\,. \tag{3.8}$$

Lewis (1964) and Cox and Lewis (1966) study cases where the distribution of $S-1$ is (1) geometric, (2) negative binomial, and (3) Poisson. If $S-1$ is geometric with probability of success p, then $E(S) = (1/p) - 1$, and (3.8) becomes

$$\lim_{t\to\infty} \mu(t) = \lambda\left[1+\frac{1}{p}-1\right] = \frac{\lambda}{p}\,. \tag{3.9}$$

From (3.9) we can see what effect p has on the limiting ROCOF. If $p=1$, so that with probability 1 there are no subsidiary failures, then the limiting ROCOF is λ. This is clear because with no subsidiary failures, the only failures are primary failures, and they occur with rate λ. If p is small, then the limiting ROCOF is large.

The main advantage of the BPP is that it can model systems where failures tend to be clustered. A disadvantage is that parameter estimation and other inferences are difficult to perform. We will discuss parameter estimation briefly in Chapter 4.

Lewis (1967) extended the BPP to allow for the primary failure process to be a NHPP. He presented a number of properties of this nonhomogeneous branching Poisson process, but he did not discuss parameter estimation or other inference for the model.

3.5 IMPERFECT REPAIR MODELS

Brown and Proschan (1983) suggested a model whereby a failed unit is either repaired to a like-new condition or to a bad-as-old condition (i.e., a condition identical to that of a functioning unit of the same age). We call these types of repair perfect and minimal, respectively, and we will call this the Brown and Proschan (BP) imperfect repair model. Specifically, under the BP imperfect repair model, a failed unit is given a perfect repair with probability p and an imperfect repair with probability $q = 1 - p$. As special cases, if $p = 0$, then we have a NHPP and if $p = 1$ we have a renewal process.

Let $F(t)$ denote the cdf of the failure-time distribution after a perfect repair, and let $r(t)$ denote the hazard function corresponding to $F(t)$. The hazard function for the time to perfect repair is then

$$\begin{aligned}
r_{\text{Perfect}}(t) &= \lim_{\Delta t \to 0} \frac{P(t < T \le t + \Delta t \,|\, T > t)}{\Delta t} \\
&= \lim_{\Delta t \to 0} \frac{1}{\Delta t} P\left(\text{a failure occurs in } (t, t+\Delta t] \text{ \& the repair is perfect}\right) \\
&= \lim_{\Delta t \to 0} \frac{1}{\Delta t} P\left(\text{a failure occurs in } (t, t+\Delta t]\right) \\
&\quad \times P\left(\text{repair is perfect} \,|\, \text{failure occurs in } (t, t+\Delta t]\right) \\
&= p\, r(t).
\end{aligned}$$

Here, T represents the time to first perfect repair. Similarly, the survival function for the time to first perfect repair is

$$\begin{aligned}
\overline{F}_{\text{Perfect}}(t) = P(T > t) &= \exp\left[-\int_0^t r_{\text{Perfect}}(x)\, dx\right] \\
&= \exp\left[-p \int_0^t r(x)\, dx\right] \\
&= \left[\overline{F}(t)\right]^p.
\end{aligned}$$

Brown and Proschan gave a number of results regarding the preservation of aging properties (e.g., if F has an increasing hazard function, so does F_{Perfect}) and monotonicity properties (e.g., the expected time to perfect repair is a monotonic function of p). They also showed that the ROCOF $\mu(t)$ for the BP imperfect repair model satisfies

$$\mu(t) = \frac{1}{p} \mu_{\text{Perfect}}(t),$$

where $\mu_{\text{Perfect}}(t)$ is the ROCOF or the renewal function for the process of perfect repairs. It is difficult to apply this result in a specific case, but the asymptotic result follows directly from Theorem 28:

$$\lim_{t \to \infty} \mu(t) = \frac{1}{p} \lim_{t \to \infty} \mu_{\text{Perfect}}(t) = \frac{1}{p\eta},$$

where η is the expected time between perfect repairs.

The Block, Borges, and Savits (1985) model (BBS model) is an extension of the BP imperfect repair model which allows for the probability of perfect repair to be time dependent. They suppose that a failed unit is given a perfect repair with probability $p(t)$, where t is the time since the most recent perfect repair, and an imperfect repair with probability $q(t) = 1 - p(t)$. Let F denote the cdf of the time to first failure. (We will assume that F is differentiable, although Block, Borges, and Savits make slightly weaker assumptions.) They showed that the perfect repairs form a renewal process that has cdf

$$F_{\text{Perfect}}(t) = \exp\left[- \int_0^t p(x)\, \overline{F}^{-1}(x)\, F'(x)\, dx \right].$$

They also derived a number of properties regarding the preservation of aging patterns that are analogous to the results found in Brown and Proschan (1983).

3.6 EXERCISES

1. Suppose that a machine fails every day exactly at noon. Is this machine deteriorating? What is the complete intensity function? What is $N(t)$?

2. Prove the four properties of the Laplace transform mentioned in the proof of Theorem 28.

3. Show that if $X_1, X_2, \ldots X_n$ are iid GAM$(2, \theta)$, then $Y = \sum_{i=1}^{n} X_i \sim$ GAM$(2n, \theta)$. What if the X_i's are GAM$(3, \theta)$? Can you generalize to GAM(κ, θ) where κ is a positive integer?

4. Show that

$$\Lambda(t) = F(t) + \int_0^t \Lambda(t-y)\, f(t)\, dy$$

by conditioning on and integrating over the time of the first failure. Then differentiate both sides to obtain

$$\mu(t) = f(t) + \int_0^t \mu(t-y)\, f(y)\, dy.$$

Hint: Use the fact that

$$\frac{d}{dt} \int_{c(t)}^{d(t)} F(t,y)\, dy = \int_{c(t)}^{d(t)} \frac{\partial}{\partial t} F(t,y)\, dy + F(t, d(t))\, d'(t) - F(t, c(t))\, c'(t).$$

(See Kreider, Kuller, Ostberg, and Perkins (1966, pp. 670 671) for a proof of this result.)

5. Suppose that we begin observing a renewal process at some point in time at which a failure did not occur. The distribution of the time to the first failure is then different from the distribution of times between

all other events. Find the cdf of the time to the first failure in terms of the cdf $F(x)$ of the times between all other failures. (A renewal process like this is called a **modified renewal process**; see Cox and Miller (1965) and Cox and Lewis (1966).)

6. Show that Theorem 28 holds for modified renewal processes.

7. Show that Cox's model with $h(t) = \lambda$, $p = 1$, $\beta_1 < 0$ and $z_1(t) = t$ leads to $E[N(\infty)] < \infty$; in other words, there will be a finite number of failures even in an infinite amount of time.

4

Analyzing Data from a Single Repairable System

In this chapter we deal with the problems of analyzing failure data from a repairable system. We begin by discussing graphical methods and how these can be used to select a model. Two of the most commonly applied models, the homogeneous Poisson process and the power law process, are discussed in later sections. In the Chapter 5 we discuss the problem of simultaneously analyzing data from several, similar systems.

4.1 GRAPHICAL METHODS

Graphical methods for displaying data from repairable systems can be used to gain insight into the data, and also to select a reasonable model. The first question one should ask when analyzing data from repairable systems is whether there is a trend in the times between failure. That is, are the times between failure getting longer? Or are they getting shorter? A simple, but powerful, graphical method is to plot the (global) failure time t_i along the horizontal axis, and the cumulative number of failures through time t_i, $N(t_i)$, on the vertical axis.

Example 4.1 For illustration, consider the failure data in Table 4.1. Suppose that each of the three systems was observed until the time of the twelfth failure. These failure times have been selected to illustrate how the graphical methods can be used to gain insight into the data. Figures 4.1 show graphs of $N(t_i)$ versus t_i. A linear relationship like the one shown in the middle figure indicates that the system remained stable over the time that data were collected. In this case a renewal process, or possibly a homogeneous Poisson process, may be an appropriate model if the times between failure are independent. If the plot shows some curvature, either concave up or concave down (or possibly concave up over some range of t and concave down over another range) then a renewal process is not an adequate model. A plot that is concave down, as in the first part of Figure 4.1, indicates reliability improvement since the times between failures are larger for large t. A concave-up plot, such as the one shown in the third part of Figure 4.1, indicates deterioration. □

Failure Number i	System Number 1	System Number 2	System Number 3
1	3	9	45
2	9	20	76
3	20	65	113
4	25	88	129
5	41	104	152
6	50	107	174
7	69	138	193
8	91	143	199
9	128	149	210
10	151	186	219
11	182	208	224
12	227	227	227

Table 4.1 Failure times for three hypothetical systems. Note that the 12 times between failures are the same for the three systems. Only the order of the times between failures are different.

Systems for which the plot of $N(t_i)$ versus t_i shows a significant amount of curvature must be modeled by a nonstationary process. Such a process must be capable of describing the occurrence of events in time. One such model, the nonhomogeneous Poisson process, is described in Chapter 2. A common functional form for the intensity function is

$$\lambda(t) = \frac{\beta}{\theta}\left(\frac{t}{\theta}\right)^{\beta-1}. \tag{4.1}$$

Because the intensity function is proportional to the global time t raised to a power, this special case of the nonhomogeneous Poisson process is usually called the **power law process**, although it has been called other names as well, most notably the Weibull process. This model is capable of modeling systems that are improving and systems that are deteriorating.

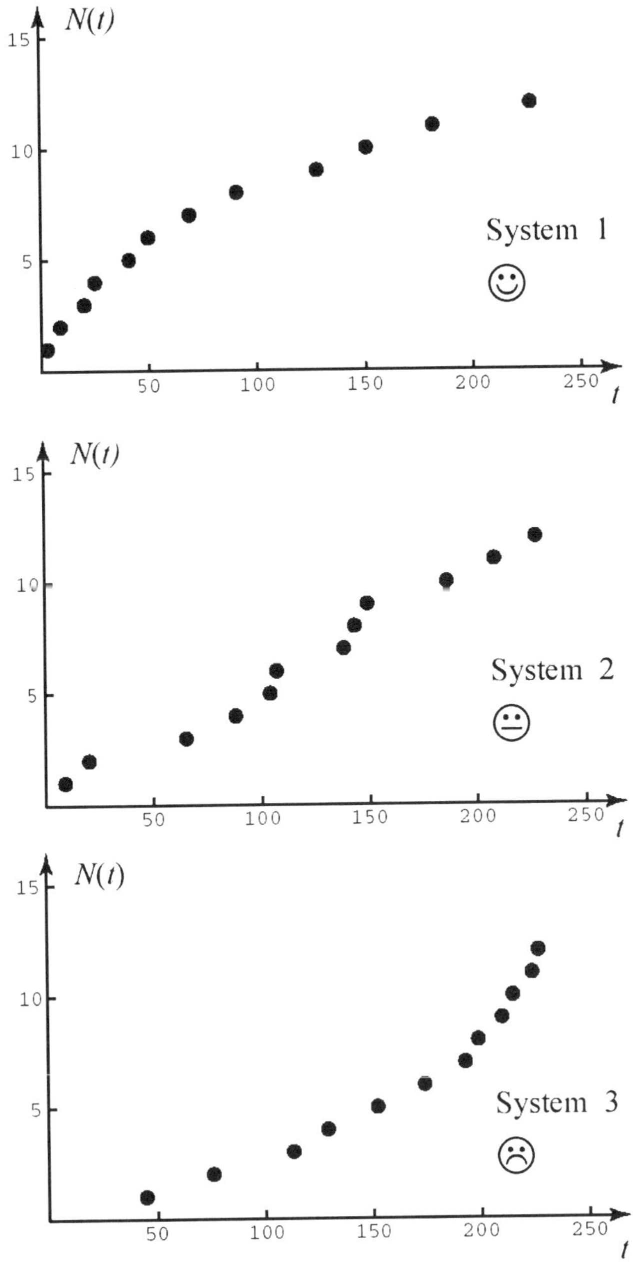

Fig. 4.1 Plots of number of failures $N(t_i) = i$ against cumulative operating time t.

4.1.1 Duane Plots

Later sections in this chapter deal with analyzing data from a power law process and include a discussion of testing for goodness-of-fit. Here we describe a rough graphical procedure that can be used to assess the adequacy of the power law process.

A Duane plot is a scatter plot of $N(t_i)/t_i$ versus t_i, $i = 1, 2, ..., n$ on double logarithmic paper. (Some implementations of the Duane plot put $t_i/N(t_i)$ on the vertical axis and t_i on the horizontal axis.) The quantity $N(t_i)/t_i$ is often called the **cumulative failure rate**.

It is widely believed that Duane plots should be roughly linear if the power law process governs the failure times of the repairable system. This claim is motivated by noting that the expected number of failures through time t is

$$E[N(t)] = \Lambda(t) = \left(\frac{t}{\theta}\right)^{\beta}$$

so

$$E\left[\frac{N(t)}{t}\right] = \frac{1}{t}\Lambda(t) = \frac{t^{\beta-1}}{\theta^{\beta}}.$$

If we take the natural logarithm of both sides, we get

$$\log E\left[\frac{N(t)}{t}\right] = (\beta - 1)\log t - \beta \log \theta,$$

which shows that $\log E[N(t)/t]$ is a linear function of $\log t$. This argument assumes that t is fixed, in which case $N(t)$ is random. On the other hand, the Duane plot is a plot of the observations of the random quantities $\log(N(T_i)/T_i)$ and $\log T_i$. Under these assumptions, the argument to N is not fixed; it is random. Also, $N(t_i)$ is not a random variable because $N(T_i) = i$. (The number of failures through the time of the ith failure must be i.) Thus

$$\frac{N(T_i)}{T_i} = \frac{i}{T_i}.$$

The expected values of the points on the Duane plot are therefore

$$\left(E(\log T_i), E\left(\log \frac{i}{T_i}\right)\right) = \left(E(\log T_i), \log i - E(\log T_i)\right), \quad i = 1, 2, \ldots, n.$$

Because $T_i \sim \text{GGAM}(i, \beta, \theta)$, the expectation of $\log T_i$ is equal to integral

$$E(\log T_i) = \int_0^{\infty} \log t \frac{1}{\Gamma(i)} \frac{\beta}{\theta} \left(\frac{t}{\theta}\right)^{i\beta-1} \exp\left[-(t/\theta)^{\beta}\right] dt.$$

If we make the change of variable $y = (t/\theta)^\beta$ in this integral then we obtain

$$
\begin{aligned}
E(\log T_i) &= \int_0^\infty \log\left(\theta y^{1/\beta}\right) \frac{1}{\Gamma(i)} \frac{\beta}{\theta} \left(\frac{\theta y^{1/\beta}}{\theta}\right)^{i\beta - 1} \exp\left[-y^\beta\right] \left(\frac{\theta}{\beta} y^{1/\beta - 1} dy\right) \\
&= \log\theta + \frac{1}{\beta\Gamma(i)} \int_0^\infty (\log y)\, y^{i-1} e^{-y}\, dy.
\end{aligned}
$$

Mathematica (Wolfram, 1996) shows that the integral in this last expression is

$$\int_0^\infty (\log y)\, y^{i-1} e^{-y}\, dy = \Gamma(i)\, \psi(i),$$

where ψ is the digamma function, which is defined by

$$\psi(z) = \frac{d}{dz} \log \Gamma(z).$$

Hence

$$
\begin{aligned}
E(\log T_i) &= \log\theta + \frac{1}{\beta\Gamma(i)} \Gamma(i)\psi(i) \\
&= \log\theta + \frac{\psi(i)}{\beta}.
\end{aligned}
$$

Thus the expected values of points on a Duane plot are

$$\left(\log\theta + \frac{\psi(i)}{\beta},\ \log i - \log\theta - \frac{\psi(i)}{\beta}\right).$$

Figure 4.2 shows the expected values for the first 20 points on a Duane plot with $\beta = 1.5$ and $\theta = 1$. This figure clearly shows nonlinearity of the points. The points on a Duane plot can even be nonmonotonic, as shown by Figure 4.3.

Duane (1964) suggested fitting a line to the points

$$\left(\log t_i, \log\left(N(t_i)/t_i\right)\right)$$

and using the slope as an estimate for $\beta - 1$ and the intercept as an estimate for $\beta \log \theta$. The efficiency of this estimator of β was studied by Molitor and Rigdon (1993), who found that this least squares estimator was almost as efficient as the MLE of β.

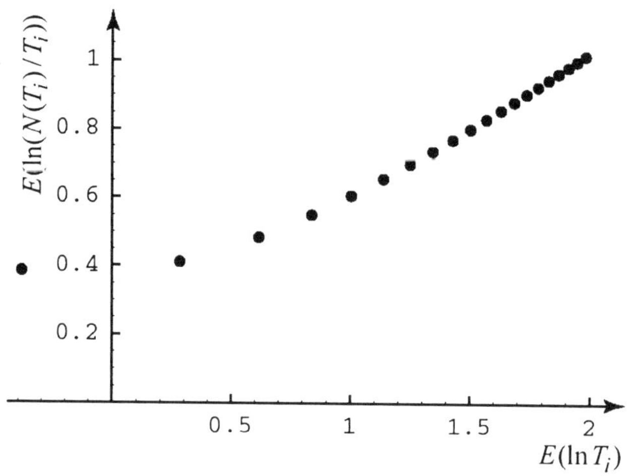

Fig. 4.2 Expected values for points on a Duane plot with $\beta = 1.5$ and $\theta = 1$.

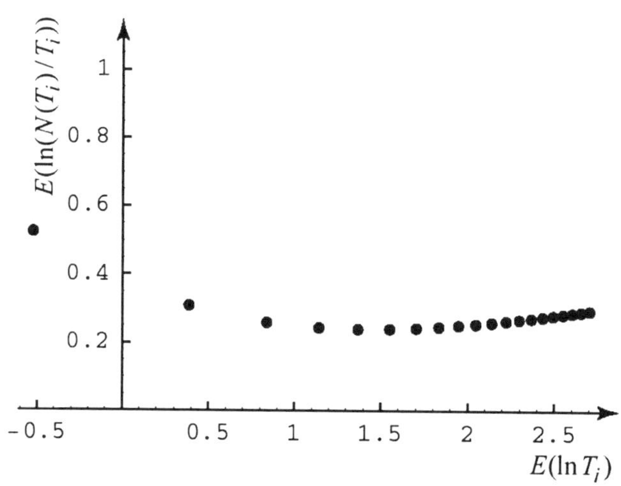

Fig. 4.3 Expected values for points on a Duane plot with $\beta = 1.1$ and $\theta = 1$.

Failure Number	System Number					
	1		2		3	
i	t_i	$N(t_i)/t_i$	t_i	$N(t_i)/t_i$	t_i	$N(t_i)/t_i$
1	3	0.333	9	0.111	45	0.022
2	9	0.222	20	0.100	76	0.026
3	20	0.150	65	0.046	113	0.027
4	25	0.160	88	0.045	129	0.031
5	41	0.122	104	0.048	152	0.033
6	50	0.120	107	0.056	174	0.034
7	69	0.101	138	0.051	193	0.036
8	91	0.088	143	0.056	199	0.040
9	128	0.070	149	0.060	210	0.043
10	151	0.066	186	0.054	219	0.047
11	182	0.060	208	0.053	224	0.049
12	227	0.053	227	0.053	227	0.053

Table 4.2 Calculations for constructing Duane plot for data in Table 4.1.

Example 4.2 Table 4.2 shows, for the three systems described above, the values of $N(t_i)/t_i$ that are needed to construct a Duane plot. Figure 4.4 shows the completed Duane plots. The downward slope of the points in the first part of Figure 4.4 indicates that there is reliability improvement for the first system. The third system shows deterioration since the points increase. For the second system, the points lie roughly along a horizontal line if we ignore the first two points. This indicates that the system remained stable over the time that the failures were observed. That is, the system experienced neither improvement nor deterioration. □

Example 4.3 Table 4.3 gives data from fatigue failure tests that were done by Butler and Rees (1974). (These data are reported in Bain and Engelhardt (1991).) To study the suitability of steel and titanium for use in aircraft manufacturing, they constructed steel and titanium specimens and subjected these specimens to stresses. Each time a crack appeared, the time was recorded and the crack was repaired. A plot of $N(t_i) = i$ against cumulative operating time t_i is shown in Figure 4.5.

Several features can be observed from these data. First, the time between the cracks in the titanium specimen are much longer than the times between cracks in the steel specimen. The Duane plot in Figure 4.6 shows an increase in $N(t)/t$ for both systems, although the increase is much steeper for the steel specimen. The points for the titanium specimen are below the points for the steel specimen, indicating the superiority of titanium.

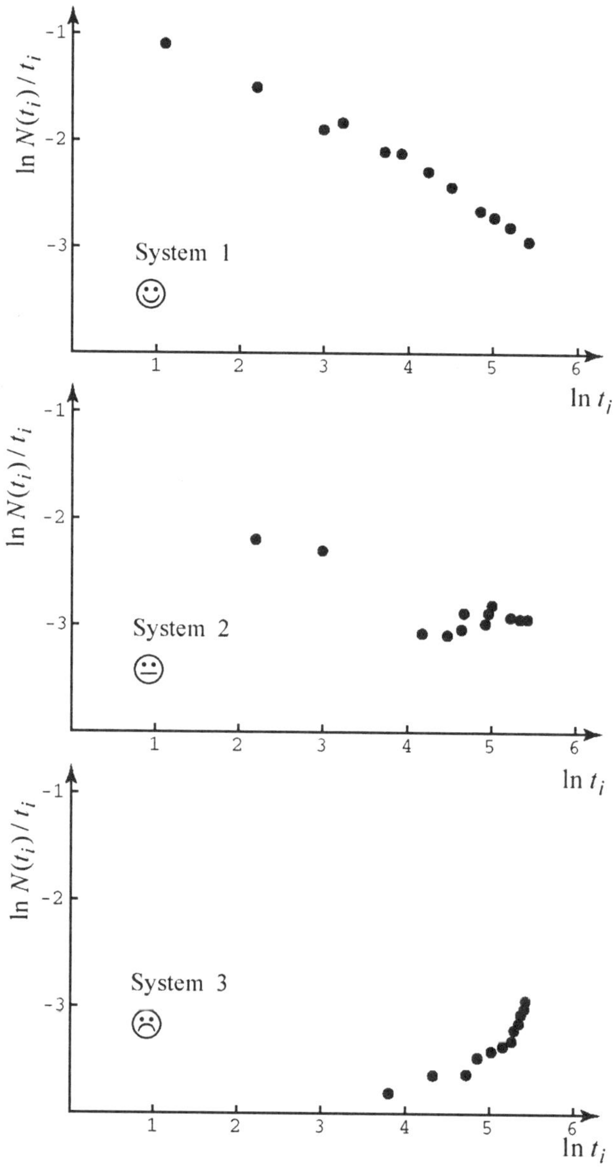

Fig. 4.4 Duane plots for Systems 1–3 of Table 4.1.

Titanium		Steel	
Failure Number i	Failure Time t_i	Failure Number i	Failure Time t_i
1	2356	1	2452
2	3498	2	3750
3	5038	3	3788
4	9501	4	3788
5	11500	5	4000
6	12000	6	4000
7	12000	7	4350
8	12000	8	4350
9	12000	9	4750
10	13000	10	5000
11	13000	11	5750
12	13500	12	5750
13	14000	13	5750
14	17000	14	6000
15	17500	15	6000
16	17500	16	6000
17	18500	17	6075
18	20505	18	6400
19	20505	19	6798
		20	7400
		21	7500

Table 4.3 Failure times for titanium and steel specimens.

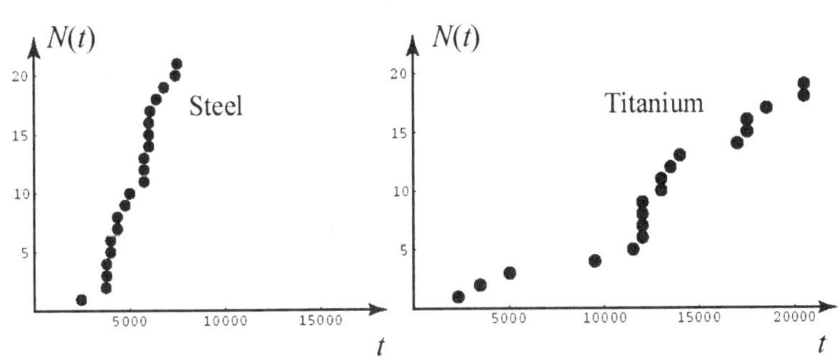

Fig. 4.5 Plot of number of failures $N(t_i) = i$ against cumulative operating time t for the Butler and Rees data.

Another aspect of this set of data is the presence of several simultaneous failures. Although one of the assumptions of the Poisson process is that

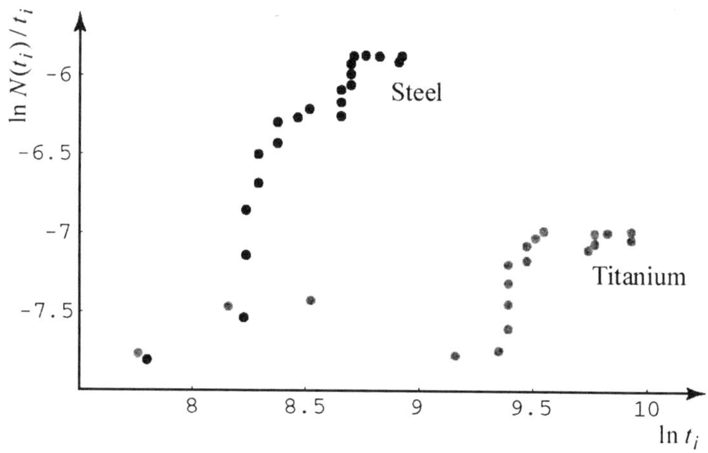

Fig. 4.6 Overlaid Duane plots for the fatigue crack failure data with steel and titanium materials.

simultaneous failures cannot occur, there are situations in practice in which simultaneous failures do occur. Notice also that the first few failure times were recorded rather precisely and the latter failure times were recorded as round numbers. It may have been the case that after the experiment was running for a while, the specimens were observed at discrete points in time. This could account for the simultaneous failures, since there would have been several cracks that occurred between the times the specimen was observed. □

4.1.2 Total Time on Test Plots

Total time on test (TTT) plots were developed by Barlow and Campo (1975) to aid in selecting a model for the lifetime distribution of nonrepairable systems. We describe this application first, and then show how this method can be applied to repairable systems.

Suppose that n identical *nonrepairable* units are simultaneously put on test and that failed units are removed from testing and not replaced. The order statistics $t_{(1)} < t_{(2)} < \cdots < t_{(n)}$ represent the failure times in the order that they were observed. The total time on test (TTT) statistic is defined as follows. First, let $x_i = t_{(i)} - t_{(i-1)}$ be the times between the observed failures. Then define

$$S_0 = 0$$

$$S_j = nx_1 + (n-1)x_2 + \cdots + +(n-j+1)x_j, \quad j \geq 1.$$

The statistic S_j represents the total time that all units have been on test at the time of the jth failure. The scaled TTT statistic is then defined by

$$u_j = \frac{S_j}{S_n}.$$

The TTT plot is then a plot of the ordered pairs $(j/n\,,\,u_j)$ for $j = 0, 1, \ldots, n$, with lines connecting the points. Langberg, Leone, and Proschan (1980) showed that with probability 1 the TTT plot converges to the function

$$\phi(u) = \frac{1}{\mu}\int_0^{F^{-1}(u)} [1 - F(t)]\, dt,$$

where F is the cdf of the distribution from which the $t_{(i)}$'s were observed, and μ is the mean of this distribution. If the distribution is exponential with mean θ, then $1 - F(t) = \exp(-t/\theta)$ and $F^{-1}(u) = -\theta\log(1-u)$. This gives

$$\begin{aligned}\phi(u) &= \frac{1}{\theta}\int_0^{-\theta\log(1-u)} \exp(-t/\theta)\, dt \\ &= \frac{1}{\theta}\left[-\theta\exp(-t/\theta)\right]_0^{-\theta\log(1-u)} \\ &= u\,, \qquad 0 \le u \le 1.\end{aligned}$$

Thus if the distribution of component lifetimes is exponential, then the TTT plot should converge to the diagonal of the unit square. The TTT plot can therefore serve as a graphical goodness-of-fit test for the exponential distribution.

Klefsjö and Kumar (1992) suggested how TTT plots can be applied to data from repairable systems. The application of TTT plots to repairable systems comes about by making note of the fact that if the power law process with parameters β and θ governs the failure times $T_1 < T_2 < \cdots < T_n$ of a repairable system, then

$$w_j = -\log\left(\frac{T_{n-j}}{T_n}\right), \qquad j = 1, 2, \ldots, n-1$$

are distributed as order statistics from the exponential distribution with mean $1/\beta$ (see Møller, 1976). Thus the TTT plot using $w_1, w_2, \ldots, w_{n-1}$ in place of $t_{(1)}, t_{(2)}, \ldots, t_{(n)}$ provides a graphical goodness-of-fit test for the power law process. (Note also that n is replaced by $n-1$ because w_n is undefined; we lose one observation by making the transformation from T_i to w_i.) If the TTT plot lies close to the diagonal line in the unit square, then the power law process may be an acceptable model. On the other hand, large deviations from the diagonal indicate that the power law process is inadequate.

Example 4.4 Consider the data from System 1 in Table 4.1. The data and the appropriate transformations are shown in Table 4.4.

i	$t_{(i)}$	w_i	S_i	u_i	$i/(n-1)$
0	0	0	0	0	0
1	3	0.221	2.430	0.034	0.091
2	9	0.408	6.507	0.091	0.182
3	20	0.573	11.663	0.162	0.273
4	25	0.914	18.98	0.264	0.364
5	41	1.191	27.31	0.380	0.455
6	50	1.513	36.39	0.507	0.545
7	69	1.711	44.95	0.626	0.636
8	91	2.206	53.77	0.748	0.727
9	128	2.429	61.06	0.850	0.818
10	151	3.228	67.51	0.940	0.909
11	182	4.326	71.84	1.000	1.000
12	227				

Table 4.4 Computations for TTT plot for System 1 in Table 4.1.

Notice that we "lose" one observation by making the transformation from the $t_{(i)}$'s to the w_i's; this leaves just 11 nonzero w_i's. The TTT plot, shown in Figure 4.7, lies close to the diagonal of the unit square, indicating that the power law process may be an adequate model for System 1. □

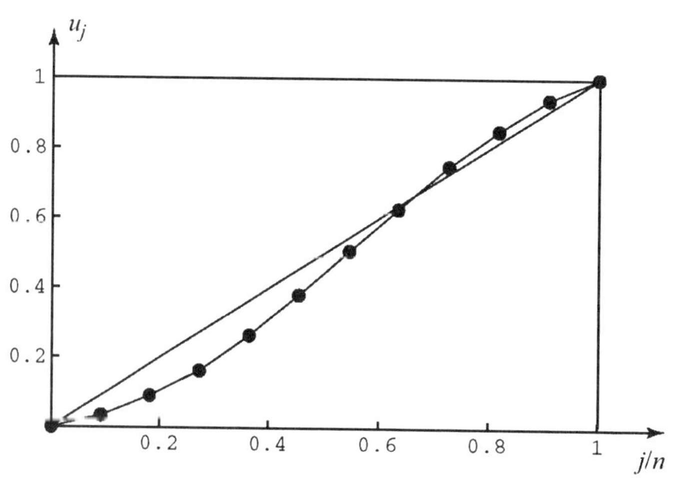

Fig. 4.7 TTT plot for System 1 in Table 4.1.

Klefsjö and Kumar (1992) suggested a way to create a goodness-of-fit test for the power law process using the TTT plot. This test will be described later in this chapter.

4.2 NONPARAMETRIC METHODS FOR ESTIMATING λ

Large data sets will often defy any simple model, such as the HPP, the NHPP, or the renewal process. For such cases a nonparametric estimate of the ROCOF may be appropriate.

4.2.1 Natural Estimates of the Intensity Function

Suppose that a repairable system has been observed over the interval $[0, a]$ where a could be a predetermined time (time truncated case) or the last failure time (failure truncated case). A simple nonparametric estimate of the ROCOF can be obtained as follows. Partition the interval $[0, a]$ into m subintervals: $[0, a_1)$, $[a_1, a_2)$, $\ldots$, $[a_{n-1}, a_n]$, where $0 < a_1 < a_2 < \cdots < a_{n-1} < a_n = a$. These intervals may be of equal length, but this is not necessary. If N_i is the number of failures in the ith interval, $i = 1, 2, \ldots, M$, then the estimate of the ROCOF is

$$\hat{\lambda}(t) = \begin{cases} N_1/(a_1 - 0), & 0 \le t < a_1 \\ \vdots & \vdots \\ N_m/(a_m - a_{m-1}), & 0 \le t < a_1 \,. \end{cases}$$

In many situations, the available data indicate only how many failures occurred in each interval, and not the individual failure times. The data set from Balaban (1978), described in Example 1.1 of Chapter 1, is of this form. Such data sets are particularly amenable to this estimate of λ.

Another form of nonparametric estimate of the intensity function λ, called a natural estimate by Ascher and Feingold (1984), is the following:

$$\hat{\lambda}_{\text{natural}}(t) = \frac{\text{Number of failures in } [t, t + \Delta t]}{\Delta t} \,.$$

This estimator of the ROCOF looks at the number of failures in a moving window of length Δt and estimates the ROCOF as the ratio of the number of failures and Δt. When Δt is small, the graph of this estimator will appear jagged, moving up when a number of failures are closely spaced, and down when the failures are sparse. The graph of $\hat{\lambda}_{\text{natural}}(t)$ will be smoother for larger values of Δt. Of course, this estimate of the ROCOF will be seriously biased when t is in the interval $[a - \Delta t, a]$, where a is the time that observation of the system stopped. This bias to the low side is caused by the fact that when $t > a - \Delta t$, the interval $[t, t + \Delta t]$ will include times beyond a. Thus, it makes sense to calculate $\hat{\lambda}_{\text{natural}}(t)$ only over the interval $[0, t - \Delta t]$.

4.2.2 Kernel Estimates

Another class of estimators of $\lambda(t)$ was suggested by Lewis and Shedler (1976). Their method is based on the results of Theorems 25 and 26. The latter of these two theorems says that conditioned on $N(t) = n$, the failure times $T_1 < T_2 < \cdots < T_n$ are distributed as order statistics from a distribution with cdf

$$G(y) = \begin{cases} 0, & y \leq 0 \\ \Lambda(y)/\Lambda(t), & 0 < y \leq t \\ 1, & y > t. \end{cases}$$

or equivalently, with pdf

$$g(y) = \begin{cases} \lambda(y)/\Lambda(t), & 0 < y \leq t \\ 0, & \text{otherwise.} \end{cases}$$

Lewis and Shedler (1976) suggested a kernel density estimate for the pdf g using the method of Rosenblatt (1956). Notice that g is not the pdf of any physical quantity, but arises only as the pdf from which the observed failure times $T_1 < T_2 < \cdots < T_n$ are order statistics. The kernel density estimate of g is

$$\hat{g}(t) = \frac{1}{n\,b(n)} \sum_{i=1}^{n} W\left(\frac{t - T_j}{b(n)}\right),$$

where $b(n)$ is a function satisfying $nb(n) \to 0$ as $n \to \infty$ and $W(u)$ is a bounded nonnegative weight function satisfying $\int_{-\infty}^{\infty} W(u)\,du = 1$.

The function $\hat{g}(t)$ serves to estimate $\lambda(y)/\Lambda(t_n)$ which contains the unknown value $\Lambda(t)$. The quantity $\Lambda(t)$, which is the expected number of failures by time t, is estimated by n. Thus, the estimate of the ROCOF is

$$\hat{\lambda}_{\text{kernel}}(t) = n\hat{g}(t) = \frac{1}{b(n)} \sum_{i=1}^{n} W\left(\frac{t - T_j}{b(n)}\right).$$

In the examples that follow, we chose for W the normal density with mean $\mu - 0$ and standard deviation σ. Thc examples that follow show that larger values of σ lead to smoother estimates of $\lambda(t)$.

This procedure assumes that the failure process is an NHPP, in which case the ROCOF and the intensity function are the same. Since we observe only a single realization of order statistics from the pdf g, we cannot use asymptotic results to justify this procedure.

4.2.3 An Estimate Assuming a Convex Intensity Function

Pang, Thompson, and Fallahi (1989) gave a nonparametric procedure for estimating the intensity function of an NHPP assuming that $\lambda(t)$ is convex, that is, concave up. This method may be applicable when the intensity function is bathtub-shaped, but in some other cases, such as the earthquake data described in the next subsection, convexity may be too strong an assumption.

The log-likelihood function for an arbitrary intensity function λ is given by

$$k(\lambda) = \sum_{i=1}^{n} \log \lambda(t_i) - \int_0^a \lambda(t)\,dt. \tag{4.2}$$

The MLE is that function λ belonging to the class C of all convex, continuous nonnegative functions defined on $[0, a]$ that maximizes (4.2). Pang, Thompson, and Fallahi proved that the function $\lambda \in C$ that maximizes the log-likelihood function is piecewise linear, with break points at the failure times $t_1 < t_2 < \cdots < t_n$ and it satisfies $\int_0^a \lambda(t)\,dt = n$. They were thus able to replace the problem of maximization over the infinite dimensional space C with a nonlinear programming problem over a finite dimensional space.

4.2.4 Examples

Example 4.5 The data set in Table 4.5 is given in the book *Small Data Sets* edited by Hand et al. (1994), although the data are originally attributed to Musa (1979). This is a good example of a large date set (despite the name of the book containing it) that is amenable to nonparametric methods. Figure 4.8 shows a plot of the cumulative number of failures against the cumulative operating time. The concave down nature of the graph suggests improvement in the software system.

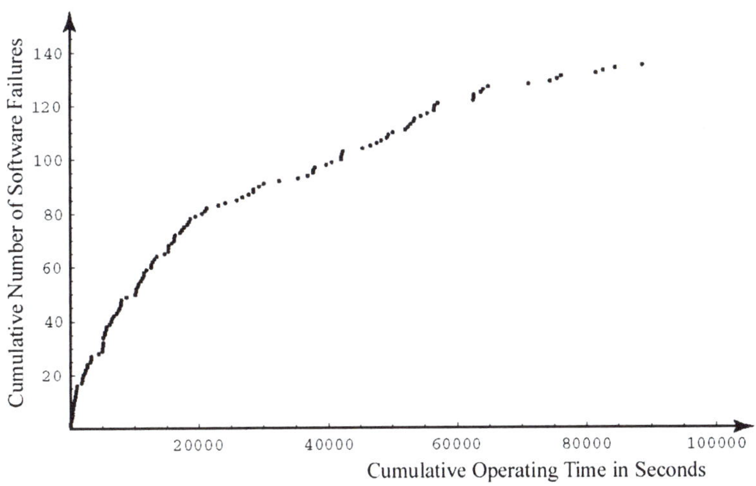

Fig. 4.8 Plots of number of software failures $N(t_i) = i$ against cumulative operating time t.

i	X_i	i	X_i	i	X_i	i	X_i	i	X_i
1	3	28	1146	55	357	82	296	109	245
2	30	29	600	56	193	83	1755	110	729
3	113	30	15	57	236	84	1064	111	1897
4	81	31	36	58	31	85	1783	112	447
5	115	32	4	59	369	86	860	113	386
6	9	33	0	60	748	87	983	114	446
7	2	34	8	61	0	88	707	115	122
8	91	35	227	62	232	89	33	116	990
9	112	36	65	63	330	90	868	117	948
10	15	37	176	64	365	91	724	118	1082
11	138	38	58	65	1222	92	2323	119	22
12	50	39	457	66	543	93	2930	120	75
13	77	40	300	67	10	94	1461	121	482
14	24	41	97	68	16	95	843	122	5509
15	108	42	263	69	529	96	12	123	100
16	88	43	452	70	379	97	261	124	10
17	670	44	255	71	44	98	1800	125	1071
18	120	45	197	72	129	99	865	126	371
19	26	46	193	73	810	100	1435	127	790
20	114	47	6	74	290	101	30	128	6150
21	325	48	79	75	300	102	143	129	3321
22	55	49	816	76	529	103	108	130	1045
23	242	50	1351	77	281	104	3110	131	648
24	68	51	148	78	160	105	1247	132	5485
25	422	52	21	79	828	106	943	133	1160
26	180	53	233	80	1011	107	700	134	1864
27	10	54	134	81	445	108	875	135	4116

Table 4.5 Computations for TTT plot for System 1 in Table 4.1.

Figure 4.9 shows the kernel estimates of the ROCOF for Musa's data on software failures. The weight function W was chosen to be the $N(0, \sigma^2)$ density, and for this data set we chose $\sigma = 5000$, $10,000$, and $20,000$. Figure 4.9 indicates that the ROCOF decreased initially from about 0.005 failures per second to 0.001 failures per second after the first 20,000 seconds of testing. After that time, the ROCOF remained roughly constant. □

Example 4.6 Table 4.6 gives another data set from *Small Data Sets* The data set gives the number of failures per week of a DEC-20 computer starting in late 1983. Let i denote the week and Y_i denote the number of failures during that week.

For this data set, we do not have the exact failure times; we just have the number of failures per week. The failures seem to be clustered, with many weeks having no failures and some weeks having over ten failures.

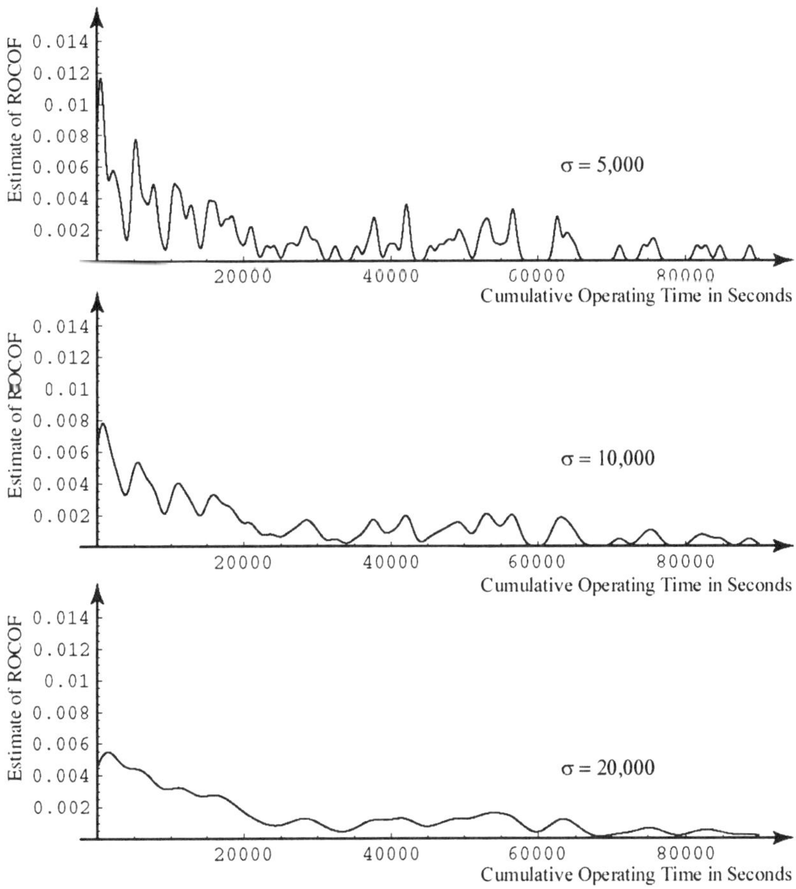

Fig. 4.9 Kernel estimates of failure intensity for software failure data.

i	V_i	i	V_i	i	V_i	i	V_i	i	V_i
1	4	27	8	53	2	79	3	105	0
2	0	28	4	54	2	80	4	106	2
3	0	29	5	55	3	81	1	107	1
4	0	30	0	56	1	82	3	108	1
5	3	31	5	57	2	83	6	109	0
6	2	32	4	58	0	84	4	110	3
7	0	33	1	59	2	85	4	111	16
8	0	34	0	60	1	86	5	112	22
9	6	35	8	61	6	87	2	113	5
10	7	36	2	62	3	88	10	114	1
11	6	37	5	63	3	89	4	115	2
12	2	38	2	64	6	90	1	116	4
13	1	39	1	65	11	91	5	117	7
14	11	40	12	66	10	92	6	118	8
15	6	41	8	67	4	93	9	119	6
16	1	42	9	68	3	94	7	120	11
17	2	43	10	69	0	95	3	121	3
18	1	44	17	70	2	96	1	122	0
19	1	45	2	71	4	97	3	123	4
20	2	46	3	72	2	98	0	124	7
21	0	47	4	73	1	99	2	125	8
22	2	48	8	74	5	100	2	126	4
23	2	49	1	75	3	101	1	127	4
24	1	50	2	76	3	102	4	128	5
25	0	51	5	77	2	103	2		
26	12	52	1	78	5	104	13		

Table 4.6 Failures per week for DEC 20 computer.

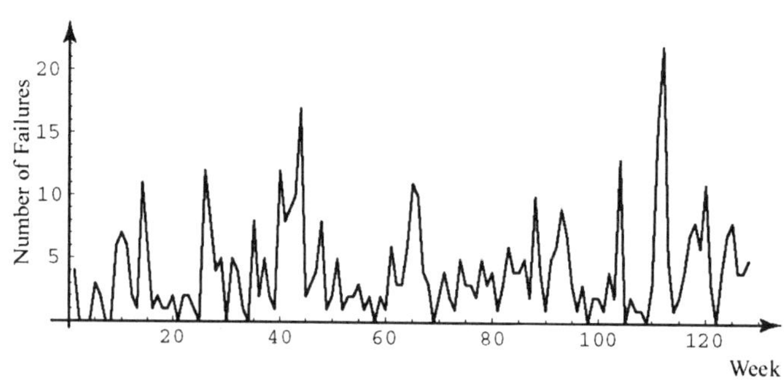

Fig. 4.10 Weekly failures for a DEC 20 computer.

Figures 4.11 and 4.12 indicate that the ROCOF remained roughly constant throughout the three-year period, although the failures do seem to be clustered. □

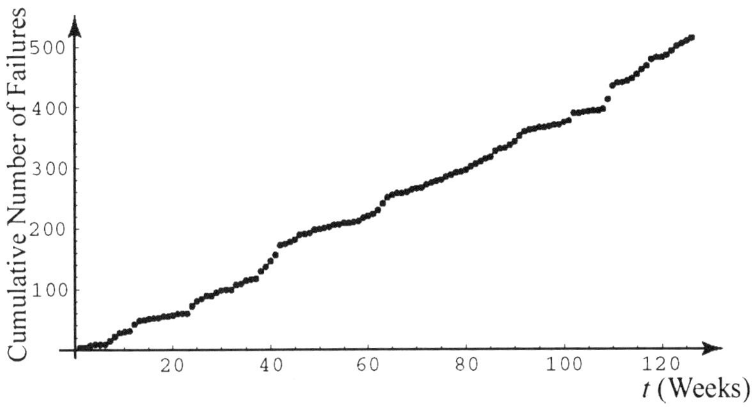

Fig. 4.11 Cumulative number of failures for DEC-20 computer.

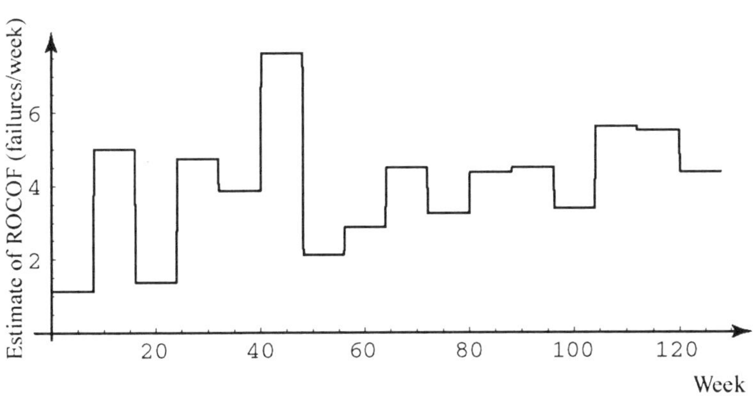

Fig. 4.12 Estimate of ROCOF for DEC-20 data. The data were partitioned into groups of eight weeks.

Example 4.7 Here is a large data set. Table 4.7 gives the first few and last few times of earthquakes of magnitude 1.0 or greater in southern California from May 10–17, 1999 (US Geological Survey, 1999,

http://quake.wr.usgs.gov/recenteqs/Quakes/quakes0.html).

There were 1364 events (earthquakes) in that seven-day period. The plot of cumulative number of events $N(t_i)$ against time t_i, shown in Figure 4.13, suggests that near the beginning of the fifth day, actually just after midnight on May 15, 1999, the rate of occurrence of earthquakes increased dramatically from less than 100/day to 1000/day. The rate of occurrence of earthquakes then gradually returned to the previous value of about 100/day. Figures 4.14 and 4.15 are graphs of the natural estimator and the kernel estimator of $\lambda(t)$. The effect of σ is evident in Figure 4.15; larger values of σ lead to a smoother estimate of the ROCOF. □

Date	Time	Hours Since Previous Event	Hours Since First Event	Magnitude
May 10, 1999	11:56:54	0	0.000	2.2
May 10, 1999	12:05:36	0.145	0.145	1.6
May 10, 1999	12:33:59	0.473	0.618	1.8
⋮	⋮	⋮	⋮	⋮
May 17, 1999	11:02:46	0.355	167.098	1.1
May 17, 1999	11:07:11	0.073	167.171	1.1
May 17, 1999	11:08:42	0.026	167.197	1.0

Table 4.7 First three and last three earthquakes over the indicated time interval.

4.3 TESTING FOR THE HOMOGENEOUS POISSON PROCESS

The homogeneous Poisson process (HPP) is the simplest statistical model for describing the occurrence of failures of a repairable system. The HPP, however, carries with it some strong assumptions that in practice are often clearly violated. The HPP implies that the system does not age in global time, that is, the system does not deteriorate and it does not exhibit reliability improvement. The HPP also implies that the system does not wear out in local time, that is, during the times between failures. If either of these assumptions does not hold, then the HPP is not the appropriate model. Graphical methods are useful for determining the suitability of the HPP. If the plot of $N(t_i)/t_i$ versus t_i is linear, or if the Duane plot has a zero slope, then the HPP may be

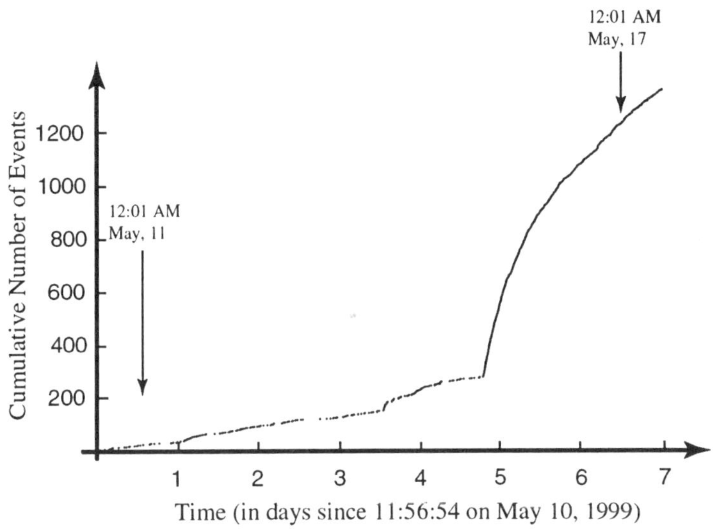

Fig. 4.13 Plots of cumulative number of events (earthquakes) $N(t_i) = i$ against cumulative time t.

an appropriate model. It would then be useful to look at the times between failures to see if they can be modeled by an exponential distribution. If they *can*, then the HPP is appropriate; if they can't, then a renewal process with a distribution besides the exponential would be appropriate.

Several statistical tests have been proposed for testing the null hypothesis that the process is an HPP against the alternative that there is a monotonic trend. Here we discuss two of these tests, the Laplace test, one of the earliest statistical tests of any kind, and the MIL-HDBK-189 test.

It is important to distinguish between the two common ways of terminating the process of data collection, since statistical inference procedures are slightly different for the two cases.

Definition 33 Failure Truncation and Time Truncation *If testing of the repairable system stops after a predetermined number of failures, say n, then the data are said to be* failure truncated. *If testing stops at a predetermined time t then the data are said to be* time truncated.

This concept of truncation is similar to that of censoring, which is often applied to the testing of nonrepairable units. Type II censoring occurs when several nonrepairable units are placed on test until a predetermined number, say r, out of the n, fail. Type I censoring occurs when several nonrepairable units are placed on test until a predetermined time. In the present case, we consider several failures on a single repairable system, rather than single failures on many nonrepairable systems.

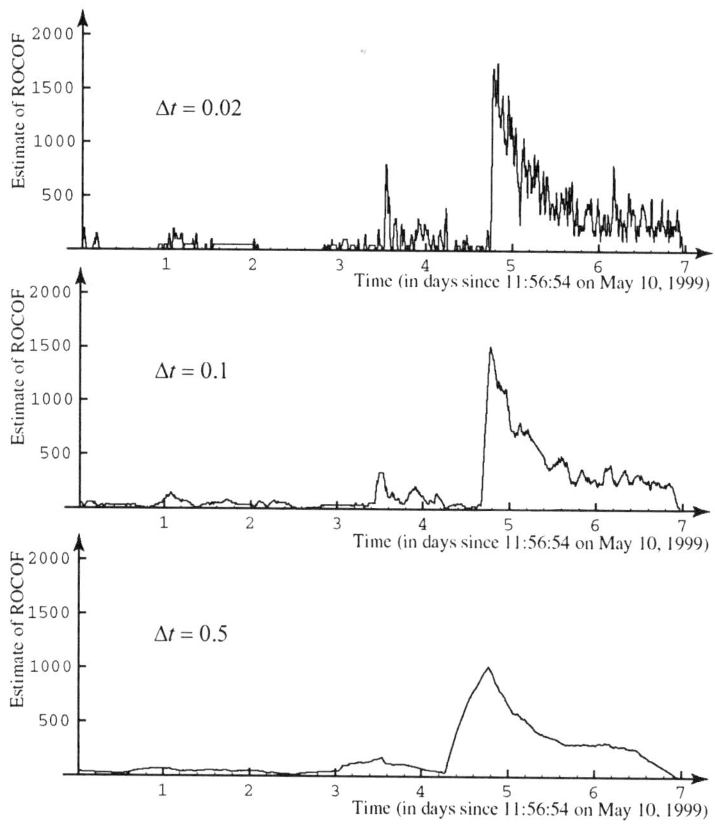

Fig. 4.14 Natural estimates of ROCOF for earthquake data.

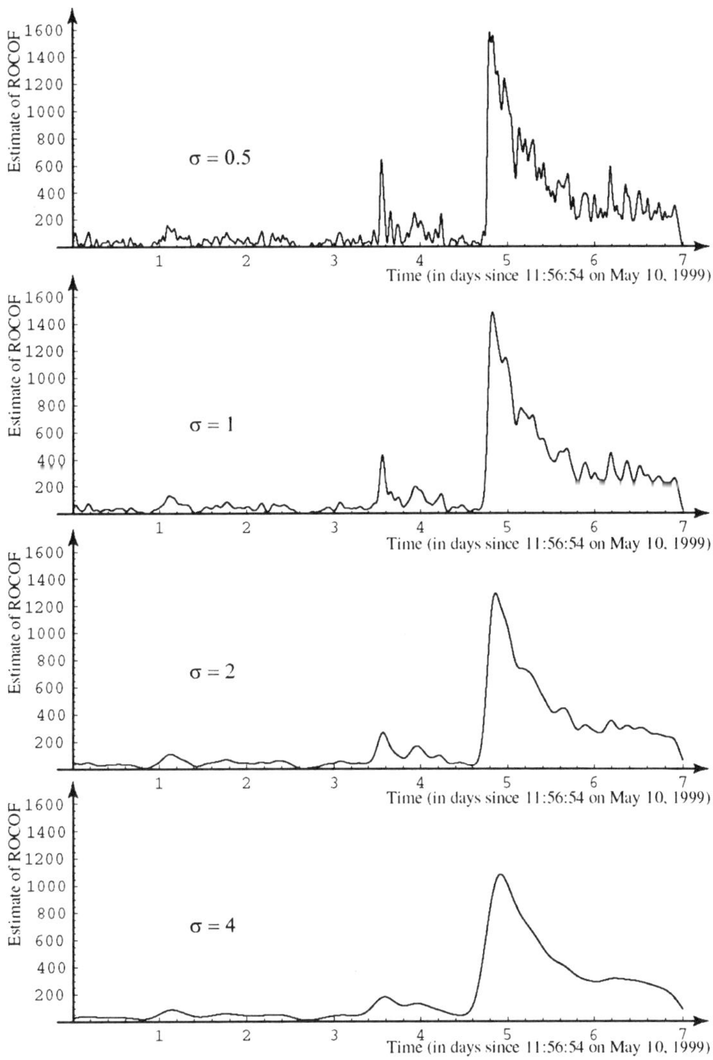

Fig. 4.15 Kernel estimates of ROCOF for earthquake data.

Suppose first that the data collection process was failure truncated. We showed in Theorem 21 that for an HPP conditioned on $T_n = t_n$, the failure times $T_1 < T_2 < \cdots < T_{n-1}$ are distributed as $n-1$ order statistics from a uniform distribution on the interval $(0, t_n)$. Thus the sum

$$S = \sum_{i=1}^{n-1} T_i$$

is distributed as the sum of $n-1$ random variables that are uniformly distributed on the interval $(0, t_n)$. This uniform distribution has mean $t_n/2$ and standard deviation $t_n/\sqrt{12}$. For moderate or large n (usually $n \geq 20$ is adequate) the standardized average

$$L = \frac{\sum_{i=1}^{n-1} T_i/(n-1) - t_n/2}{t_n/\sqrt{12(n-1)}}$$

is approximately normally distributed with mean 0 and standard deviation 1. The null hypothesis

$$H_0 : \text{The process is an HPP}$$

is rejected if

$$L < -z_{\alpha/2} \text{ or } L > z_{\alpha/2},$$

where $z_{\alpha/2}$ is the value that leaves a probability of $\alpha/2$ is the right tail of the standard normal distribution. Large values of L mean that the sum of the failure times are larger than expected, indicating that the failure times tend to bunch up toward the end of the interval $(0, t_n)$; this in turn suggests that the system in *deteriorating*. On the other hand, small values of L indicate reliability *improvement*. This test, which is due to Laplace, was shown by Cox and Lewis (1966) to be optimal against the alternative hypothesis that the failure process is a nonhomogeneous Poisson process with intensity function

$$\lambda(t) = \exp(\alpha + \beta t), \quad t > 0.$$

Example 4.8 For illustration, consider the data from system 3 in Table 4.1. The Duane plot shown in Figure 4.4 indicates reliability deterioration, so we would expect the homogeneous Poisson process to be an inadequate mode. The test statistic for testing the null hypothesis that the failure times are governed by a homogeneous Poisson process is

$$L = \frac{\frac{1734}{11} - \frac{227}{2}}{227/\sqrt{12 \times (12-1)}} = 2.23.$$

In this case, the null hypothesis would be rejected for $\alpha = 0.05$ since $|2.23| > z_{0.025} = 1.96$. The P-value for this test is 0.0257. □

If the data are time truncated, that is, the testing was stopped at a predetermined time t, then conditioned on $N = n$ the random variables $T_1 < T_2 < \cdots < T_n$ are distributed as n order statistics from a distribution that is uniform on the interval $(0, t)$. In this case, the test statistic for Laplace's test is

$$L = \frac{\sum_{i=1}^{n} T_i/n - t/2}{t/\sqrt{12n}}.$$

The other test we discuss here is the test described in MIL-HDBK 189. (The rationale for this test, and the associated notation will become clearer once we study the power law process.) For failure truncated data, this test is based on the statistic $2n/\hat{\beta}$, where

$$\hat{\beta} = \frac{n}{\sum_{i=1}^{n-1} \log(t_n/t_i)}.$$

The null hypothesis of an HPP is rejected if

$$2n/\hat{\beta} < \chi^2_{1-\alpha/2}(2(n-1)) \quad \text{or} \quad 2n/\hat{\beta} > \chi^2_{\alpha/2}(2(n-1)),$$

where $\chi^2_\gamma(\nu)$ is the value that leaves a probability of γ in the right tail of the chi-square distribution with ν degrees of freedom. This result is based on the result of Theorem 31.

Example 4.9 Let's consider again the data of system 3 from Table 4.1. For these data

$$\hat{\beta} = \frac{n}{\sum_{i=1}^{n-1} \log(t_n/t_i)} = \frac{12}{5.2375} = 2.2912.$$

The MIL-HDBK 189 test statistic for testing whether the process can be modeled by an HPP is thus

$$\frac{2n}{\hat{\beta}} = \frac{2 \times 12}{2.2912} = 10.475.$$

From the chi-square table, Table A.1, we find $\chi^2_{0.975}(22) = 10.982$ and $\chi^2_{0.025}(22) = 36.781$. Since $2n/\hat{\beta} = 10.475$ is less than 10.982 we would reject the null hypothesis that the homogeneous Poisson process is an adequate model. The P-value for this test is $2 \times 0.01854 = .03708$, giving evidence against the HPP model. Notice that the two tests (Laplace's test and the MIL-HDBK 189 test) reached the same conclusions at the $\alpha = 0.05$ level, although their P-values (0.0257 and 0.0371) are different. Having different P-values is not unusual because the two tests have specific alternative hypotheses for which each is optimal. This example also illustrates the advantage of reporting the P-value rather than just the accept/reject decision. Seeing the P-value, readers can make their own judgment as to the evidence against the null hypothesis. □

When the data are time truncated the test statistic is

$$\hat{\beta} = \frac{N}{\sum_{i=1}^{N} \log(t/t_i)}$$

and we reject the null hypothesis that the HPP is an adequate model if

$$2n/\hat{\beta} < \chi^2_{1-\alpha/2}(2n) \qquad \text{or} \qquad 2n/\hat{\beta} > \chi^2_{\alpha/2}(2n)\,.$$

Bain and Engelhardt (1991) showed that the MIL-HDBK 189 test is optimal if the alternative is that the failure process is a nonhomogeneous Poisson process with intensity function

$$\lambda(t) = \frac{\beta}{\theta}\left(\frac{t}{\theta}\right)^{\beta-1},$$

that is, if the true failure process is the power law process.

Bain, Engelhardt, and Wright (1985) studied the power of several tests for which the null hypothesis is that the process is an HPP against the alternative that the intensity is increasing. They compared the Laplace test (using the exact and approximate distributions), the MIL-HDBK 189 test, a likelihood ratio test due to Boswell (1966), nonparametric tests based on Kendall's tau and Spearman's ρ, and a test based on the F distribution proposed by Barlow, Bartholomew, Bremner, and Brunk (1972). Using simulation to estimate the power of each test, Bain, Engelhardt, and Wright (1985) found that the MIL-HDBK 189 test had power at least 87% of any of the others for all situations they studied. The Laplacc test had the largest power when the intensity is a step function.

4.4 INFERENCE FOR THE HOMOGENEOUS POISSON PROCESS

We consider in this section the HPP, which as we explained in Chapter 3, is characterized by exponentially distributed times between failures. Since the HPP is also a renewal process, or a same-as-new model, the HPP *cannot* be used to model systems that deteriorate or show improvement. The HPP implies that the condition of the system is the same at any point in time, and this condition is independent of the previous pattern of failures. In particular, this means that after repair a system is in exactly the same condition as a brand new system. The homogeneous Poisson process should thus never be applied without first displaying the data graphically, and without giving a thorough consideration of the implications of the model. Laplace's test or the MIL-HDBK 189 test, which were discussed in the previous section, should be applied before assuming the HPP.

We assume now that the failure process is an HPP with intensity

$$\lambda(t) = 1/\mu, \qquad t > 0.$$

The times between failures are then independent and identically distributed EXP(μ) random variables. If the data are failure truncated then we observe the failure times $0 < t_1 < t_2 < \cdots < t_n$. In this case the maximum likelihood estimator for μ is

$$\hat{\mu} = t_n/n. \tag{4.3}$$

Since the quantity $2t_n/\mu$ has a chi-square distribution with $2n$ degrees of freedom, a $100 \times (1 - \alpha)\%$ confidence interval for μ is

$$\frac{2t_n}{\chi^2_{\alpha/2}(2n)} < \mu < \frac{2t_n}{\chi^2_{1-\alpha/2}(2n)}. \tag{4.4}$$

The maximum likelihood estimator for $\lambda(t_n)$ is

$$\overline{\lambda}(t_n) = 1/\hat{\mu} = n/t_n. \tag{4.5}$$

Example 4.10 For illustration, we consider the System 2 data from Table 4.1. For this data set, there seems to be no deterioration or improvement, although the failures do seem to be evenly spaced. If the homogeneous Poisson process is assumed, then the maximum likelihood estimator for μ is

$$\hat{\mu} = 227/12 = 18.917.$$

From the chi-square table (Table A.1) we find that $\chi^2_{0.975}(24) = 12.401$ and $\chi^2_{0.025}(24) = 39.364$. The confidence interval for μ is thus

$$\frac{2 \times 227}{39.364} < \mu < \frac{2 \times 227}{12.401}$$

or

$$11.533 < \mu < 36.61.$$

This is a wide interval because the sample size is so small. It is, however, good to know when there is insufficient information to estimate a parameter with precision. □

Suppose now that the process is time truncated, that is, the testing is stopped at the predetermined time t. In this case, the number of observed failures N is random and has a Poisson distribution with mean t/μ. In addition, N is a sufficient statistic for μ, so the failure times $t_1 < t_2 < \cdots < t_N$ do not enter into the expressions for the MLE of μ or for the confidence limits on μ. The likelihood function is

$$L(\mu) = c\left(1/\mu^N\right)\exp\left(-t/\mu\right),$$

where c is a constant. The MLE is thus

$$\hat{\mu} = N/t.$$

Confidence intervals for μ can be based directly on the Poisson distribution, in which case the true confidence is usually slightly greater than the specified level. Confidence intervals for μ can also be based on a normal approximation to the Poisson distribution.

Pearson and Hartley(1966) showed that if $N = c$ events are observed, then a $100\,(1-\alpha)\,\%$ confidence interval for μ is

$$\left(\frac{1}{2}\chi^2_{1-\alpha/2}(2c)\,,\ \frac{1}{2}\chi^2_{1-\alpha/2}(2(c+1))\right).$$

Since this formula involves only the cdf of the χ^2 distribution, a chi-square table is sufficient for computing confidence intervals. To simplify matters we give in Table A.3 a table for obtaining a confidence interval for μ given c. This table yields conservative confidence intervals in the sense that the probability that an interval covers the true mean t/μ is α or greater. The example below illustrates the use of this table.

If the number of failures is large, say $n = 20$ or more, then the normal approximation to the Poisson distribution can be applied. If the true mean λ is large, then

$$z = \frac{N-\lambda}{\sqrt{\lambda}} \tag{4.6}$$

is distributed approximately as a standard normal random variable. If $z_{\alpha/2}$ is the point that leaves an area of $\alpha/2$ in the upper tail of a normal distribution, then

$$P\left(-z_{\alpha/2} < \frac{N-\lambda}{\sqrt{\lambda}} < z_{\alpha/2}\right) \approx 1-a.$$

If $N = n$ failures are observed, then the two inequalities will just be broken (i.e., the inequalities will become equalities) whenever

$$\frac{n-\lambda}{\sqrt{\lambda}} = z_{\alpha/2}.$$

This leads to

$$(n - \lambda)^2 = \lambda z_{\alpha/2}^2,$$

which, in turn, leads to the quadratic equation

$$\lambda^2 - \lambda(2n + z_{\alpha/2}^2) + n^2 = 0.$$

The solutions to this quadratic equation are

$$\begin{aligned}\lambda &= \frac{2n + z_{\alpha/2}^2 \pm \sqrt{(2n + z_{\alpha/2}^2)^2 - 4n^2}}{2} \\ &= \left(n + z_{\alpha/2}^2/2\right) \pm z_{\alpha/2}\sqrt{n + z_{\alpha/2}^2/4}.\end{aligned}$$

This is a confidence interval for the parameter $\lambda = t/\mu$. A confidence interval for μ is then

$$\begin{aligned}&\frac{t}{\left(n + z_{\alpha/2}^2/2\right) + z_{\alpha/2}\sqrt{n + z_{\alpha/2}^2/4}} < \mu \\ &\qquad < \frac{t}{\left(n + z_{\alpha/2}^2/2\right) - z_{\alpha/2}\sqrt{n + z_{\alpha/2}^2/4}}. \qquad (4.7)\end{aligned}$$

Example 4.11 Suppose that a repairable system was operated for $t = 3000$ hours and experienced $n = 28$ failures. Suppose further that the times between failures do not tend to increase or decrease, so that the HPP may be a reasonable model. In this case the MLE of μ is

$$\hat{\mu} = 3000/28 = 107.14.$$

From Table A.3 we see that a 95% confidence interval for $\lambda = 3000/\mu$ is

$$18.606 < \lambda < 40.468.$$

This means that

$$18.606 < t/\mu < 40.468$$

so the confidence interval for μ is

$$\frac{3000}{40.468} < \mu < \frac{3000}{18.606}$$

or,

$$74.13 < \mu < 161.24.$$

If we apply the normal approximation from (4.6), we get

$$28 + 1.96^2/2 - 1.96\sqrt{28 + 1.96^2/4} = 29.92 \pm 10.55$$

or

$$19.37 < \lambda < 40.47$$

as the 95% confidence interval for $\lambda = 3000/\mu$. The confidence interval for μ is thus

$$74.13 < \mu < 154.85.$$

In this example, where we had $n = 28$ failures, the normal approximation worked very well. □

4.5 INFERENCE FOR THE POWER LAW PROCESS: FAILURE TRUNCATED CASE

As we mentioned previously, the power law process is a nonhomogeneous Poisson process with intensity function that has the form

$$\lambda(t) = \frac{\beta}{\theta}\left(\frac{t}{\theta}\right)^{\beta-1}. \tag{4.8}$$

The β parameter affects how the system deteriorates or improves over time. If $\beta > 1$, then the intensity function $\lambda(t)$ is increasing, and the failures tend to occur more frequently. If $\beta < 1$, then $\lambda(t)$ is decreasing, and the system is improving. Finally, if $\beta = 1$, then the power law process reduces to the simpler homogeneous Poisson process with intensity $1/\theta$. The θ parameter is a scale parameter. There are several reasons why the power law process is a widely used model for repairable systems. Duane (1964) demonstrated that many systems developed at General Electric seemed to follow a model closely related to the power law process. In what became known as a Duane plot, he suggested plotting $N(t_i)/t_i$ against t_i on double logarithmic paper. (The Duane plot is discussed in Section 4.1.1.) He found that for many systems such a plot was nearly linear and suggested fitting a line to the data. Crow (1974) later discussed the relationship between the power law process and the linearity on a Duane plot. Another reason for the popularity of the power law process is that statistical inference procedures are well developed and existing tables can often be used. It is also flexible enough to model repairable systems that are deteriorating or improving.

4.5.1 Point Estimation for β and θ

Suppose that a repairable system is observed until n failures occur, so we observe the failure times $0 < t_1 < t_2 < \cdots < t_n$. Theorem 23 gives the joint pdf of a failure truncated NHPP as

$$f(t_1, t_2, \ldots, t_n) = \left(\prod_{i=1}^{n} \lambda(t_i)\right) \exp\left[-\int_0^{t_n} \lambda(x)\, dx\right], \quad 0 < t_1 < t_2 < \cdots < t_n.$$

For the case when $\lambda(t) = (\beta/\theta)(t/\theta)^{\beta-1}$, we have

$$\begin{aligned} f(t_1, t_2, \ldots, t_n) &= \left(\prod_{i=1}^{n} \frac{\beta}{\theta}\left(\frac{t_i}{\theta}\right)^{\beta-1}\right) \exp\left[-\int_0^{t_n} \frac{\beta}{\theta}\left(\frac{x}{\theta}\right)^{\beta-1} dx\right] \\ &= \frac{\beta^n}{\theta^{n\beta}}\left(\prod_{i=1}^{n} t_i\right)^{\beta-1} \exp\left[-\left(\frac{t_n}{\theta}\right)^{\beta}\right], \quad 0 < t_1 < t_2 < \cdots < t_n. \end{aligned}$$

To get the MLE's, we take the logarithm of this joint density and set the first partial derivatives (with respect to θ and β) equal to zero. The log-likelihood function is

$$\ell(\theta, \beta|\mathbf{t}) = n \log \beta - n\beta \log \theta + (\beta - 1)\sum_{i=1}^{n} \log t_i - \left(\frac{t_n}{\theta}\right)^{\beta}.$$

Differentiating with respect to θ and β, and setting the results equal to zero, yields the two equations

$$0 = \frac{\partial \ell}{\partial \theta} = -\frac{n\beta}{\theta} + \frac{\beta}{\theta}\left(\frac{t_n}{\theta}\right)^{\beta}$$

and

$$0 = \frac{\partial \ell}{\partial \beta} = \frac{n}{\beta} - n \log \theta + \sum_{i=1}^{n} \log t_i - \left(\frac{t_n}{\theta}\right)^{\beta} \log\left(\frac{t_n}{\theta}\right).$$

The first equation simplifies to

$$0 = -n + \left(\frac{t_n}{\theta}\right)^{\beta},$$

which can be solved for θ (in terms of β) to obtain

$$\theta = \frac{t_n}{n^{1/\beta}}.$$

Substituting back into the first equation yields

$$0 = \frac{n}{\beta} - n \log \frac{t_n}{n^{1/\beta}} + \sum_{i=1}^{n} \log t_i - \left(\frac{t_n n^{1/\beta}}{t_n}\right)^{\beta} \log\left(\frac{t_n n^{1/\beta}}{t_n}\right),$$

which simplifies to

$$0 = \frac{n}{\beta} - n \log t_n + \sum_{i=1}^{n} \log t_i + \frac{n}{\beta} \log n - n \log n^{1/\beta}.$$

The last two terms sum to zero. After noting that

$$\sum_{i=1}^{n} \log t_i - n \log t_n = \sum_{i=1}^{n} \log t_i - \sum_{i=1}^{n} \log n = \sum_{i=1}^{n} \log(t_i/t_n)$$

we get

$$0 = \frac{n}{\beta} + \sum_{i=1}^{n} \log(t_i/t_n) = \frac{n}{\beta} - \sum_{i=1}^{n} \log(t_n/t_i).$$

Solving for β yields

$$\beta = \frac{n}{\sum_{i=1}^{n} \log(t_n/t_i)}.$$

Since this gives β in terms of the observed failure times only, it is the MLE of β. Note that the last term in the sum is $\log(t_n/t_n) = \log 1 = 0$, so the MLE of β can be written as

$$\hat{\beta} = \frac{n}{\sum_{i=1}^{n-1} \log(t_n/t_i)}.$$

Once the MLE of β is computed, it can be substituted back into the expression for θ, giving the MLE for θ as

$$\hat{\theta} = \frac{t_n}{n^{1/\hat{\beta}}}.$$

The MLE's for the parameters β and θ are therefore

$$\hat{\beta} = \frac{n}{\sum_{i=1}^{n} \log(t_n/t_i)} \tag{4.9}$$

and

$$\hat{\theta} = \frac{t_n}{n^{1/\hat{\beta}}}. \tag{4.10}$$

The estimator $\hat{\beta}$ is biased. An unbiased estimator is

$$\bar{\beta} = \frac{n-2}{n}\hat{\beta} = \frac{n-2}{\sum_{i=1}^{n-1} \log(t_n/t_i)}. \tag{4.11}$$

4.5.2 Interval Estimation and Tests of Hypotheses

A useful result for deriving confidence intervals and tests of hypotheses about β is that $2n\beta/\hat{\beta}$ has a chi-square distribution with $2(n-1)$ degrees of freedom. To derive this important result we need the following lemma.

Lemma 30 *If U has a uniform distribution on the interval (0,1), then $X = -\theta \log U$ has an exponential distribution with mean θ.*

Proof. The cdf of $X = -\theta \log U$ is

$$\begin{aligned} F_X(x) &= P(X \le x) \\ &= P(-X/\theta \ge -x/\theta) \\ &= P\left(\exp(-X/\theta) \ge \exp(-x/\theta)\right) \\ &= P\left(U \ge \exp(-x/\theta)\right) \\ &= 1 - \exp(-x/\theta), \qquad x > 0. \end{aligned}$$

This is the cdf of the exponential distribution with mean θ. ■

We are now ready to prove the main result.

Theorem 31 *For failure truncated data $2n\beta/\hat{\beta}$ has a chi-square distribution with $2(n-1)$ degrees of freedom.*

Proof. We can write the expression $2n\beta/\hat{\beta}$ as

$$\frac{2n\beta}{\hat{\beta}} = 2\beta \sum_{i=1}^{n} \log(t_n/t_i) = -2\beta \sum_{i=1}^{n} \log(t_i/t_n).$$

Applying Theorem 21 from Chapter 2 we conclude that conditioned on $T_n = t_n$ the random variables $T_1 < T_2 < \cdots < T_{n-1}$ are distributed as $n-1$ order statistics from the distribution with cdf

$$G(y) = \begin{cases} 0, & y \le 0 \\ \Lambda(y)/\Lambda(t_n), & 0 < y < t_n \\ 1, & y \ge t_n . \end{cases}$$

For the power law process, we have $\Lambda(t) = (t/\theta)^\beta$, so

$$\frac{\Lambda(y)}{\Lambda(t_n)} = \frac{(y/\theta)^\beta}{(t_n/\theta)^\beta} = \left(\frac{y}{t_n}\right)^\beta .$$

In this case we have

$$G(y) = \begin{cases} 0 & y \le 0 \\ (y/t_n)^\beta & 0 < y < t_n \\ 1 & y \ge t_n . \end{cases}$$

Letting Y be a random variable with cdf G, we have on the one hand

$$\left(\frac{y}{t_n}\right)^{\beta} = G(y) = P(Y \leq y)$$

for $0 < y < t_n$. On the other hand,

$$P(Y \leq y) = P\left(Y/t_n \leq y/t_n\right) = P\left((Y/t_n)^{\beta} \leq (y/t_n)^{\beta}\right).$$

This implies that

$$P\left((Y/t_n)^{\beta} \leq (y/t_n)^{\beta}\right) = (y/t_n)^{\beta}$$

for $0 < y < t_n$, or $0 < y/t_n < 1$. This means that the random variable $(Y/t_n)^{\beta}$ has a uniform distribution over the interval $(0,1)$. Therefore the quantities $(T_i/t_n)^{\beta}$, $i = 1, 2, \ldots, n-1$ are distributed as $n-1$ order statistics from a uniform distribution on the interval $(0,1)$. By Lemma 30 the sum

$$\sum_{i=1}^{n} -\log(t_i/t_n)^{\beta} = -\beta \sum_{i=1}^{n} \log(t_i/t_n)$$

is thus distributed as the sum of $n-1$ exponential random variables, each with mean 1. By Theorem 7, the sum of $n-1$ exponential random variables has a gamma distribution with parameters $\kappa = n-1$ and $\theta = 1$. Finally, twice a gamma distribution with parameters $n-1$ and 1 yields a chi-square distribution with $2(n-1)$ degrees of freedom. Thus

$$-2\beta \sum_{i=1}^{n} \log(t_i/t_n) = \frac{2n\beta}{\hat{\beta}}$$

has a $\chi^2(2(n-1))$ distribution. ■

To get a $100 \times (1-\alpha)\%$ two-sided confidence interval for β we would look up in Table A.1 the values $\chi^2_{1-\alpha/2}(2(n-1))$ and $\chi^2_{\alpha/2}(2(n-1))$, the points that leave an area of $\alpha/2$ in the left and right tails, respectively, of the chi-square distribution with $2(n-1)$ degrees of freedom. We can then write

$$P\left(\chi^2_{1-\alpha/2}(2(n-1)) < \frac{2n\beta}{\hat{\beta}} < \chi^2_{\alpha/2}(2(n-1))\right) = 1-\alpha$$

so the confidence interval for β is

$$\frac{\chi^2_{1-\alpha/2}(2(n-1))\hat{\beta}}{2n} < \beta < \frac{\chi^2_{\alpha/2}(2(n-1))\hat{\beta}}{2n}. \tag{4.12}$$

This interval is interpreted like any other confidence interval. Before data are collected, the probability is $1-\alpha$ that an interval calculated in this way will include the true value of β. Once the data are collected, no probability statements can be made since there are no longer any random quantities in the expression for the confidence interval. The confidence is in the *method* that produced the interval, rather than in the *interval* itself.

It is often of interest to see whether the confidence interval for β includes or excludes the value $\beta = 1$. When $\beta = 1$ the intensity function reduces to

$$\lambda(t) = (1/\theta)(t/\theta)^{1-1} = 1/\theta,$$

which is independent of t. Thus the power law process reduces to the homogeneous Poisson process when $\beta = 1$.

We can test the hypothesis $H_0 : \beta = 1$ against one of the alternatives $H_1 : \beta \neq 1$, $H_1 : \beta < 1$, or $H_1 : \beta > 1$. (The discussion here explains the rationale behind the MIL-HDBK 189 test that we covered in a previous section.) If the alternative is the two-sided alternative $H_0 : \beta \neq 1$, then since $2n\beta/\hat{\beta}$ has a χ^2 distribution with $2(n-1)$ degrees of freedom, the rule is to reject H_0 if

$$2n\beta/\hat{\beta} < \chi^2_{1-\alpha/2}(2(n-1))$$

or if

$$2n\beta/\hat{\beta} > \chi^2_{\alpha/2}(2(n-1)).$$

After some manipulation, this becomes

$$\hat{\beta} < \frac{2n}{\chi^2_{\alpha/2}(2(n-1))} \quad \text{or} \quad \hat{\beta} > \frac{2n}{\chi^2_{1-\alpha/2}(2(n-1))}. \tag{4.13}$$

For testing $H_0 : \beta \geq 1$ versus $H_1 : \beta < 1$, that is, a test of whether the system is remaining stable or possibly deteriorating against the alternative that it is improving, the rule is to reject H_0 if

$$\hat{\beta} < \frac{2n}{\chi^2_{\alpha/2}(2(n-1))}.$$

For testing $H_0 : \beta \leq 1$ versus $H_1 : \beta > 1$, that is, a test of whether the system is remaining stable or possibly improving against the alternative that it is deteriorating, the rule is to reject H_0 if

$$\hat{\beta} > \frac{2n}{\chi^2_{1-\alpha/2}(2(n-1))}.$$

The P-value can also be used to assess the strength of evidence against the null hypothesis. For the two-sided alternative the P-value would be double the tail area beyond $2n/\hat{\beta}$. The smaller the P-value, the stronger the evidence against the null hypothesis.

Confidence intervals and tests of hypotheses for θ can be constructed from the statistic $W = (\hat{\theta}/\theta)^{\hat{\beta}}$. The statistic W is a pivotal quantity; that is, its distribution does not depend on any unknown parameters. The percentage points of the distribution of W for various values of n were determined by Finkelstein (1976) who used simulation. The tables in Finkelstein's paper can be used to calculate confidence intervals for θ. Such confidence intervals for θ are often extremely wide. It is usually not necessary, however, to estimate θ with high precision, since this parameter has no simple interpretation.

Failure Number i	**Failure Time** t_i	$N(t_i)/t_i$
1	10	0.1000
2	55	0.0364
3	166	0.0181
4	205	0.0195
5	341	0.0147
6	488	0.0123
7	567	0.0123
8	731	0.0109
9	1308	0.0069
10	2050	0.0049
11	2453	0.0045
12	3115	0.0039
13	4017	0.0032
14	4596	0.0030

Table 4.8 Failure times for an aircraft generator.

Example 4.12 For illustration, consider the failure times of an aircraft generator shown in Table 4.8. These failure times were read from a figure in Duane (1964). We assume that testing was stopped after the fourteenth failure. A plot of cumulative operating time t_i on the horizontal axis against cumulative number of failures $N(t_i)$ on the vertical axis, given in Figure 4.16, shows a definite concave down relationship. Table 4.8 also gives the values of t_i and $N(t_i)/t_i$, $i = 1, 2, \ldots, 14$ for these data. The Duane plot is also shown in Figure 4.16. The maximum likelihood estimates of β and θ are

$$\hat{\beta} = \frac{14}{\sum_{i=1}^{14-1} \log(4596/t_i)} = \frac{14}{28.97} = 0.483$$

and

$$\hat{\theta} = \frac{4596}{14^{1/0.483}} = \frac{4596}{236.013} = 19.474.$$

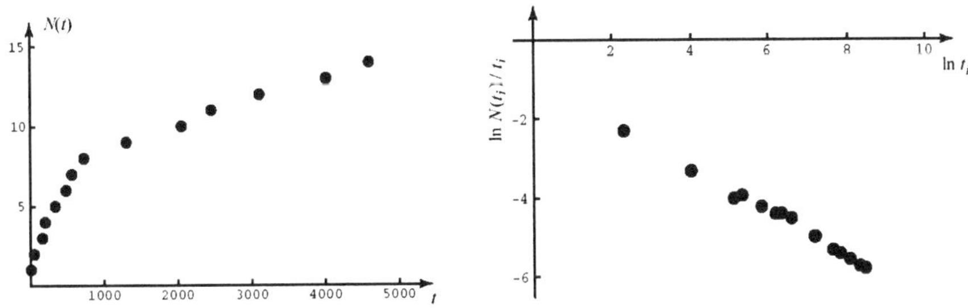

Fig. 4.16 Plot of number of failures $N(t_i) = i$ against cumulative operating time t_i, and the corresponding Duane plot.

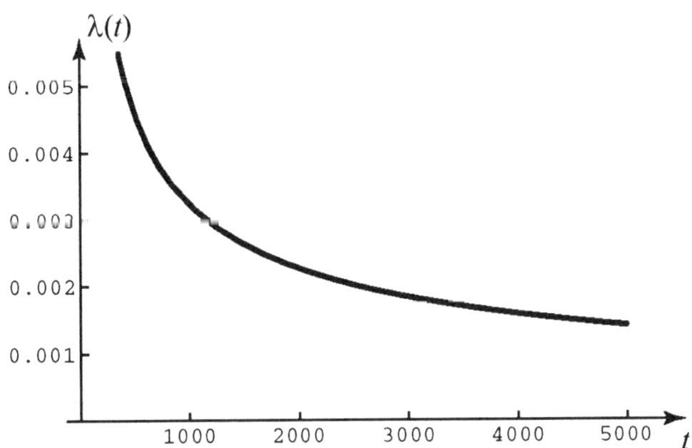

Fig. 4.17 Estimate of intensity function $\lambda(t)$ for aircraft generator data.

Figure 4.17 shows a plot of the estimated intensity function

$$\hat{\lambda}(t) = (\hat{\beta}/\hat{\theta})(t/\hat{\theta})^{\hat{\beta}-1} = \frac{0.483}{19.474}\left(\frac{t}{19.474}\right)^{0.483-1} = 0.115\, t^{-0.517}.$$

An unbiased estimate for β is $\bar{\beta} = (12/14)\hat{\beta} = 0.414$. To obtain a 95% confidence interval for β we determine from Table A.1 the values $\chi^2_{0.975}(2(n-1))$ and $\chi^2_{0.025}(2(n-1))$, that is, the values that leave 0.025 in the left and right tails of the $\chi^2(2(n-1))$ distribution. For $n = 14$, the degrees of freedom are $2 \times (14-1) = 26$, and we obtain $\chi^2_{0.975}(26) = 13.844$ and $\chi^2_{0.025}(26) = 41.923$. Using (4.12) we find that the lower confidence limit is

$$\frac{\chi^2_{0.975}(26)\hat{\beta}}{2n} = \frac{13.844 \times 0.483}{2 \times 14} = 0.239$$

and the upper confidence limit is

$$\frac{\chi^2_{0.025}(26)\hat{\beta}}{2n} = \frac{41.923 \times 0.483}{2 \times 14} = 0.723.$$

The 95% confidence interval for β is thus

$$0.239 < \beta < 0.723.$$

This interval excludes $\beta = 1$, indicating reliability improvement.

A hypothesis test with $\alpha = 0.05$ for $H_0 : \beta = 1$ against $H_0 : \beta \neq 1$ would have a critical region of the form

$$\hat{\beta} < \frac{2n}{\chi^2_{\alpha/2}(2(n-1))} = \frac{28}{41.92} = 0.668$$

or

$$\hat{\beta} > \frac{2n}{\chi^2_{1-\alpha/2}(2(n-1))} = \frac{28}{13.84} = 2.02.$$

Since $\hat{\beta} = 0.483 < 0.668$ we would reject the null hypothesis and conclude that there is evidence that β is not equal to 1. The P-value for this test is less than 0.005.

We can use Table 1 from Finkelstein (1976) to obtain a 90% confidence interval for θ. The confidence interval is

$$\frac{\hat{\theta}}{a^{1/\hat{\beta}}} < \theta < \frac{\hat{\theta}}{b^{1/\hat{\beta}}},$$

where $\hat{\theta}$ and $\hat{\beta}$ are the MLE's of θ and β, respectively, and a and b are obtained from Finkelstein's table. In our case, we have $n = 14$, so we obtain $a = 0.43$ and $b = 5.9$. Thus, we obtain

$$\frac{19.474}{9^{1/0.483}} < \theta < \frac{19.474}{0.36^{1/0.483}}$$

or

$$0.21 < \theta < 161.$$

Such a wide interval tells us that we have little information about the parameter θ. □

4.5.3 Estimation of the Intensity Function

Frequently the value of the intensity function when testing stops is an important quantity. For a system experiencing reliability growth during the development stage, this quantity represents the rate, in failures per time unit, that the system would have if it were put into service with no further improvements. For this reason, an estimate of this value is often used to decide when sufficient testing and development have been made. For a system that is deteriorating, the estimate of the value of the intensity function can be used to decide when to remove a system from use, or when to overhaul it.

For failure truncated data the maximum likelihood estimate of the value of the intensity at time t_n (the time the testing stops) is

$$\hat{\lambda} = \frac{\hat{\beta}}{\hat{\theta}}\left(\frac{t_n}{\hat{\theta}}\right)^{\hat{\beta}-1} = \frac{\hat{\beta}}{t_n/n^{1/\hat{\beta}}}\left(\frac{t_n}{t_n/n^{1/\hat{\beta}}}\right)^{\hat{\beta}-1} = \frac{n\hat{\beta}}{t_n}. \tag{4.14}$$

This is actually the MLE conditioned on the time of the last failure, that is, on $T_n = t_n$. Note that this expression depends only on n, $\hat{\beta}$, and t_n, and not on $\hat{\theta}$.

Rigdon and Basu (1988) studied this estimator of $\lambda(t_n)$ and other estimators of the form $c_n\hat{\beta}/t_n$. We found that the bias of the estimator $c_n\hat{\beta}/t_n$ is

$$\text{Bias} = \frac{\beta}{\theta}\frac{\Gamma(n-1/\beta)}{\Gamma(n)}\left[\frac{nc_n}{n-2} - (n-1/\beta)\right]$$

and the mean squared error (MSE) is

$$\text{MSE} = \left(\frac{\beta}{\theta}\right)^2 \frac{\Gamma(n-2/\beta)}{\Gamma(n)(n-2)(n-3)}$$

$$\times\left[(n+1-2/\beta)(n-2/\beta)(n-2)(n-3) - 2c_n(n-2/\beta)(n-3) + c_n^2 n^2\right].$$

For a fixed value of β the value

$$c_n = \frac{(n-1/\beta)(n-2)}{n}$$

makes the bias equal to zero, and the value

$$c_n = \frac{(n-2/\beta)(n-3)}{n}$$

minimizes the MSE. Since $\beta = 1$ is the boundary between increasing and decreasing intensities, and since a good estimator should perform well in both cases, we suggested using $\beta = 1$ in the above expressions. In Rigdon and Basu (1988) we studied the MSEs of the two estimators

$$\hat{\lambda}_{UB} = \frac{(n-1)(n-2)\hat{\beta}}{nt_n}$$

and

$$\hat{\lambda}_{MMSE} = \frac{(n-2)(n-3)\hat{\beta}}{nt_n}.$$

The first estimator has the property of being unbiased when $\beta = 1$, and the second has minimum mean squared error within the class $\{c_n\hat{\beta}/t_n\}$ when $\beta = 1$. We found that $\hat{\lambda}_{MMSE}$ was a good estimator for a wide range of β's.

A table for obtaining confidence intervals for $\lambda(t_n)$ is given in Table A.4. A $100 \times (1-\alpha)\%$ confidence interval is obtained by looking up ω_1 and ω_2, the $\alpha/2$ and $1-\alpha/2$ quantiles of the distribution of $n\hat{\beta}/t_n$. The confidence interval is then

$$\omega_1\hat{\lambda}(t_n) < \lambda(t_n) < \omega_2\hat{\lambda}(t_n).$$

This confidence interval is an unconditional confidence interval and is interpreted in the following manner. When viewed before testing begins, a confidence interval calculated in this way will include the true value of $\lambda(t_n)$ with probability $1-\alpha$.

Example 4.13 For the aircraft generator we had $n = 14$, $t_{14} = 4596$, and $\hat{\beta} = 0.483$ so

$$\hat{\lambda}(4596) = \frac{14 \times 0.483}{4596} = 0.00147.$$

From Table A.4 we find that for $n = 14$ and for 95% confidence, $\omega_1 = 0.3850$ and $\omega_2 = 1.7924$. Therefore, a 95% confidence interval for the value of the intensity at time $t = 4596$ is

$$0.3850 \times 0.00147 < \hat{\lambda}(4596) < 1.7924 \times 0.00147$$

or

$$0.00057 < \hat{\lambda}(4596) < 0.00263.$$

If reliability growth of the generator were to stop at time $t = 4596$, the subsequent failure process would be a Poisson process with constant intensity. This last result gives a confidence interval for this intensity. □

4.6 GOODNESS-OF-FIT TESTS

The Duane plot is an informal graphical method for assessing the fit of the power law process, although it has its limitations as indicated in Section 4.1.1. The adequacy of the power law process can also be tested by applying a formal goodness-of-fit test. For goodness-of-fit tests like the one described here, the null hypothesis is that the power law process is the correct model, and the alternative hypothesis is that the power law process is not the correct model. Because of the formulation of the hypotheses, we can never have evidence that the power law process is the *correct* model. We can only have a lack of evidence that the model is inadequate. There *is* a difference. The old saying is "Absence of evidence is not evidence of absence." Unless the true model is grossly different from the hypothesized model, small samples usually do not provide enough information to indicate a departure from the assumed model. Small samples are thus unlikely to lead to rejection of the null hypothesis. On the other hand, large samples will often lead to rejection even if the data seem to fit the power law process fairly closely. Keep these points in mind when doing any goodness-of-fit test.

Exact goodness-of-fit tests can be constructed by making an appropriate transformation of the failure times. Actually, there are two transformations that lead to goodness-of-fit tests. The more common is called the ratio-power transformation, which is defined by

$$\hat{R}_i = (t_i/t_n)^{\bar{\beta}}, \qquad i = 1, 2, \cdots, n-1,$$

where $\bar{\beta}$ is the unbiased estimator

$$\bar{\beta} = \frac{n-2}{n}\hat{\beta} = \frac{n-2}{\sum_{i=1}^{n-1} \log(t_n/t_i)}.$$

The rationale for this transformation is as follows: If $\Lambda(t)$ is the expected number of failures before time t, then

$$R_i = \frac{\Lambda(t_i)}{\Lambda(t_n)}, \qquad i = 1, 2, \cdots, n$$

are distributed as $n-1$ order statistics from a uniform distribution over the interval $[0, 1]$. For the power law process $\Lambda(t) = (t/\theta)^\beta$, so

$$R_i = \frac{\Lambda(t_i)}{\Lambda(t_n)} = \frac{(t_i/\theta)^\beta}{(t_n/\theta)^\beta} = (t_i/t_n)^\beta, \qquad i = 1, 2, \cdots, n-1.$$

Thus if β were known, then any goodness-of-fit test for the uniform distribution, such as the Cramér-von Mises test, could be applied to the R_i's. However, in most cases, the parameter β is unknown and must be estimated from the observed failure data. If the unbiased estimate $\bar{\beta}$ is used in place of β, then the R_i become the $\hat{R}_i$ defined above. The test statistic for the Cramér-von Mises test

$$C_R^2 = \frac{1}{12(n-1)} + \sum_{i=1}^{n-1} \left(\hat{R}_i - \frac{2i-1}{2(n-1)} \right)^2 \tag{4.15}$$

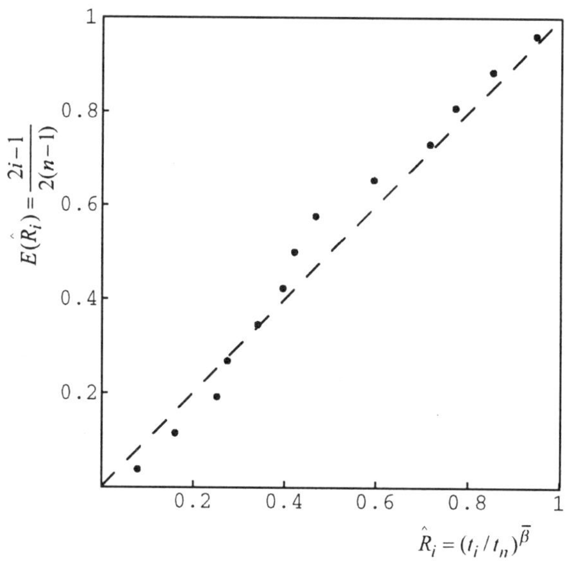

Fig. 4.18 Graphical illustration of Cramer-von Mises test.

can then be applied. The $\hat{R}_i$'s are distributed, approximately, as order statistics from a uniform distribution on the interval $[0, 1]$. The quantity

$$\frac{2i-1}{2(n-1)}$$

is the expectation of the ith order statistic from a uniform distribution on $[0, 1]$. The sum in the expression for C_R^2 is thus the sum of the squared deviations of the $\hat{R}_i$ from their expectation. Therefore, large values of C_R^2 indicate evidence of a departure from the power law process. Table A.6 gives critical values for C_R^2 for various values of M. For failure truncated data, be sure to enter Table A.6 with $M = n - 1$.

Example 4.14 Table 4.9 shows the computations involved in getting the test statistic C_R^2 for the data on the aircraft generator. Figure 4.18 shows a scatter plot of $\hat{R}_i$, on the horizontal axis, against its expectation

$$E\left(\hat{R}_i\right) = \frac{2i-1}{2(n-1)}$$

on the vertical axis. A scatter plot with points lying close to the diagonal line through the origin indicates a good fit of the model. The unbiased estimate for β is $\bar{\beta} = 0.414$. The sum of the squares of the entries in the last column is

$$\sum_{i=1}^{n}\left(\hat{R}_i - \frac{2i-1}{2(n-1)}\right)^2 = 0.03452$$

so the test statistics is

$$C_R^2 = \frac{1}{12 \times (14-1)} + 0.03452 = 0.04093.$$

For $\alpha = 0.05$, the critical value obtained form Table A.6 is 0.215. In this case we would not reject the null hypothesis that the power law process is an adequate model. The P-value for this test is greater than 0.20. □

i	t_i	$\hat{R}_i = (t_i/t_n)^{\bar{\beta}}$	$\frac{2i-1}{2(n-1)}$	$\hat{R}_i - \frac{2i-1}{2(n-1)}$
1	10	0.0790	0.0385	0.0405
2	55	0.1600	0.1154	0.0446
3	166	0.2528	0.1923	0.0605
4	205	0.2759	0.2692	0.0066
5	341	0.3406	0.3462	−0.0056
6	488	0.3951	0.4231	−0.0280
7	567	0.4204	0.5000	−0.0796
8	731	0.4670	0.5769	−0.1099
9	1308	0.5943	0.6538	−0.0596
10	2050	0.7158	0.7308	−0.0150
11	2453	0.7710	0.8077	−0.0367
12	3115	0.8512	0.8846	−0.0334
13	4017	0.9458	0.9615	−0.0158
14	4596	1.0000	1.0385	−0.0385

Table 4.9 Calculations for goodness-of-fit test.

The other transformation that can be used to develop a goodness-of-fit test for the power law process is the log-ratio transformation

$$U_i = \log(t_n/t_{n-i}), \qquad i = 1, 2, \ldots, n-1.$$

If the power law process with parameters β and θ is the true model, then given $T_n = t_n$, the random variables $T_1, T_2, \ldots, T_{n-1}$ are distributed as $n-1$ order statistics from the distribution with cdf

$$G(y) = \begin{cases} 0, & y < 0 \\ (y/t_n)^{\beta} & 0 \le y \le t_n \\ 1 & y > t_n. \end{cases}$$

Thus the random variables

$$\left(\frac{T_1}{t_n}\right)^\beta, \left(\frac{T_2}{t_n}\right)^\beta, \ldots, \left(\frac{T_{n-1}}{t_n}\right)^\beta$$

are distributed as $n-1$ order statistics from the UNIF(0,1) distribution. By Theorem 30 we conclude that

$$-\log\left(\frac{T_{n-1}}{t_n}\right)^\beta, -\log\left(\frac{T_{n-2}}{t_n}\right)^\beta, \ldots, -\log\left(\frac{T_2}{t_n}\right)^\beta, -\log\left(\frac{T_1}{t_n}\right)^\beta$$

or equivalently,

$$-\beta\log\frac{T_{n-1}}{t_n},\ -\beta\log\frac{T_{n-2}}{t_n},\ \ldots,\ -\beta\log\frac{T_2}{t_n},\ -\beta\log\frac{T_1}{t_n}$$

are distributed as $n-1$ order statistics from the EXP(1) distribution. Finally, division by β leads us to the conclusion that $U_1 < U_2 < \cdots < U_{n-1}$ are distributed as $n-1$ order statistics from an exponential distribution with mean $1/\beta$.

Once the transformation is made, any goodness-of-fit test for the exponential distribution with unknown mean (because the parameter β is usually unknown) can be used to test the adequacy of the power law process. The following tests have been developed to test for exponentiality:

1. Lilliefors' test
2. Kuiper's V test
3. Cramér-von Mises C^2 test
4. Watson's U^2 test
5. Anderson-Darling A^2 test
6. Shapiro-Wilk W test
7. Stephens' W^* test.

Here we will illustrate Lilliefors' (1969) test. See Shapiro and Wilk (1972), Stephens (1974), and Stephens (1978) for details and examples of the other tests. Notice that this transformation leads to the random variables $U_1 < U_2 < \cdots < U_{n-1}$ whose distribution depends only on β. This transformation "gets rid of" the nuisance parameter θ, thus reducing the problem from one involving two unknown parameters (β and θ) to one involving just one unknown parameter (β).

Lilliefors' test for exponentiality is a Kolmogorov-type test. The test statistic is the largest absolute difference between the estimated cumulative distribution function and the empirical distribution function. The empirical distribution function is the step function

$$S_U(u) = \frac{\text{number of } U_i\text{'s } \leq u}{n-1},$$

which can be written

$$S_U(u) = \begin{cases} 0, & \text{if } u < u_1, \\ j/(n-1), & \text{if } u_j \le u < u_{j+1}, \quad j = 1, 2, \cdots, n-2 \\ 1, & \text{if } u \ge u_{n-1} \,. \end{cases}$$

Recall that if there are n failures of the repairable system, there will be just $n-1$ of the U_j's. To get the estimated cumulative distribution we must first estimate the mean of the exponential distribution by

$$\hat{\mu} = \frac{\sum_{j=1}^{n-1} U_j}{n-1} \,.$$

Note that this exponential distribution occurs only because of the transformation made on the original failure times, and has no physical interpretation. The estimated cumulative distribution function is then

$$F_U(u) = 1 - \exp\left(-u/\hat{\mu}\right), \qquad u > 0.$$

The statistic T for Lilliefors' test of exponentiality is the largest absolute deviation between $F_U(u)$ and $S_U(u)$ which will always occur at a jump point in the step function. The quantities $A_j^L = |F_U(u_j) - S_U(u_j)|$ measure the absolute deviations at the left end points of the steps, and the quantities $A_j^R = |F_U(u_j) - S_U(u_{j-1})|$ measure the absolute deviations at the right end points. This largest deviation can be found by computing

$$T_1 = \max_{1 \le j \le n-1} |F_U(u_j) - S_U(u_j)| = \max_{1 \le j \le n-1} A_j^L \tag{4.16}$$

and

$$T_2 = \max_{1 \le j \le n-1} |F_U(u_j) - S_U(u_{j-1})| \max_{1 \le j \le n-1} A_j^R \,, \tag{4.17}$$

where u_0 is defined to be zero. The test statistic for testing whether $U_1 < U_2 < \cdots < U_{n-1}$ make up a random sample from an exponential distribution with mean $1/\beta$, that is, a test of whether $T_1 < T_2 < \cdots < T_n$ were generated by a power law process with parameters β and θ, is

$$T = \max\left(T_1, T_2\right).$$

The larger of T_1 and T_2 is then the maximum (actually, the supremum) absolute deviation between $F_U(u)$ and $S_U(u)$. Table A.7 gives critical values for the test statistic T. The null hypothesis that the power law process is the true model is rejected for values of T larger than those values in Table A.7.

Example 4.15 Table 4.10 shows the computations required to apply Lilliefors' test to the data in Table 4.8. The last two columns show the quantities

$|F_U(u_j) - S_U(u_j)|$ and $|F_U(u_j) - S_U(u_{j-1})|$. The largest entries in these two columns are $T_1 = 0.1001$ and $T_2 = 0.1771$. This makes the test statistic

$$T = \max(T_1, T_2) = 0.1771.$$

i	t_i	$U_j = \log(t_n/t_j)$	j	$F_U^*(u_j)$	$S_U(u_j)$	A_j^L	A_j^R
1	10	6.1304	13	0.9361	1.0000	0.0639	0.0130
2	55	4.4256	12	0.8627	0.9231	0.0604	0.0165
3	166	3.3210	11	0.7746	0.8462	0.0716	0.0054
4	205	3.1099	10	0.7522	0.7692	0.0170	0.0599
5	341	2.6011	9	0.6887	0.6923	0.0036	0.0733
6	488	2.2426	8	0.6344	0.6154	0.0190	0.0959
7	567	2.0926	7	0.6089	0.5385	0.0704	0.1474
8	731	1.8385	6	0.5617	0.4615	0.1001	0.1771
9	1308	1.2567	5	0.4309	0.3846	0.0463	0.1233
10	2050	0.8073	4	0.3039	0.3077	0.0038	0.0731
11	2453	0.6279	3	0.2455	0.2308	0.0147	0.0916
12	3115	0.3890	2	0.1601	0.1538	0.0063	0.0832
13	4017	0.1347	1	0.0586	0.0769	0.0183	0.0586
14	4569						

Table 4.10 Calculations for Lilliefors' test for aircraft generator data.

Table A.7 gives critical values for the goodness-of-fit test using this test statistic. Since we have $m = n - 1 = 13$ U_j's, we enter Table A.7 with $m = 13$ and we obtain, using interpolation, that the critical value for a test of size $\alpha = 0.05$ is 0.287. Since 0.1771 is not larger than 0.287, we have no reason to believe that the power law process is an inadequate model. Figure 4.19 shows the empirical cumulative distribution function $S_U(u)$ and the estimated cumulative distribution function $F_U^*(u)$. □

All of the goodness-of-fit tests mentioned previously have the same shortcoming. When the true process is a renewal process and the distribution of times between failure has a small coefficient of variation, then the failure times will be evenly spaced. Consequently the $\hat{R}_i^2$'s and the U_i's will also be evenly spaced. the tests suggested above are poor at detecting this "super-uniformity." Rigdon (1989) investigated the power of these tests and found them almost powerless at detecting renewal processes with small coefficients of variation. To increase the sensitivity of these tests we suggest a further transformation of the failure times. This transformation is called Durbin's modification and is defined by

$$D_j = (n + 1 - j)(U_j - U_{j-1}), \qquad j = 1, 2, \cdots, n - 1,$$

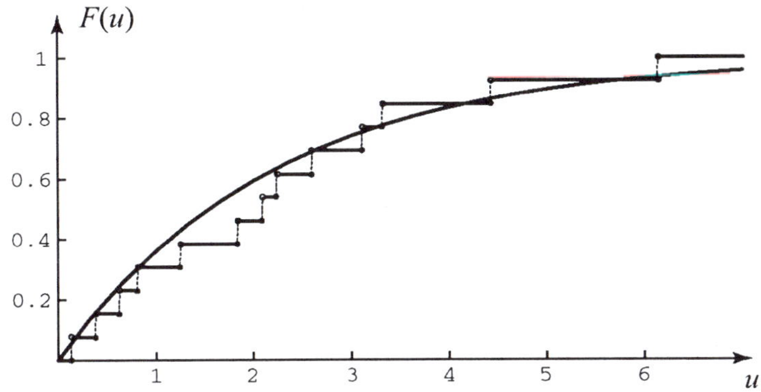

Fig. 4.19 Lilliefors' test for exponentiality.

where U_0 is defined to be zero. If the U_j's are order statistics from an exponential distribution, and they *are* if the power law process is the true model, then the D_j's are independent exponential random variables whose mean is the same as that of the U_j's; that is, the mean is $1/\beta$. Once this transformation is made, we can apply a goodness-of-fit test for the exponential distribution.

j	U_i	$D_j = (n-j)(U_j - U_{j-1})$
1	0.1347	1.7505
2	0.3890	3.0517
3	0.6279	2.6281
4	0.8073	1.7947
5	1.2567	4.0441
6	1.8385	4.6547
7	2.0926	1.7784
8	2.2426	0.9003
9	2.6011	1.7922
10	3.1099	2.0355
11	3.3210	0.6331
12	4.4256	2.2093
13	6.1304	1.7047

Table 4.10 Calculations for goodness-of-fit test using Durbin's modification.

Table 4.10 shows the normalized spacings $D_1, D_2, \ldots, D_{n-1}$ for the data on the aircraft generator. The calculations involved in determining Lilliefors' test statistic for the normalized spacings are shown in Table 4.11. The average of the D_j's is

$$\hat{\mu} = \frac{1}{n-1}\sum_{j=1}^{n-1} D_j = 2.229.$$

Here $T_1 = 0.3038$ and $T_2 = 0.3807$. The critical value for $\alpha = 0.05$ is 0.2877, so in this case we would reject the hypothesis that the power law process is an adequate model. Since the null hypothesis was rejected when Durbin's modification was applied but not otherwise, we conclude that the failure times are spaced more evenly than would have occurred by chance.

i	D_j	$F_D^*(d_j)$	$S_D(d_j)$	$\|F_D^*(d_j) - S_D(d_j)\|$	$\|F_D^*(d_j) - S_D(d_{j-1})\|$
1	0.6331	0.2472	0.0769	0.1703	0.2472
2	0.9003	0.3323	0.1538	0.1784	0.2554
3	1.7047	0.5346	0.2308	0.3038	0.3807
4	1.7505	0.5440	0.3077	0.2363	0.3133
5	1.7784	0.5497	0.3846	0.1651	0.2420
6	1.7922	0.5525	0.4615	0.0909	0.1679
7	1.7947	0.5530	0.5385	0.0145	0.0914
8	2.0355	0.5988	0.6154	0.0166	0.0603
9	2.2093	0.6289	0.6923	0.0635	0.0135
10	2.6281	0.6924	0.7692	0.0768	0.0001
11	3.0517	0.7457	0.8462	0.1005	0.0236
12	4.0441	0.8370	0.9231	0.0860	0.0091
13	4.6547	0.8761	1.0000	0.1239	0.0470

Table 4.11 Calculations for Lilliefors' test using Durbin's modification.

Klefsjö and Kumar (1992) suggested a way to use the TTT (total time on test) plot to form a goodness of fit test for the power law process. The key idea is that the area in the unit square under the TTT plot should be roughly 1/2 if the power law process is the correct model. They showed that this area can be calculated as follows. Define the statistic V to be

$$V = \sum_{i=1}^{n} u_j,$$

where $u_j = (n-j)(w_j - w_{j-1})$ and $w_j = -\log(T_{n-j}/T_n)$. The area under the TTT plot is then

$$A = \frac{V + 1/2}{n-1}.$$

When the null hypothesis of the power law process is true, we would expect the area under the TTT plot to be 1/2, so we would then expect

$$\frac{1}{2} = \frac{V + 1/2}{n - 1},$$

that is, $V = n/2 - 1$. Barlow, Bartholomew, Bremner and Brunk (1972) showed that the statistic V is asymptotically normal with mean $n/2 - 1$ and variance $(n-2)/12$. A test of the power law process can then be based on the statistic

$$z = \frac{V - (n/2 - 1)}{\sqrt{(n-2)/12}},$$

which has approximately a standard normal distribution.

Example 4.16 Consider the data from System 1 in Table 4.1. This same data set was studied in Example 4.1 where we presented the TTT plot (Figure 4.7). The V statistic is $V = 4.60$ and so $z = -0.44$. For a test with $\alpha = 0.95$, we would reject the null hypothesis that the power law process is the model governing the failure process if $|z| > 1.96$. In this case we would fail to reject the null hypothesis. Hence, for the data on this system, there is no evidence that the power law process is an inadequate model. □

Klefsjö and Kumar (1992) point out that their test is virtually powerless against a renewal process alternative. They therefore suggest using Durbin's modification to detect a renewal process.

4.6 STATISTICAL INFERENCE FOR THE TIME TRUNCATED CASE

If testing stops at a predetermined time t, then the data are said to be time truncated. In this case, the number of failures N is random and the time that testing stops is fixed. (In the failure truncated case, just the reverse is true; the number of failures is fixed and the time that testing stops is random.) This difference in the data collection strategy causes the statistical inference procedures to be slightly different. Many of the procedures, however, are similar to the failure truncated case.

4.6.1 Point Estimation for β and θ

Let $T_1 < T_2 < \cdots < T_N < t$ denote the observed failure times before time t. Since the number of failures N is random, we observe in the time truncated case a random number of random variables. For the time truncated case that we are considering here, the likelihood must take into account the randomness of N as well as the randomness of the failure times $t_1 < t_2 < \cdots < t_N$. (It is possible that we observe no failures before time t. In this case $N = 0$ and the MLE's of β and θ do not exist.) The likelihood equation for the time truncated can be derived as follows: The joint density of $(N, T_1, \ldots, T_N)$ is

$$f(n, t_1, \ldots, t_n) = \begin{cases} f_N(n) f(t_1, \ldots, t_n | n), & n \geq 1 \\ f_N(0), & n = 0. \end{cases}$$

The random variable N has a Poisson distribution with mean $(t/\theta)^\beta$, so

$$f_N(n) = \frac{\left[(t/\theta)^\beta\right]^n \exp\left[-(t/\theta)^\beta\right]}{n!}, \qquad n = 0, 1, \ldots.$$

The conditional distribution of $T_1 < T_2 < \cdots < T_N$ given $N = n$ was given in Theorem 26 of Chapter 2. These random variables are distributed as n order statistics from the distribution with cdf

$$G(y) = \begin{cases} 0 & y < 0 \\ \Lambda(y)/\Lambda(t) & 0 \leq y \leq t \\ 1 & y > t. \end{cases}$$

For the power law process this is

$$G(y) = \frac{(y/\theta)^\beta}{(t/\theta)^\beta} = \left(\frac{y}{t}\right)^\beta, \qquad 0 \leq y \leq t.$$

The density corresponding to G would then be

$$g(y) = \frac{\beta}{t}\left(\frac{y}{t}\right)^{\beta-1}, \qquad 0 \leq y \leq t.$$

Given $N = n$, the distribution of $T_1 < T_2 < \cdots < T_n$ is therefore

$$\begin{aligned} f(t_1, t_2, \ldots, t_n | n) &= n! \prod_{i=1}^{n} G'(t_i) \\ &= n! \prod_{i=1}^{n} \frac{\beta}{t}\left(\frac{t_i}{t}\right)^{\beta-1}, \qquad 0 < t_1 < t_2 < \cdots < t_n < t. \end{aligned}$$

The joint density of N and $T_1 < T_2 < \cdots < T_N$ is thus

$$\begin{aligned} f(n, t_1, t_2, \ldots, t_n) &= \frac{(t/\theta)^{n\beta} \exp\left[-(t/\theta)^\beta\right]}{n!} n! \prod_{i=1}^{n} \frac{\beta}{t}\left(\frac{t_i}{t}\right)^{\beta-1} \\ &= \frac{t^{n\beta}}{\theta^{n\beta}} \frac{\beta^n}{t^n} \left[\prod_{i=1}^{n}\left(\frac{t_i}{t}\right)\right]^{\beta-1} \exp\left[-(t/\theta)^\beta\right] \\ &= \frac{\beta^n}{\theta^{n\beta}} \left(\prod_{i=1}^{n} t_i\right)^{\beta-1} \exp\left[-(t/\theta)^\beta\right], \qquad (4.18) \\ & \qquad n \geq 1,\ 0 < t_1 < t_2 < \cdots < t_n < t \end{aligned}$$

and

$$f(0) = \exp\left[-(t/\theta)^{\beta}\right], \qquad n = 0.$$

This is nearly identical to the likelihood function from the failure truncated case; this is remarkable given the completely different approaches to their derivations. The log-likelihood function is thus

$$\ell(\theta, \beta | n, \mathbf{t}) = n \log \beta - n \log \theta + (\beta - 1) \sum_{i=1}^{n} \log t_i - n(\beta - 1) \log \theta - \left(\frac{t}{\theta}\right)^{\beta}.$$

The derivation of the MLE's is nearly the same as for the failure truncated case. If at least one failure occurs before time t then the maximum likelihood estimators (MLE's) exist and are equal to

$$\hat{\beta} = \frac{N}{\sum_{i=1}^{N} \log(t/t_i)} \tag{4.19}$$

and

$$\hat{\theta} = \frac{t}{N^{1/\hat{\beta}}}. \tag{4.20}$$

Compare these expressions in (4.19) and (4.20) with the corresponding expressions for the failure truncated case that are found in (4.9) and (4.10). The MLE's are similar, the major changes being the replacement of N for n, and t for t_n. Note also, however, that the sum in the denominator for $\hat{\beta}$ goes up to $i = N$, rather than just up to $i = n - 1$.

It can be shown that if $n \geq 1$ then

$$E\left(\frac{N-1}{N}\hat{\beta} \,\middle|\, N = n\right) = \beta.$$

Thus the estimator

$$\bar{\beta} = \frac{N-1}{N}\hat{\beta} = \hat{\beta} = \frac{N-1}{\sum_{i=1}^{N} \log(t/t_i)}$$

is called the *conditionally unbiased* estimator for β.

4.7.2 Interval Estimation and Tests of Hypotheses

Confidence intervals and tests of hypotheses for β are based on the following result.

Theorem 32 *For a power law process truncated at time t, conditioned on $N = n$, the quantity $2n\beta/\hat{\beta}$ has a chi-square distribution with $2n$ degrees of freedom.*

Proof. From Theorem 22 of Chapter 2, we conclude that conditioned on $N = n$, the random variables $T_1 < T_2 < \cdots < T_n < t$ are distributed as n order statistics from the distribution with cdf

$$G(y) = \begin{cases} 0, & y \leq 0 \\ (y/t)^\beta, & 0 < y < t \\ 1, & y \geq t\,. \end{cases}$$

Let Y be a random variable with cdf G. Then for $0 < y < t$

$$\left(\frac{y}{t}\right)^\beta = G(y) = P(Y \leq y) = P\left(Y/t \leq y/t\right) = P\left((Y/t)^\beta \leq (y/t)^\beta\right).$$

This implies that the random variable $(Y/t)^\beta$ is uniformly distributed over the interval $[0,1]$. Consequently, $(T_i/t)^\beta,\ i = 1, 2, \ldots, n$ are distributed as n order statistics from the uniform distribution over the interval $[0,1]$. Applying Lemma 30 again we can say that

$$\sum_{i=1}^{n}\left(-\log(t_i/t)^\beta\right) = -\beta\sum_{i=1}^{n}\log(t_i/t)$$

has a GAM$(n,1)$ distribution. Arguing as before, we conclude that

$$\frac{2n\beta}{\hat\beta} = 2n\beta\left(\frac{n}{\sum_{i=1}^{n}\log(t/t_i)}\right)^{-1} = 2\beta\sum_{i=1}^{n}\log(t/t_i) = -2\beta\sum_{i=1}^{n}\log(t_i/t)$$

has a chi-square distribution with $2n$ degrees of freedom. ■

Thus a $100(1-\alpha)\%$ confidence interval for β is given by

$$\frac{\chi^2_{1-\alpha/2}(2n)\hat\beta}{2n} < \beta < \frac{\chi^2_{\alpha/2}(2n)\hat\beta}{2n}\,. \tag{4.21}$$

Since the power law process reduces to the homogeneous Poisson process when $\beta = 1$, it is often of interest to see if $\beta = 1$ is in the confidence interval.

The result that $2n\beta/\hat\beta$ has a chi-square distribution with $2n$ degrees of freedom can also be used to construct a size α test of

$$H_0 : \beta = \beta_0 \quad \text{versus} \quad H_1 : \beta \neq \beta_0.$$

The rule is to reject H_0 if

$$2n\beta_0/\hat\beta < \chi^2_{1-\alpha/2} \quad \text{or} \quad 2n\beta_0/\hat\beta > \chi^2_{\alpha/2},$$

that is, if

$$\hat\beta < \frac{2n\beta_0}{\chi^2_{\alpha/2}} \quad \text{or} \quad \hat\beta > \frac{2n\beta_0}{\chi^2_{1-\alpha/2}}\,. \tag{4.22}$$

Again, it is often useful to test $H_0 : \beta = 1$ versus $H_1 : \beta \neq 1$ since the power law process reduces to the homogeneous Poisson process when $\beta = 1$. One-sided tests, that is tests with a one-sided alternative such as $H_1 : \beta \leq 1$ or $H_1 : \beta \geq 1$ can also be constructed.

There appears to be no method of obtaining exact confidence intervals for θ alone when the data are time truncated. Bain and Engelhardt (1980) have suggested an approximate method that uses the tables of Finkelstein (1976). The parameter has no simple interpretation, so confidence intervals for θ are usually not computed.

4.6.3 Estimation of the Intensity Function

The MLE of the intensity function at time t is

$$\hat{u}(t) = \frac{\hat{\beta}}{\hat{\theta}} \left(\frac{t}{\hat{\theta}} \right)^{\hat{\beta}-1} = \frac{N\hat{\beta}}{t}. \tag{4.23}$$

This quantity represents the rate, in failures per time unit, that failures are occurring at time t. Rigdon and Basu (1990b) studied estimators in the class $\left\{ c_N \hat{\beta}/t \right\}$ of estimators of $\lambda(t)$. Based on the results from the failure truncated case, the estimator

$$\hat{u}_{MMSE} = \frac{(N-2)(N-3)\hat{\beta}}{Nt}$$

was suggested. For a wide range of β this estimator has high efficiency relative to the MLE shown above in (4.23).

Confidence intervals for $\lambda(t)$ are easily found, by looking up the values ρ_1 and ρ_2 in Table A.5. The confidence interval is then

$$\rho_1 \hat{u}(t) < \lambda(t) < \rho_2 \hat{u}(t).$$

Example 4.17 To illustrate the statistical inference for the time truncated case, let's consider the failure times given in Table 4.12. These failure data, which were given by Crow (1990) were time truncated at time $t = 2000$. The MLE's of β and θ are

$$\hat{\beta} = \frac{9}{\sum_{i=1}^{9} \log(t_n/t)} = \frac{9}{23.9806} = 0.375$$

and

$$\hat{\theta} = \frac{2000}{9^{1/\hat{\beta}}} = 5.773.$$

A conditionally unbiased estimator for β is

$$\bar{\beta} = \frac{8}{9}\hat{\beta} = 0.334.$$

i	t_i	$\log(2000/t_i)$
1	1.2	7.419
2	55.6	3.583
3	72.7	3.315
4	111.9	2.883
5	121.9	2.798
6	303.6	1.885
7	326.9	1.811
8	1568.4	0.243
9	1913.5	0.044

Table 4.12 Data from a process time truncated at $t = 2000$.

To get a 95% confidence interval for β we look up $\chi^2_{0.975}(2 \times 9) = 8.231$ and $\chi^2_{0.025}(2 \times 9) = 31.526$. The confidence interval for β is thus

$$\frac{0.375 \times 8.231}{18} < \beta < \frac{0.375 \times 31.526}{18},$$

which is equivalent to

$$0.171 < \beta < 0.657.$$

Note that this interval excludes 1, so there is evidence of system reliability improvement.

For a size α test of the hypotheses $H_0 : \beta = 1$ versus $H_1 : \beta \neq 1$ we would reject the null hypothesis if

$$\hat{\beta} < \frac{18}{31.526} = 0.571 \qquad \text{or} \qquad \hat{\beta} > \frac{18}{8.231} = 2.187.$$

Therefore, we would reject the null hypothesis in this example.

The point estimate of the intensity function at time $t = 2000$ (the time that testing stopped) is

$$\hat{u}(2000) = \frac{9 \times 0.375}{2000} = 0.00169.$$

A confidence interval for $\lambda(2000)$ is obtained by looking up the values $\rho_1 = 0.3044$ and $\rho_2 = 2.4832$ in Table A.5. The confidence interval is then

$$0.3044 \times 0.00169 < \lambda(2000) < 2.4832 \times 0.00169,$$

which is equivalent to

$$0.00051 < \lambda(2000) < 0.00420.$$

This is a confidence interval for the value of the intensity function at time $t = 2000$. □

4.7.4 Goodness-of-Fit Tests

Goodness-of-fit tests for the time truncated case can be constructed, and are similar to tests for the failure truncated case. Two transformations can be applied to the original failure times. The ratio-power transformation is

$$\hat{R}_i = \left(\frac{t_i}{t}\right)^{\bar{\beta}}.$$

The Cramér-von Mises test statistics then becomes

$$C_R^2 = \frac{1}{12n} + \sum_{i=1}^{n}\left(\hat{R}_i - \frac{2i-1}{2n}\right)^2. \tag{4.24}$$

Table A.6 gives critical values for the test statistic C_R^2 for various values of M. If the data are time truncated, we enter Table A.6 with $M = n$. (Recall that for failure truncated data, we entered the table with $M = n - 1$.) A value of C_R^2 larger than the tabulated entry leads to rejection of the null hypothesis

H_0 : the failure times were governed by a power law process.

The rationale for this transformation is the same as it was for the failure truncated case. Here again, a plot of $\hat{R}_i$ on the horizontal axis, against its expectation

$$E\left(\hat{R}_i\right) = \frac{2i-1}{2n}$$

on the vertical axis can show where there are deviations from the power law process. Such a plot is expected to be approximately linear along the 45° line that goes through the origin.

The other transformation that leads to a goodness-of-fit test for the power law process is the log-ratio transformation

$$W_j = \log\left(\frac{t}{t_{n-j+1}}\right), \qquad j = 1, 2, \ldots n.$$

If the power law process is the correct model, then the W_j's are distributed as n order statistics from an exponential distribution with mean $1/\beta$. An estimate for $1/\beta$ is given by

$$\hat{\mu}_T = \sum_{j=1}^{n} W_j/n = \frac{1}{\hat{\beta}}.$$

Thus, once this transformation is made, a goodness-of-fit test for testing exponentiality can be applied. Several of these tests were mentioned in the previous section on failure truncated data. Here we discuss Lilliefors' Kolmogorov-type test. The test statistic in the time truncated case, as it was for the failure truncated case is the largest difference between the empirical distribution function for the w's

$$S_W(w) = \begin{cases} 0, & w < w_1 \\ j/n, & w_j \leq w < w_{j+1} \\ 1, & w \geq w_n \end{cases}$$

and the estimated cumulative distribution function (assuming the W's are exponentially distributed)

$$F_W^*(w) = 1 - \exp\left[-w/\mu_T\right], \qquad w > 0.$$

The largest (actually the supremum) difference between $S_W(w)$ and $F_W^*(w)$ can be found by taking the larger of

$$T_1 = \max_{1 \leq j \leq n} |F_W^*(w_j) - S_W(w_j)| = \max_{1 \leq j \leq n} A_j^L \tag{4.25}$$

and

$$T_2 = \max_{1 \leq j \leq n} |F_W^*(w_j) - S_W(w_{j-1})| = \max_{1 \leq j \leq n} A_j^R. \tag{4.26}$$

The test statistic, $T = \max(T_1, T_2)$ is then compared to the critical value found in Table A.7. Test statistics larger than the critical value lead to rejection of the null hypothesis that the power law process is the correct model.

We noted in the previous section that these goodness-of-fit tests are almost powerless for detecting a renewal process, or any failure process where the failure times are more evenly spaced than expected. The same story applies in the time truncated case as well. For this situation, we suggest using Durbin's modification

$$D_j = (n + 1 - j)(W_j - W_{j-1}), \qquad j = 1, 2, \ldots, n,$$

where W_0 is defined to be 0. Assuming the power law process governs the failure times, the D_j's are distributed independently and exponentially with mean $1/\beta$. Lilliefors' test, as it was described in the previous section, could then be applied to the D_j's.

Example 4.18 For the data in Table 4.12, Lilliefors' (1969) goodness-of-fit test statistics (without applying Durbin's modification) are

$$T_1 = 0.150, \quad T_2 = 0.271, \quad \text{and} \quad T = \max(T_1, T_2) = 0.271.$$

The critical value 0.340 for a size $\alpha = 0.05$ test is obtained from Table A.7. Since our test statistic $T = 0.271$ does not exceed the tabulated value, we have no evidence to reject the null hypothesis that the power law process is an adequate model. This example was done without applying Durbin's modification; we leave it as an exercise to perform the goodness-of-fit test using Durbin's modification.

i	t_i	j	W_j	$F_W^*(w_j)$	$S_W(w_j)$	A_j^L	A_j^R
1	1.2	9	7.419	1.000	0.938	0.062	0.049
2	55.6	8	3.583	0.889	0.739	0.150	0.039
3	72.7	7	3.315	0.778	0.712	0.066	0.045
4	111.9	6	2.883	0.667	0.661	0.006	0.105
5	121.9	5	2.798	0.556	0.650	0.094	0.106
6	303.6	4	1.885	0.444	0.507	0.063	0.164
7	326.9	3	1.811	0.333	0.493	0.160	0.271
8	1568.4	2	0.243	0.222	0.087	0.135	0.024
9	1913.5	1	0.044	0.111	0.016	0.095	0.016
t	2000.0	0	0.000	0.000			

Table 4.13 Computations for Lilliefors' test for time truncated data.

The Cramér-von Mises test statistic is

$$C_R^2 = \frac{1}{12 \times 9} + \sum_{i=1}^{n} \left(\hat{R}_i - \frac{2i-1}{2 \times 9} \right)^2 = 0.0093 + 0.804 = 0.0897.$$

The critical value for the Cramér-von Mises test with $\alpha = 0.05$ is 0.167, so again, we fail to reject the null hypothesis. □

i	t_i	$\hat{R}_i = (t_i/t)^{\hat{\beta}}$	$(2i-1)/2n$	$\hat{R}_i - (2i-1)/2n$
1	1.2	0.084	1/18	0.028
2	55.6	0.303	3/18	0.136
3	72.7	0.331	5/18	0.053
4	111.9	0.382	7/18	−0.007
5	121.9	0.393	9/18	−0.107
6	303.6	0.533	11/18	−0.078
7	326.9	0.546	13/18	−0.176
8	1568.4	0.922	15/18	−0.089
9	1913.5	0.985	17/18	0.041

Table 4.14 Calculations for Cramèr-von Mises test.

4.7 THE EFFECT OF ASSUMING AN HPP WHEN THE TRUE PROCESS IS A POWER LAW PROCESS

The two previous sections discussed inference for the homogeneous Poisson process and the power law process. How does one go about selecting between these two models? What are the effects if we choose the simpler model when the process is really a power law process? We studied this latter question in a paper in the *Journal of Quality Technology* (Rigdon and Basu, 1990). In that paper, we studied the effect of model misspecification on estimation of the intensity at the conclusion of testing; that is, at time t_n if the data are failure truncated, and at time t if the data are time truncated.

If the homogeneous Poisson process were assumed, then the estimate of the intensity would be

$$\bar{u} = \frac{n}{t_n}$$

if the data were failure truncated, and

$$\bar{u} = \frac{N}{t}$$

if the data were time truncated. On the other hand, if the power law process were assumed, then the estimates would be

$$\hat{u}(t_n) = \frac{n\hat{\beta}}{t_n}$$

for failure truncated data, and

$$\hat{u}(t) = \frac{N\hat{\beta}}{t}$$

for time truncated data.

The mean squared error (MSE) of an estimator is defined to be

$$\text{MSE} = E\left[(\text{estimator} - \text{true parameter})^2\right].$$

The **efficiency** of the estimator $\bar{u}$ relative to $\hat{u}$ is the ratio of MSE's

$$\text{Efficiency} = \frac{\text{MSE of } \hat{u}}{\text{MSE of } \bar{u}}.$$

An efficiency greater than one indicates that the estimator $\hat{u}$ has a larger MSE than $\bar{u}$, in which case $\bar{u}$ is the better estimator. On the other hand, if the efficiency is less than one, then the MSE of $\hat{u}$ has a smaller MSE than $\bar{u}$, in which case $\hat{u}$ is the better estimator.

In Rigdon and Basu (1990) we derived expressions for the MSE's for both estimators, $\bar{u}$ and $\hat{u}$, and we computed the efficiency for $n = 5, 10, 20, 40$, and

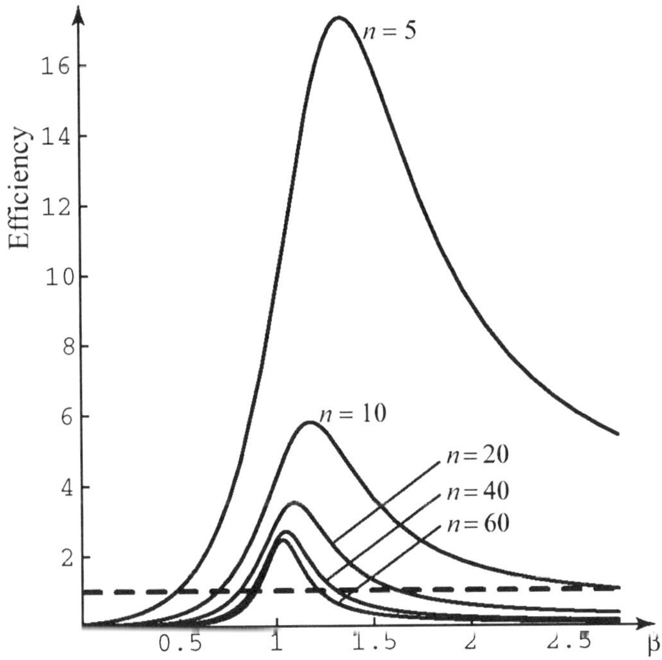

Fig. 4.20 Efficiency of $\overline{u} = n/t_n$ relative to $\hat{u} = n\hat{\beta}/t_n$.

60 (failure truncated case), and $(t/\theta)^\beta = 5, 10, 20, 40$, and 60 (time truncated case). Note that, in the failure truncated case, $(t/\theta)^\beta$ is the expected number of failures before the truncation time t. For both estimators, $\bar{u}$ and $\hat{u}$, the MSE is a multiple of θ. Thus when we compute the ratio of MSE's, the θ's cancel, so the efficiency is *independent* of θ. We can then plot the efficiency as a function of β. Figure 4.20 shows the efficiencies for $n = 5, 10, 20, 40$, and 60 when the data are failure truncated. This figure indicates that for small sample sizes, say ten or less, the homogeneous Poisson process estimator $\bar{u}$ is better than the power law process estimator $\hat{u}$ for a wide range of β. Similar results apply for the time truncated case.

We also considered the effect of model misspecification on interval estimation of the intensity function. Figure 4.21 shows the probability that a confidence interval for $\lambda(t_n)$, calculated under the assumption of a homogeneous Poisson process, includes the true value of $\lambda(t_n)$. This probability is shown as a function of β for $n =$ 5, 10, 20 and 40. For small sample sizes n the probability that the confidence interval will include the true value of the intensity is greater than $\frac{1}{2}$ for a side range of β.

The conclusion we draw from this analysis is that for small sample sizes (certainly for $n \leq 5$, and depending of the true value of β, also for $n \leq 10$) it is better to assume the homogeneous Poisson process even if some deterioration

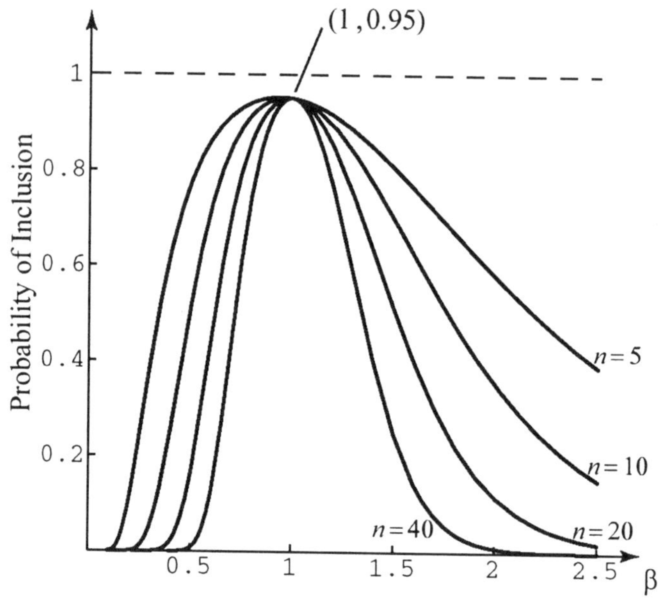

Fig. 4.21 Probability that a confidence interval calculated under the HPP assumption includes the true value of $u(t_n)$.

or improvement is expected. When the sample size is small, there is a lot of variability in the estimates of β and θ, and hence in the estimate of the intensity. In this case, it us usually better to have a good estimate of the wrong thing (the value of the intensity under the incorrect homogeneous Poisson process model) than a poor estimate of the right thing (the value of the intensity under the correct power law process model).

4.8 BAYESIAN ESTIMATION

The classical approach to statistical inference treats parameters as fixed but unknown values. In contrast, the Bayesian approach treats parameters as unobserved random variables. We give here a brief summary of the Bayesian approach.

Suppose that γ is some unknown parameter, and that we can observe random variables $X_1, X_2, \ldots, X_n$ having joint probability density function

$$f(x_1, x_2, \ldots, x_n|\gamma), \tag{4.27}$$

which depends on the unknown parameter γ. (The parameter γ and the random variables $X_1, X_2, \ldots, X_n$ could be vectors. For the power law process, we think of $X_1, X_2, \ldots, X_n$ as being the observed failure times $T_1 < T_2 <$

$\cdots < T_n$, and we think of the parameter γ as being the vector (θ, β).) Before observing any data, we select for γ a probability distribution (a joint distribution if γ is a vector) that reflects our degree of belief in the value of the parameter γ. This is called the **prior distribution** and we will denote it by $p(\gamma)$. Once we observe the data $x_1, x_2, \ldots, x_n$, we compute the **posterior distribution**, that is, the distribution of the unknown parameter given the observed data $x_1, x_2, \ldots, x_n$, by applying Bayes' Theorem:

$$p(\gamma | x_1, x_2, \ldots, x_n) = \frac{p(\gamma) f(x_1, x_2, \ldots, x_n | \gamma)}{\displaystyle\int p(\gamma) f(x_1, x_2, \ldots, x_n | \gamma)\, d\gamma},$$

where f is the joint probability density function given in (4.27). Notice that the denominator is a function of $x_1, x_2, \ldots, x_n$ but not of γ. The function $f(x_1, x_2, \ldots, x_n | \gamma)$ is usually called the likelihood function. The above result can then be summarized by saying that the posterior distribution is proportional to the prior times the likelihood. This posterior distribution then reflects what is known about the unknown parameter γ *after* observing the data $x_1, x_2, \ldots, x_n$.

If γ is a scalar, then the point estimate for γ is usually taken to be the posterior mean. This is obtained by integrating γ times the posterior pdf, that is,

$$\hat{\gamma} = \int \gamma\, p(\gamma | x_1, x_2, \ldots, x_n)\, d\gamma = \frac{\displaystyle\int \gamma\, p(\gamma)\, f(x_1, x_2, \ldots, x_n | \gamma)\, d\gamma}{\displaystyle\int p(\gamma)\, f(x_1, x_2, \ldots, x_n | \gamma)\, d\gamma}. \tag{4.28}$$

If γ is a vector, say $\gamma = (\gamma_1, \ldots, \gamma_p)$, then the point estimate of γ_i is

$$\hat{\gamma}_i = \underbrace{\int \cdots \int}_{p \text{ integrals}} \gamma_i\, p(\gamma_1, \ldots, \gamma_p | x_1, x_2, \ldots, x_n)\, d\gamma_p \cdots d\gamma_1$$

$$= \frac{\displaystyle\int \cdots \int \gamma_i\, p(\gamma_1, \ldots, \gamma_p)\, f(x_1, x_2, \ldots, x_n | \gamma_1, \ldots, \gamma_p)\; d\gamma_p \cdots d\gamma_1}{\displaystyle\int \cdots \int p(\gamma_1, \ldots, \gamma_p)\, f(x_1, x_2, \ldots, x_n | \gamma_1, \ldots, \gamma_p)\; d\gamma_p \cdots d\gamma_1}. \tag{4.29}$$

If the sample size is large, then under certain regularity conditions the posterior distribution is approximately normal. This fact can be used to obtain the approximate $100 \times (1 - \alpha)\%$ confidence interval for γ

$$\hat{\gamma} \pm z_{\alpha/2} \sqrt{V(\gamma_i | x_1, x_2, \ldots, x_n)},$$

where $V(\gamma_i | x_1, x_2, \ldots, x_n)$ denotes the posterior variance of γ_i, that is,

$$V(\gamma_i | x_1, x_2, \ldots, x_n) = E(\gamma_i^2 | x_1, x_2, \ldots, x_n) - \hat{\gamma}_i^2.$$

Compare the Bayesian approach with the classical approach to statistics where parameters are treated as fixed and nonrandom (but unknown) values. Many statisticians and engineers see the Bayesian paradigm as the only way to approach statistical inference. Others see the subjective selection of the prior distribution as violating the objectivity of the data analysis. We caution the reader about the controversy surrounding the Bayesian approach. While we do not go into the debate here, we refer the reader to other references that support or denounce Bayesian analysis of reliability. A recent exchange in *IEEE Transactions on Reliability* featured articles entitled "Bayes Is for the Birds" (Evans, 1989), "Comment on Bayes Is for the Birds" (Buynoski, 1990), "Bayes Is for Eagles" (Kaplan, 1990), and "Bayesianism Is for Turkeys" (Dougherty, 1991). The book *Bayesian Reliability Analysis* by Martz and Waller (1982) discusses thoroughly the Bayesian approach to reliability analysis. Easterling (1972) gives a strong argument against using Bayesian inference for the analysis of reliability data.

Besides the controversial aspect of the Bayesian approach, we should also point out that computations required to obtain point and interval estimates for the parameters are often complicated integrals that cannot be evaluated analytically. In these cases, the integrals (often *multiple* integrals) must be evaluated numerically. Software is often needed to do the required computations.

Finally, although the expressions for point and interval estimates of the model parameters often look complicated (for example, ratios of double integrals) the approach to Bayesian inference is quite straightforward:

1. Choose a prior distribution $p(\gamma)$ that reflects the prior information about the parameter.

2. Collect the data.

3. Base the inference on the posterior distribution.

Often prior information is based on reliable sources of information, such as similar product models having known failure characteristics. In cases like this, the prior information would prove valuable in terms of reducing the testing time or increasing the accuracy of estimates. Practitioners must avoid letting wishful thinking guide their choice of a prior distribution. This can lead to a situation where the data are inconsistent with the proposed prior distribution. In a case like this, the prior will have an unduly strong influence on the posterior, which can lead to misleading conclusions.

4.8.1 Bayesian Inference for the Parameters of the HPP

We treat separately the failure truncated and time truncated cases. Throughout this section, we assume that the HPP is a reasonable model for the repairable system.

4.8.1.1 Failure Truncated Case Suppose that we observe the failure times $T_1 < T_2 < \cdots < T_n$ from a system whose failure process is modeled by an HPP. Then the times between failure, $X_i = T_i - T_{i-1}$ (where $T_0 \equiv 0$) are iid exponential random variables with mean $1/\lambda$. The likelihood function is

$$p(\mathbf{x} \mid \lambda) = \prod_{i=1}^{n} \lambda \exp\left[-\lambda x_i\right] = \lambda^n \exp\left[-\lambda \sum_{i=1}^{n} x_i\right] = \lambda^n \exp\left[-\lambda t_n\right], \quad \lambda > 0.$$

The conjugate prior for the intensity λ is the gamma, with pdf

$$p(\lambda) = \frac{b^a \lambda^{a-1}}{\Gamma(a)} e^{-b\lambda}, \qquad \lambda > 0. \tag{4.30}$$

The parameters a and b can be chosen by relating the prior mean

$$\mu = \frac{a}{b}$$

and the prior variance

$$\sigma^2 = \frac{a}{b^2}$$

Solving for a and b yields

$$a = \mu^2/\sigma^2$$

and

$$b = \mu/\sigma^2.$$

Thus, the posterior distribution of λ given $\mathbf{x}$, or equivalently, given the sufficient statistic t_n, is

$$\begin{aligned} p(\lambda \mid \mathbf{x}) &= c\left(\frac{b^a \lambda^{a-1}}{\Gamma(a)} \exp\left[-b\lambda\right]\right)\left(\lambda^n \exp\left[-\lambda t_n\right]\right) \\ &= c_1 \lambda^{a+n-1} \exp\left[-\left(b + t_n\right)\lambda\right]. \end{aligned}$$

This indicates that the posterior distribution of λ is $\text{GAM}(a+n, b+t_n)$. Thus the posterior mean is

$$\hat{\lambda}_B = E\left(\lambda \mid \mathbf{x}\right) = \frac{a+n}{b+t_n}.$$

A noninformative prior for the parameter λ is $p(\lambda) = 1/\lambda$. We leave it as an exercise to work out the details for this noninformative prior.

4.8.1.2 Time Truncated Case Suppose that a single system is placed on test until time t, where t is predetermined. The number of failures N is a sufficient statistic for the parameter λ and, conditioned on λ, the random variable N has a Poisson distribution with mean λ. The likelihood function is

$$p(n \mid \lambda) = \frac{(\lambda t)^n \, e^{-\lambda t}}{n!}, \qquad n = 0, 1, 2, \ldots .$$

The conjugate prior for the Poisson parameter λ is the gamma, with pdf

$$p(\lambda) = \frac{b^a \lambda^{a-1}}{\Gamma(a)} e^{-b\lambda}, \qquad \lambda > 0. \tag{4.31}$$

The parameters a and b can be chosen be relating the prior mean variance as described in the previous subsection. The posterior distribution is therefore

$$\begin{aligned} p(\lambda \mid n) &= cp(\lambda)p(n \mid \lambda) \\ &= \frac{b^a \lambda^{a-1}}{\Gamma(a)} e^{-b\lambda} \frac{(\lambda t)^n \, e^{-\lambda t}}{n!} \\ &= c' \lambda^{a+n-1} \exp\left[-\left(b+t\right)\lambda\right], \qquad \lambda > 0. \end{aligned}$$

Thus, conditioned on $N = n$, the distribution of λ is $\text{GAM}(a+n, b+t)$. The posterior mean is therefore

$$\hat{\lambda}_B = E\left(\lambda \mid \mathbf{x}\right) = \frac{a+n}{b+t}$$

and the posterior variance is

$$V\left(\lambda \mid \mathbf{x}\right) = \frac{a+n}{(b+t)^2} \, .$$

Example 4.19 Suppose that a model of photocopy machines has historically had a failure intensity of 0.05 failures per day of operation. (Mention of a failure intensity that is a fixed number implies an HPP.) A new copy machine is designed that has mostly the same features and many of the same components. Because the designs are so similar, it may be reasonable to use the prior information about the previous model to specify a prior distribution. Suppose that a gamma prior distribution is chosen with mean $\mu = 0.05$ and $\sigma = 0.015$. This leads to prior parameters of approximately $a = 11.111$ and $b = 222.222$. The prior pdf is shown in Figure 4.22

Suppose that in $t = 200$ days of testing, there were $n = 21$ failures. The posterior distribution is thus a $\text{GAM}(11.111 + 22,\ 222.222 + 200)$, or a $\text{GAM}(33.111, 422.222)$ distribution; its pdf is shown in Figure 4.23. The posterior mean is therefore

$$\hat{\lambda}_B = \frac{a+n}{b+t} = \frac{33.111}{422.222} = 0.0784$$

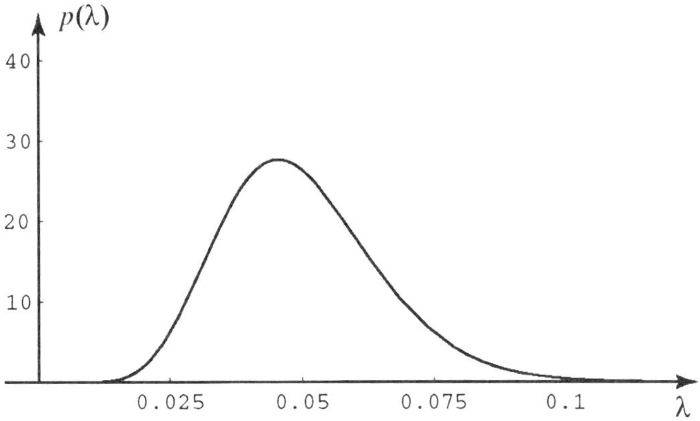

Fig. 4.22 Prior pdf for intensity λ.

and the posterior variance is

$$V(\lambda \mid n) = \frac{a+n}{(b+t)^2} = \frac{33.111}{(422.222)^2} \approx 0.0001857.$$

Since the posterior standard deviation is $\sqrt{V(t \mid n)} \approx \sqrt{0.0001857} \approx 0.01363$, an approximate 95% posterior probability interval is

$$\hat{\lambda} \pm 1.96\sqrt{V(t \mid n)} \approx 0.0784 \pm 1.96 \times 0.01363$$

or, in interval notation, $(0.0517, 0.1052)$. □

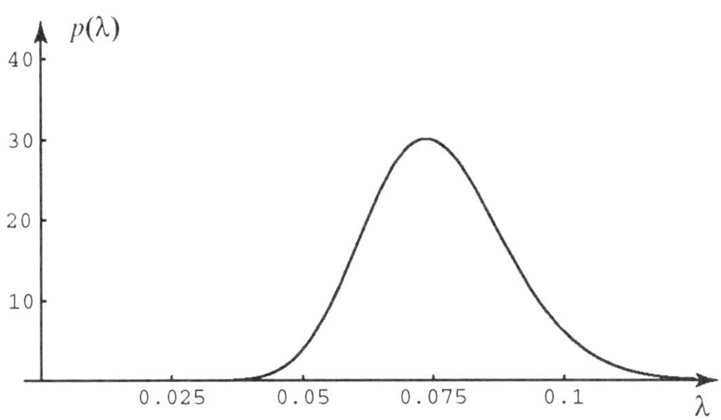

Fig. 4.23 Posterior pdf of λ given 21 failures in 200 days of testing.

4.8.2 Bayesian Inference for Predicting the Number of Failures from the HPP

Suppose now that we wish to predict the number of failures in the future time interval $(t_n, t_n + \tau)$. The number M of failures in this interval is a discrete random variable having (conditioned on λ) a Poisson distribution with mean λ. The joint posterior distribution for (λ, M) is

$$\begin{aligned} p(\lambda, m \,|\, \mathbf{t}) &= p(\lambda \,|\, \mathbf{t})\, P\,(M = m \,|\, \mathbf{t}) \\ &= c\, p\,(\lambda)\ \lambda^n \exp\left(-\lambda t_n\right)\ \frac{(\lambda\tau)^m \exp\left(-\lambda\tau\right)}{m!}\,. \end{aligned}$$

The marginal of M, conditioned on $\mathbf{t}$, is obtained by integrating over λ:

$$p(m \,|\, \mathbf{t}) = c \int_0^\infty p\,(\lambda)\ \lambda^n \exp\left(-\lambda t_n\right)\ \frac{(\lambda\tau)^m \exp\left(-\lambda\tau\right)}{m!}\, d\lambda.$$

With the informative gamma prior from (4.31), this becomes

$$\begin{aligned} p(m \,|\, \mathbf{t}) &= c \int_0^\infty \left(\frac{b^a \lambda^{a-1}}{\Gamma(a)} e^{-b\lambda}\right)\ \lambda^n \exp\left(-\lambda t_n\right)\ \frac{(\lambda\tau)^m \exp\left(-\lambda\tau\right)}{m!}\, d\lambda \\ &= \frac{c\, b^a}{\Gamma(a)} \frac{\tau^m}{m!} \int_0^\infty \lambda^{a+n+m-1} \exp\left[-\left(b + t_n + \tau\right)\lambda\right]\, d\lambda \\ &= \frac{c\, b^a}{\Gamma(a)} \frac{\tau^m}{m!} \frac{\Gamma(a+n+m)}{(b+t_n+\tau)^{a+n+m}} \\ &\quad \times \int_0^\infty \frac{\lambda^{a+n+m-1}(b+t_n+\tau)^{a+n+m}}{\Gamma(a+n+m)} \exp\left[-\left(b + t_n + \tau\right)\lambda\right]\, d\lambda \\ &= \frac{c\, b^a}{\Gamma(a)} \frac{\tau^m}{m!} \frac{\Gamma(a+n+m)}{(b+t_n+\tau)^{a+n+m}}, \qquad m = 0, 1, 2, \ldots\,. \end{aligned}$$

The integrand in the second last line is the gamma density, so the value of the integrand is 1. If a is an integer, then the conditional distribution of M given $\mathbf{t}$ is a negative binomial distribution with probability mass function

$$p(m \,|\, \mathbf{t}) = \binom{m+n+a-1}{m} \left(\frac{b+t_n}{b+t_n+\tau}\right)^{a+n} \left(\frac{\tau}{b+t_n+\tau}\right)^m, \qquad m = 0, 1, 2, \ldots\,. \tag{4.32}$$

With the noninformative prior $p(\lambda) = 1/\lambda$, the marginal distribution of M given $\mathbf{t}$ is

$$\begin{aligned} p(m \,|\, \mathbf{t}) &= c \int_0^\infty \frac{1}{\lambda}\, \lambda^n \exp\left(-\lambda t_n\right)\ \frac{(\lambda\tau)^m \exp\left(-\lambda\tau\right)}{m!}\, d\lambda \\ &= \frac{c\tau^m}{m!} \int_0^\infty \lambda^{n+m-1} \exp\left[-\left(t_n + \tau\right)\lambda\right]\, d\lambda \\ &= \binom{m+n-1}{m} \left(\frac{t_n}{t_n+\tau}\right)^n \left(\frac{\tau}{t_n+\tau}\right)^m \qquad m = 0, 1, 2, \ldots\,. \end{aligned}$$

Thus, when the prior for λ is the noninformative prior $p(\lambda)$, the marginal of M given $\mathbf{t}$ has a negative binomial distribution.

Example 4.20 For illustration, we will consider one of the aircraft air-conditioning systems in the data set presented in Proschan (1963). The data for this system are shown in Table 4.15.

Failure Number i	Failure Time t_i	Failure Number i	Failure Time t_i	Failure Number i	Failure Time t_i
1	90	11	649	21	1576
2	100	12	733	22	1620
3	160	13	777	23	1643
4	346	14	836	24	1705
5	407	15	965	25	1835
6	456	16	983	26	2043
7	470	17	1008	27	2113
8	494	18	1164	28	2214
9	550	19	1474	29	2422
10	570	20	1550		

Table 4.15 Failure times from aircraft air-conditioning unit number 3.

This system experienced $n = 29$ failures in 2422 hours of operating time. The system showed no deterioration or reliability improvement, so the HPP seems to be an adequate model for the failure process. Suppose we wish to predict the number of failures in the next $\tau = 500$ hours using the noninformative prior for λ.

Given the data $\mathbf{t}$, the marginal distribution of M is negative binomial with probability mass function

$$P(M=m \mid \mathbf{t}) = \binom{28+m}{m} \left(\frac{2422}{2422+500}\right)^{29} \left(\frac{500}{2422+500}\right)^{m}, \quad m=0,1,2,\ldots.$$

Figure 4.24 shows these probabilities for $m = 0, 1, \ldots$. The most likely number of failures is $m = 5$ with probability 0.1507. A point prediction for the number of failures is the mean of the marginal posterior distribution. Using the result from Bain and Engelhardt (1992) on the mean of the negative binomial distribution, we find that the marginal posterior mean of M is given by

$$\hat{m} = \frac{n\tau}{t_n} = \frac{29 \times 500}{2422} \approx 5.987.$$

Since $P(M \leq 11 \mid \mathbf{t}) = 0.9683$, a conservative lower 95% Bayesian prediction interval for M is $[0, 11]$. □

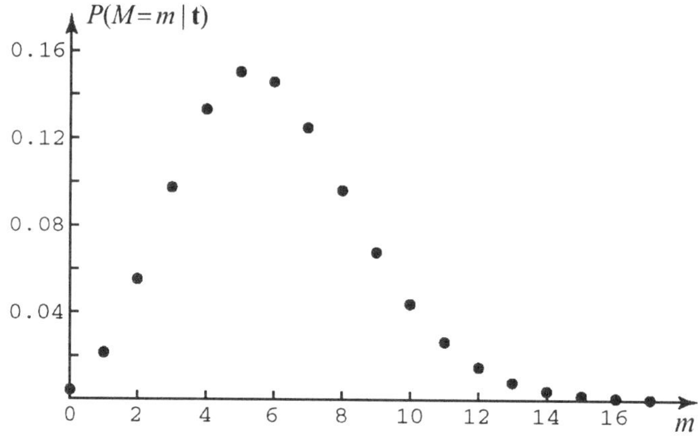

Fig. 4.24 Posterior probabilities of observing exactly m failures in the next $\tau = 500$ hours.

One advantage of the Bayesian paradigm is that it often yields tractable solutions to problems that, when viewed from the classical point of view, are intractable. In the case of predicting the number of failures of a repairable system, the uncertainty in the estimate for λ must be combined with the variability in the distribution of M. (Even if we knew λ with certainty, there would still be variability in the value of the random variable M.) In the classical paradigm, it is difficult to combine these two sources of uncertainty. In the Bayesian paradigm, the analysis is straightforward.

For users who are not inclined to use the Bayesian approach, we suggest using a noninformative prior and viewing the Bayesian approach as a recipe for a classical procedure. Using simulation, Beiser and Rigdon (1992) studied the coverage rates for this procedure and found that the actual coverage rates are usually slightly above the nominal rates. For example, when $n = 50$, $\tau = 8$, $\lambda = 1$, the coverage rate for a 95% prediction interval is 95.9%. For further results of this simulation study, see Beiser and Rigdon (1992).

4.8.3 Bayesian Inference for the Parameters of the Power Law Process

For the power law process, the unknown parameters are θ and β, and the observed data are the failure times $t_1 < t_2 < \cdots < t_n$. We will denote the observed failure times by the vector $\mathbf{t}$. Guida, Calabria, and Pulcini (1989), Calabria, Guida, and Pulcini (1990), and Bar-Lev, Lavi, and Reiser (1992) studied the Bayesian approach to analyzing data from a single repairable system that is modeled by the power law process. We now describe this Bayesian approach.

If we denote the prior distribution by $p(\theta, \beta)$ then the posterior distribution is

$$p(\theta, \beta \mid \mathbf{t}) = c\, p(\theta, \beta)\, f(\mathbf{t} \mid \theta, \beta). \tag{4.33}$$

Here $f(\mathbf{t} \mid \theta, \beta)$ is the likelihood function. The value of the normalizing constant c is

$$c = \left\{ \int_0^\infty \int_0^\infty p(\theta, \beta)\, f(\mathbf{t} \mid \theta, \beta)\, d\theta\, d\beta \right\}^{-1}.$$

We discuss two situations. First, we suppose that there is some prior information about the parameters θ and β. In this case, an informative prior distribution is appropriate. Second, we suppose that there is no, or at most very little, information about θ and β.

4.8.3.1 An Informative Prior Distribution The parameter β has a clear interpretation as a measure of improvement or deterioration, so it may be a good idea to begin the assessment of the prior information regarding β. Let's denote the prior distribution for β by $g(\beta)$. The interpretation of the parameter θ is less clear, so we will obtain the prior for θ in a roundabout way. Instead of selecting a joint prior for (θ, β) directly, we choose a time T and specify a prior distribution for

$$\eta = (T/\theta)^\beta.$$

The quantity η is the expected number of failures before time T. As Guida et al. (1989) suggested, we can select the gamma prior distribution for η

$$h(\eta) = \frac{b^a}{\Gamma(a)} \eta^{a-1} \exp(-b\eta), \qquad \eta > 0. \tag{4.34}$$

The prior parameters a and b can be selected by relating them to the mean μ and variance σ^2 of the prior information about η. That is, we can set

$$a = \mu^2/\sigma^2$$

and

$$b = \mu/\sigma^2.$$

If we assume that the prior knowledge about η and β is independent, then the joint prior distribution for (η, β) is

$$p(\eta, \beta) = g(\beta) h(\eta).$$

Making the change of variable $\eta = (T/\theta)^\beta$, or $\theta = T/\eta^{1/\beta}$, we find that the joint prior density is

$$p(\theta, \beta) = g(\beta) \frac{\beta b^a}{\theta \Gamma(a)} \left(\frac{T}{\theta}\right)^{\beta a} \exp\left[-b\left(\frac{T}{\theta}\right)^\beta\right], \qquad \beta > 0,\ \theta > 0. \tag{4.35}$$

Note that with this joint prior, the parameters θ and β are not independent a priori. If we define

$$P_n = \prod_{i=1}^{n} t_i$$

then, with this informative prior distribution, the posterior density is

$$\begin{aligned}
p(\theta, \beta|\mathbf{t}) &= c\, p(\theta, \beta)\, f(\mathbf{t}|\theta, \beta) \\
&= c\, g(\beta) \frac{\beta b^a}{\theta \Gamma(a)} \left(\frac{T}{\theta}\right)^{\beta a} \exp\left[-b\left(\frac{T}{\theta}\right)^{\beta}\right] \\
&\quad \times \left(\frac{\beta}{\theta}\right)^{n} \left(\frac{P_n}{\theta^n}\right)^{\beta-1} \exp\left[-\left(\frac{t_n}{\theta}\right)^{\beta}\right] \\
&= c \left(\frac{\beta}{\theta}\right)^{n} \left(\frac{P_n}{\theta^n}\right)^{\beta-1} g(\beta) \frac{\beta b^a}{\theta \Gamma(a)} \left(\frac{T}{\theta}\right)^{\beta a} \\
&\quad \times \exp\left[-\left(\frac{t_n}{\theta}\right)^{\beta} - b\left(\frac{T}{\theta}\right)^{\beta}\right],
\end{aligned}$$

where c is the normalizing constant chosen to make the joint posterior integrate to one; that is,

$$\begin{aligned}
c = \Bigg\{ \int_0^\infty \int_0^\infty & \left(\frac{\beta}{\theta}\right)^{n} \left(\frac{P_n}{\theta^n}\right)^{\beta-1} g(\beta) \frac{\beta b^a}{\theta \Gamma(a)} \left(\frac{T}{\theta}\right)^{\beta a} \\
& \times \exp\left[-\left(\frac{t_n}{\theta}\right)^{\beta} - b\left(\frac{T}{\theta}\right)^{\beta}\right] d\theta\, d\beta \Bigg\}^{-1}.
\end{aligned}$$

The point estimates for θ and β are then the posterior means

$$\begin{aligned}
\hat{\theta} = c \int_0^\infty \int_0^\infty & \theta \left(\frac{\beta}{\theta}\right)^{n} \left(\frac{P_n}{\theta^n}\right)^{\beta-1} g(\beta) \frac{\beta b^a}{\theta \Gamma(a)} \left(\frac{T}{\theta}\right)^{\beta a} \\
& \times \exp\left[-\left(\frac{t_n}{\theta}\right)^{\beta} - b\left(\frac{T}{\theta}\right)^{\beta}\right] d\theta\, d\beta
\end{aligned} \tag{4.36}$$

and

$$\begin{aligned}
\hat{\beta} = c \int_0^\infty \int_0^\infty & \beta \left(\frac{\beta}{\theta}\right)^{n} \left(\frac{P_n}{\theta^n}\right)^{\beta-1} g(\beta) \frac{\beta b^a}{\theta \Gamma(a)} \left(\frac{T}{\theta}\right)^{\beta a} \\
& \times \exp\left[-\left(\frac{t_n}{\theta}\right)^{\beta} - b\left(\frac{T}{\theta}\right)^{\beta}\right] d\theta\, d\beta.
\end{aligned} \tag{4.37}$$

While these expressions may seem very complicated, they are easily evaluated on a computer. Many mathematical software packages, such as *Mathematica, Maple, Matlab,* and *Gauss,* can do double integration in a straightforward manner, requiring the user to input only the integrand and the limits.

Example 4.21 For illustration, let us consider the $n = 14$ failures given in Table 4.8. First we consider the selection of the prior distribution for β; this should, of course, be done before any observations are taken. In this example, reliability improvement is expected, so we will take the prior distribution to be triangular over the interval $0.3 < \beta < 1.2$; that is,

$$g(\beta) = \begin{cases} \dfrac{4\,(x-0.3)}{(1.2-0.3)^2}\,, & 0.3 \leq \beta \leq 0.75 \\ \dfrac{-4\,(x-1.2)}{(1.2-0.3)^2}\,, & 0.75 < \beta < 0.1 \\ 0\,, & \text{otherwise;} \end{cases}$$

see Figure 4.25.

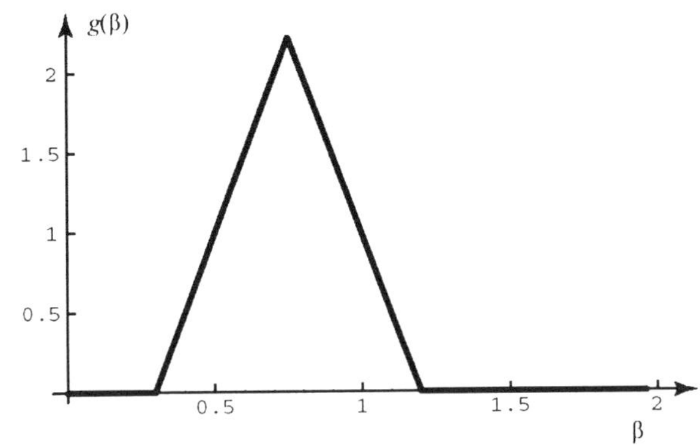

Fig. 4.25 The triangular prior distribution for β.

To obtain a joint prior for (θ, β) we follow the procedure suggested previously. Suppose that the time $T = 1000$ was chosen, and that expert opinion based on experience with other systems gave the following prior information about the number of failures before time $T = 1000$:

$$\mu = 9.0 \text{ and } \sigma = 3.0.$$

The parameters a and b of the prior distribution are then

$$a = \frac{\mu^2}{\sigma^2} = 9.0$$

and

$$b = \frac{\mu}{\sigma^2} = 1.0.$$

The joint prior density for (θ, β) is thus

$$p(\theta, \beta) = g(\beta) \frac{\beta}{8!\theta} \left(\frac{1000}{\theta}\right)^{9\beta - 1} \exp\left[-\left(\frac{1000}{\theta}\right)^{\beta}\right],$$
$$\theta > 0, 0.3 < \beta < 1.2.$$

Figure 4.26 gives a surface plot and a contour plot for this prior density.

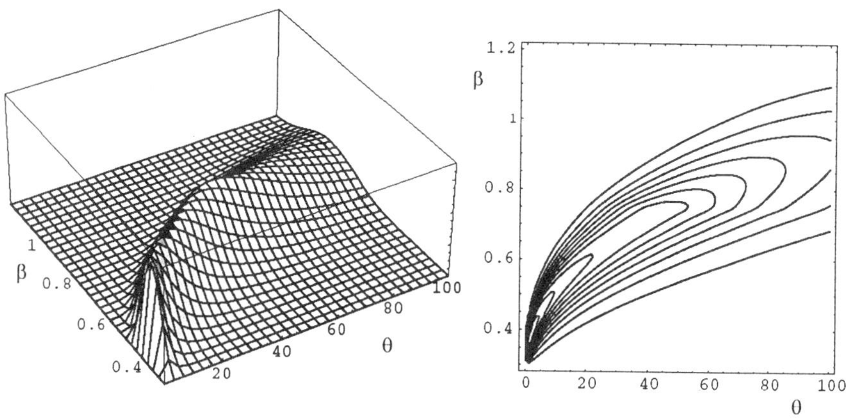

Fig. 4.26 Three-dimensional and contour plots for prior distribution of θ and β for the aircraft generator data.

The likelihood function and the posterior density, computed from (4.35), are shown in Figures 4.27 and 4.28. These graphs of the posterior distribution show a sharp ridge of plausible values for θ and β. The marginal posterior distributions, obtained by integrating (numerically) over the other variable are shown in Figure 4.29. The posterior means computed from (4.36) and (4.37), are

$$\hat{\beta} = 0.5615 \quad \text{and} \quad \hat{\theta} = 41.18.$$

□

4.8.3.2 A Noninformative Prior Distribution For the situation in which there is little or no prior information about the parameters θ and β Guida et al. (1989) suggest the use of a noninformative prior. In many cases the noninformative prior distribution is improper; that is, it does not integrate to 1. For

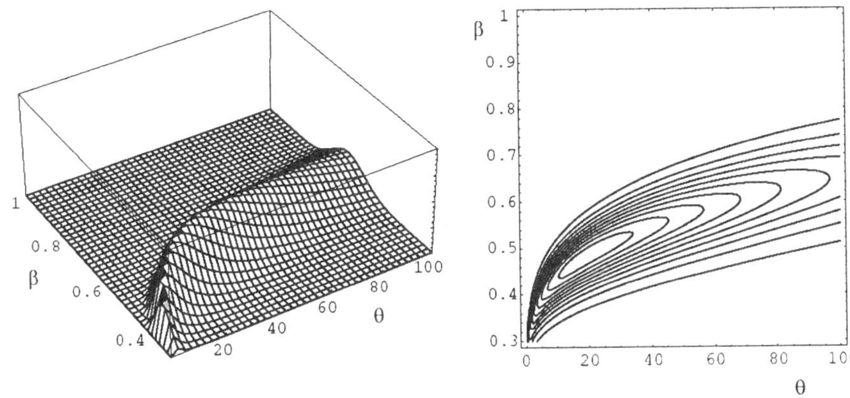

Fig. 4.27 Three-dimensional and contour plots for the likelihood function for θ and β for the aircraft generator data.

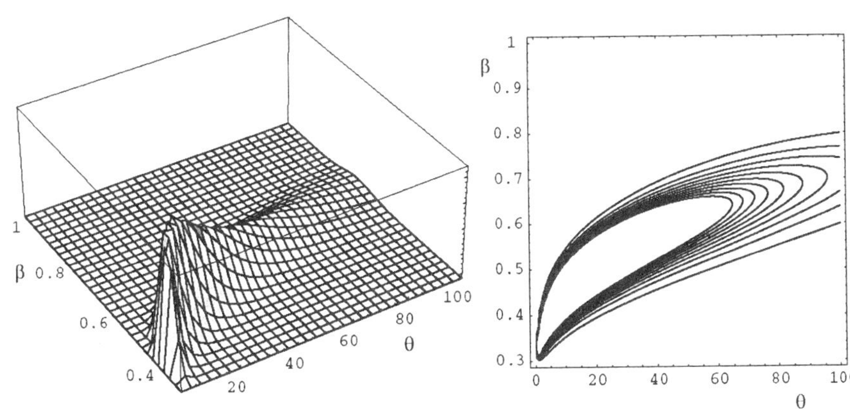

Fig. 4.28 Three-dimensional and contour plots for the posterior distribution of θ and β for the aircraft generator data.

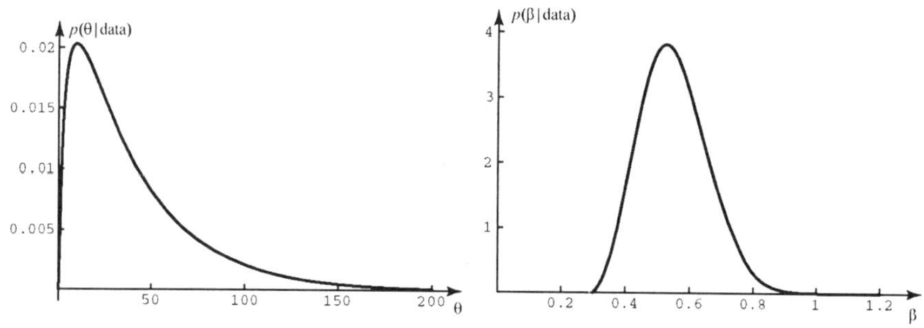

Fig. 4.29 Marginal posterior distributions for θ and β.

the power law process the noninformative prior pdf is

$$p(\theta, \beta) = \frac{1}{\theta\beta}, \qquad \theta, \beta > 0. \tag{4.38}$$

The joint posterior is then

$$\begin{aligned} p(\theta, \beta \,|\, \mathbf{t}) &= c\, p(\theta, \beta)\, L\,(\theta, \beta) \\ &= c\, \beta^{n-1}\, P_n^{\beta-1}\, \theta^{-n\beta-1}\, \exp\left[-\left(\frac{t_n}{\theta}\right)^{\beta}\right]. \end{aligned}$$

The marginal posterior of β is therefore

$$\begin{aligned} p\,(\beta \,|\, \mathbf{t}) &= c \int_0^\infty \beta^{n-1}\, P_n^{\beta-1}\, \theta^{-n\beta-1}\, \exp\left[-\left(\frac{t_n}{\theta}\right)^{\beta}\right] d\theta \\ &= c_1\, \beta^{n-2} \left(\frac{P_n}{t_n^n}\right)^{\beta}. \end{aligned} \tag{4.39}$$

Bar-Lev et al. (1992) showed the posterior distribution of $2n\beta/\hat{\beta}$ (where $\hat{\beta}$ is the MLE) is $\chi^2\,(2(n-1))$. Compare this result to Theorem 31; in Theorem 31, β was considered fixed (but unknown) and the random variable was $\hat{\beta}$, whereas in the current situation $\hat{\beta}$ is fixed because we are conditioning on the observed data $\mathbf{t}$, and β is random.

The posterior mean of β works out a little nicer than it did for the case of the informative prior:

$$\hat{\beta}_B = E\,(\beta \,|\, \mathbf{t}) = \frac{n-1}{\sum_{i=1}^{n-1} \log(t_n/t_i)}.$$

Notice the relationship of this posterior mean to the MLE $\hat{\beta}$ and the unbiased estimator for β

$$\bar{\beta} = \frac{n-2}{\sum_{i=1}^{n-1} \log(t_n/t_i)} = \frac{n-2}{n}\hat{\beta}.$$

Since

$$\frac{n-2}{\sum_{i=1}^{n-1} \log(t_n/t_i)} < \frac{n-1}{\sum_{i=1}^{n-1} \log(t_n/t_i)} < \frac{n}{\sum_{i=1}^{n-1} \log(t_n/t_i)}$$

we have

$$\bar{\beta} < \hat{\beta}_B < \hat{\beta}.$$

Thus for the noninformative prior given in (4.38) the posterior mean is always between the unbiased estimator and the MLE.

The posterior mean of θ is given by

$$\begin{aligned}\hat{\theta}_B &= E(\theta \,|\, \mathbf{t}) \\ &= \frac{t_n \left(\sum_{i=1}^{n} \log(t_n/t_i) \right)^{n-1}}{\Gamma(n)\Gamma(n-1)} \int_0^{\infty} \beta^{n-2} \left(\prod_{i=1}^{n} \frac{t_i}{t_n} \right)^{\beta} \Gamma(n - 1/\beta)\, d\beta. \qquad (4.40)\end{aligned}$$

We have discussed an informative prior distribution and a noninformative prior distribution for the parameters θ and θ. Guida et al. (1989) discuss a few other prior distributions and give some results for them.

4.8.4 Bayesian Inference for Predicting the Number of Failures

In this section we use this Bayesian approach to get the posterior probability of observing exactly $M = m$ failures in a future interval of length τ, given that the process was observed through the time of the nth failure. We identify two situations. The first situation applies when the power law process governs the failure process up to the time of the nth failure and thereafter the homogeneous Poisson process governs the failures. This model applies when a unit experiences reliability improvement during the development stage and, at the time of the nth failure is put into service with no further improvements. The second situation arises when the same power law process governs the failure time both before and after the nth failure.

4.8.4.1 Reliability Improvement Stops after the nth Failure Suppose now that while a system undergoes development, the power law process with intensity $\lambda(t)$ is the model that governs the failure times, and that after the nth failure the system is put into use with no further improvements. Under these conditions the homogeneous Poisson process may be an adequate model for the failure process after time t_n and the intensity function would then be constant at the value $\lambda(t_n)$. The distribution for the number of failures M in the interval $(t_n, t_n + \tau)$, given θ and β, is a Poisson random variable with mean

$$\mu = \tau\lambda(t_n) = \tau\frac{\beta}{\theta}\left(\frac{t_n}{\theta}\right)^{\beta-1}.$$

Let $p(\theta, \beta)$ denote the prior distribution for the parameters θ and β. The posterior distribution for M given the observed data $\mathbf{t}$ is given by

$$P(M = m \mid \mathbf{t}) = \frac{c\tau_n^m}{m!}\int_0^\infty\int_0^\infty p(\theta, \beta)\frac{\beta^{n+m}t_n^{m(\beta-1)}P_n^\beta}{\theta^{(n+m)\beta}} \times \exp\left\{-\left(\frac{t_n}{\theta}\right)^\beta\left(1+\frac{\tau\beta}{t_n}\right)\right\}\,d\theta\,d\beta,$$

where

$$P_n = \prod_{i=1}^n t_i$$

and c is a normalizing constant.

4.9 INFERENCE FOR A MODULATED POWER LAW PROCESS

The modulated power law process (MPLP) is a generalization of both the power law process and the gamma renewal process (i.e., a renewal process with gamma distributed times between failure). The MPLP arises out of the assumption that failures occur at every κth shock, where (unobservable) shocks occur according to a power law process with parameters θ and β. There are therefore three parameters, θ, β, and κ that must be estimated.

4.9.1 Maximum Likelihood Estimation of θ, β, and κ

For the failure truncated case, the joint pdf of $T_1 < T_2 < \cdots < T_n$ is

$$f(t_1, t_2, \ldots, t_n \mid \theta, \beta, \kappa) = \left\{\prod_{i=1}^n \lambda(t_i)\left[\Lambda(t_i) - \Lambda(t_{i-1})\right]^{\kappa-1}\right\} \times\frac{\exp\left[-\Lambda(t_n)\right]}{\left[\Gamma(\kappa)\right]^n}, \quad 0 < t_1 < t_2 < \cdots < t_n, \tag{4.41}$$

where $\lambda(t) = (\beta/\theta)(t/\theta)^{\beta-1}$ and $\Lambda(t) = \int_0^t \lambda(x)\,dx = (t/\theta)^\beta$ are, respectively, the intensity function and mean function for the process of shocks. The joint pdf in (4.41) is a density as long as κ is a positive real number. Thus it is not necessary to assume that κ is a positive integer even though the description of the process would imply that it is.

The MLE's of θ, β, and κ can be obtained by applying an iterative method to solve the likelihood equations. The log-likelihood is

$$\begin{aligned}\ell(\theta,\beta,\kappa) = &-\left(\frac{t_n}{\theta}\right)^{\beta} + n\log\beta - n\log\Gamma(\kappa) - n\beta\kappa\log\theta \\ &+ (\beta-1)\sum_{i=1}^{n}\log t_i + (\kappa-1)\sum_{i=1}^{n}\log\left(t_i^{\beta} - t_i^{\beta-1}\right).\end{aligned}$$

The partial derivatives with respect to θ, β, and κ are

$$\frac{\partial\ell}{\partial\theta} = \frac{\beta}{\theta}\left(\frac{t_n}{\theta}\right)^{\beta} - \frac{n\beta\kappa}{\theta}$$

$$\begin{aligned}\frac{\partial\ell}{\partial\beta} = &-\left(\frac{t_n}{\theta}\right)^{\beta}\log\left(\frac{t_n}{\theta}\right) + \frac{n}{\beta} - n\kappa\log\theta + \sum_{i=1}^{n}\log t_i \\ &+ (\kappa-1)\sum_{i=1}^{n}\frac{t_i^{\beta}\log t_i - t_{i-1}^{\beta}\log t_{i-1}}{t_i^{\beta} - t_{i-1}^{\beta}}\end{aligned}$$

$$\frac{\partial\ell}{\partial\kappa} = -n\psi(\kappa) - n\beta\log\theta + \sum_{i=1}^{n}\log\left(t_i^{\beta} - t_{i-1}^{\beta}\right).$$

Here ψ denotes the digamma function $\psi(x) = \Gamma'(x)/\Gamma(x)$ and t_0 is defined to be 0. The nonlinear equation

$$\begin{bmatrix}\dfrac{\partial\ell}{\partial\theta}\\[2ex] \dfrac{\partial\ell}{\partial\beta}\\[2ex] \dfrac{\partial\ell}{\partial\kappa}\end{bmatrix} = \begin{bmatrix}0\\[2ex]0\\[2ex]0\end{bmatrix}$$

can be solved by an iterative method such as Newton-Raphson. This method involves an initial estimate $[\theta,\beta,\kappa]' = [\theta_0,\beta_0,\kappa_0]'$ and iterating

$$\begin{bmatrix}\theta_{i+1}\\[2ex]\beta_{i+1}\\[2ex]\kappa_{i+1}\end{bmatrix} = \begin{bmatrix}\theta_i\\[2ex]\beta_i\\[2ex]\kappa_i\end{bmatrix} - \left[\ \mathbf{J}(\theta_i,\beta_i,\kappa_i)\ \right]^{-1}\begin{bmatrix}\dfrac{\partial\ell}{\partial\theta}(\theta_i,\beta_i,\kappa_i)\\[2ex]\dfrac{\partial\ell}{\partial\beta}(\theta_i,\beta_i,\kappa_i)\\[2ex]\dfrac{\partial\ell}{\partial\kappa}(\theta_i,\beta_i,\kappa_i)\end{bmatrix},$$

where $\mathbf{J}(\theta, \beta, \kappa)$ is the matrix of second partial derivatives

$$\mathbf{J}(\theta, \beta, \kappa) = \begin{bmatrix} \dfrac{\partial^2 \ell}{\partial \theta^2} & \dfrac{\partial^2 \ell}{\partial \theta\, \partial \beta} & \dfrac{\partial^2 \ell}{\partial \theta\, \partial \kappa} \\ \dfrac{\partial^2 \ell}{\partial \theta\, \partial \beta} & \dfrac{\partial^2 \ell}{\partial \beta^2} & \dfrac{\partial^2 \ell}{\partial \beta\, \partial \kappa} \\ \dfrac{\partial^2 \ell}{\partial \theta\, \partial \kappa} & \dfrac{\partial^2 \ell}{\partial \beta\, \partial \kappa} & \dfrac{\partial^2 \ell}{\partial \kappa^2} \end{bmatrix}.$$

The entries in $\mathbf{J}$ are

$$\frac{\partial^2 \ell}{\partial \theta^2} = \frac{\beta}{\theta^2}\left[n\kappa - (\beta+1)\left(\frac{t_n}{\theta}\right)^{\beta}\right]$$

$$\frac{\partial^2 \ell}{\partial \theta\, \partial \beta} = \frac{\beta}{\theta}\left(\frac{t_n}{\theta}\right)^{\beta} \log\left(\frac{t_n}{\theta}\right) + \frac{1}{\theta}\left(\frac{t_n}{\theta}\right)^{\beta} - \frac{n\kappa}{\theta}$$

$$\frac{\partial^2 \ell}{\partial \theta\, \partial \kappa} = -\frac{n\beta}{\theta}$$

$$\begin{aligned}\frac{\partial^2 \ell}{\partial \beta^2} = & -\left(\frac{t_n}{\beta}\right)^{\beta}\left[\log\left(\frac{t_n}{\theta}\right)\right]^2 - \frac{n}{\beta^2} \\ & + (\kappa - 1)\sum_{i=1}^{n} \frac{t_i^{\beta}(\log t_i)^2 - t_{i-1}^{\beta}(\log t_{i-1})^2}{t_i^{\beta} - t_{i-1}^{\beta}} \\ & - (\kappa-1)\sum_{i=1}^{n} \frac{\left(t_i^{\beta}\log t_i - t_{i-1}^{\beta}\log t_{i-1}\right)^2}{\left(t_i^{\beta} - t_{i-1}^{\beta}\right)^2}\end{aligned}$$

$$\frac{\partial^2 \ell}{\partial \beta\, \partial \kappa} = -n\log\theta + \sum_{i=1}^{n} \frac{t_i^{\beta}\log t_i - t_{i-1}^{\beta}\log t_{i-1}}{t_i^{\beta} - t_{i-1}^{\beta}}$$

$$\frac{\partial^2 \ell}{\partial \kappa^2} = n\psi'(\kappa).$$

The method of Newton-Raphson is sensitive to the starting point $[\theta_0, \beta_0, \kappa_0]'$; if this starting value is not sufficiently close to the maximum of the log likelihood function, then the iterative method is likely to diverge or jump to a negative estimate of one of the parameters. Black and Rigdon (1996) found that several iterations of the Nelder-Mead simplex algorithm (Nelder and Mead, 1965), described in Buchanan and Turner (1992), can be used to get a good starting point for the Newton-Raphson method. The Nelder-Mead simplex algorithm is a derivative free method for approximating the maximum of a function; it nearly always leads to convergence, but convergence is usually slow, often requiring several thousand iterations. The Newton-Raphson method will usually converge if the starting value is determined by several iteration of the Nelder-Mead algorithm.

4.9.2 Hypothesis Tests for the Modulated Power Law Process

Tests of hypothesis regarding the parameters of the modulated power law process can be based on the likelihood ratio statistic. To test the null hypothesis $H_0 : (\theta, \beta, \kappa) \in S$ against the alternative $H_a : (\theta, \beta, \kappa) \notin S$, we compute

$$\mathbf{LR} = \frac{\max\limits_{(\theta,\beta,\kappa)\in S} L(\theta, \beta, \kappa)}{\max\limits_{(\theta,\beta,\kappa)} L(\theta, \beta, \kappa)} .$$

The denominator is just the maximum of the likelihood function, which is achieved at the MLE $\left(\hat{\theta}, \hat{\beta}, \hat{\kappa}\right)$; that is, the denominator is $L\left(\hat{\theta}, \hat{\beta}, \hat{\kappa}\right)$. The numerator is obtained by maximizing L subject to the restriction that $(\theta, \beta, \kappa) \in S$. The method to achieve this depends on the particular null hypothesis being tested. The statistic $-2\log \mathbf{LR}$ is distributed asymptotically as $\chi^2(r)$ where r is the number of restrictions on the parameters in the condition $(\theta, \beta, \kappa) \in S$.

4.9.2.1 Testing $H_0 : \kappa = 1$ If the null hypothesis $H_0 : \kappa = 1$ is true, then the failure process is a power law process with parameters θ and β. Thus, under H_0, the values of θ and β that maximize the likelihood function are the MLE's of the parameters of the power law process:

$$\hat{\beta}_{\mathrm{PLP}} = \frac{n}{\sum\limits_{i=1}^{n-1} \log \dfrac{t_n}{t_i}}$$

$$\hat{\theta}_{\mathrm{PLP}} = \frac{t_n}{n^{1/\hat{\beta}}} .$$

The likelihood ratio is therefore

$$\mathbf{LR}_{\mathrm{PLP}} = \frac{L(\hat{\theta}_{\mathrm{PLP}}, \hat{\beta}_{\mathrm{PLP}}, 1)}{L(\hat{\theta}, \hat{\beta}, \hat{\kappa})} .$$

Since the null hypothesis contains a single restriction on a parameter, the asymptotic distribution is therefore $\chi^2(1)$, so the rule is to reject $H_0 : \kappa = 1$ if $-2\log \mathbf{LR}_{\mathrm{PLP}} > \chi^2_\alpha(1)$.

4.9.2.2 Testing $H_0 : \beta = 1$ Under this null hypothesis, the process is a gamma renewal process (i.e., a renewal process for which the times between failures are iid $\mathrm{GAM}(\kappa, \theta)$ random variables). The numerator in the likelihood ratio statistic requires that we first find the MLE's of θ and κ. The likelihood function is the product of $\mathrm{GAM}(\kappa, \theta)$ densities. Differentiating the log likelihood function and setting the result equal to zero leads to the equation

$$\log\kappa - \Gamma'(\kappa)/\Gamma(\kappa) - \log(\overline{x}/\widetilde{x}) = 0, \tag{4.42}$$

where $\overline{x}$ and $\widetilde{x}$ are, respectively, the mean and median of the times between failure. Once an estimate $\hat{\kappa}_{\text{GRP}}$ for the solution to (4.42) is obtained (using some iterated method, such as Newton's method), we can obtain the MLE for θ from

$$\hat{\theta}_{\text{GRP}} = \frac{\overline{x}}{\hat{\kappa}_{\text{GRP}}}.$$

Given $\hat{\theta}_2$ and $\hat{\kappa}_2$ we can then substitute these values into the likelihood function to obtain

$$\mathbf{LR}_{\text{GRP}} = \frac{L(\hat{\theta}_{\text{GRP}}, 1, \hat{\kappa}_{\text{GRP}})}{L(\hat{\theta}, \hat{\beta}, \hat{\kappa})}.$$

Using the asymptotic result for the distribution of $-2\log\Lambda_{\text{GRP}}$, we would reject $H_0 : \beta = 1$ if $-2\log\mathbf{LR}_{\text{GRP}} > \chi^2_\alpha(1)$.

4.9.2.3 Testing $H_0 : \kappa = \beta = 1$ Under the null hypothesis $H_0 : \kappa = \beta = 1$, the process is a homogeneous Poisson process with intensity $1/\theta$; that is, the times between failures are iid EXP(θ) random variables. The estimate of θ, under the HPP assumption is

$$\hat{\theta}_{\text{HPP}} = \frac{t_n}{n}.$$

The likelihood ratio statistic is therefore

$$\mathbf{LR}_{\text{HPP}} = \frac{L(\hat{\theta}_{\text{GRP}}, 1, 1)}{L(\hat{\theta}, \hat{\beta}, \hat{\kappa})}.$$

Since there are now two restrictions on the parameters in the null hypothesis, the asymptotic chi-square distribution has 2 degrees of freedom. Thus the rule is to reject the null hypothesis if $-2\log\mathbf{LR}_{\text{HPP}} > \chi^2_\alpha(2)$.

4.9.3 Confidence Intervals for Parameters

The MLE $\left[\hat{\theta}, \hat{\beta}, \hat{\kappa}\right]'$ of $[\theta, \beta, \kappa]'$ has an asymptotic multivariate normal distribution with mean $[\theta, \beta, \kappa]'$ and covariance matrix $\left[\mathbf{J}(\hat{\theta}, \hat{\beta}, \hat{\kappa})\right]^{-1}$. Approximate $100 \times (1 - \alpha)\%$ confidence intervals for the parameters are therefore

$$\hat{\theta} \pm z_{\alpha/2}\sqrt{(1,1) \text{ entry in } \left[\mathbf{J}(\hat{\theta}, \hat{\beta}, \hat{\kappa})\right]^{-1}}$$

$$\hat{\beta} \pm z_{\alpha/2}\sqrt{(2,2) \text{ entry in } \left[\mathbf{J}(\hat{\theta}, \hat{\beta}, \hat{\kappa})\right]^{-1}}$$

$$\hat{\kappa} \pm z_{\alpha/2}\sqrt{(3,3) \text{ entry in } \left[\mathbf{J}(\hat{\theta}, \hat{\beta}, \hat{\kappa})\right]^{-1}}.$$

Black and Rigdon (1996) investigated the coverage rates of these confidence intervals based on the asymptotic normal distribution and found that the coverage rate for θ was usually below the nominal rate. For example, even for sample sizes as large as 30, the coverage rates for a 95% confidence interval for θ were actually only between 85% and 90%. They suggested instead that a confidence interval be derived for the logarithm of θ. The interval for $\log\theta$ can then be transformed to an interval for θ; the result is

$$\hat{\theta}\exp\left\{-z_{\alpha/2}\sqrt{(1,1)\text{ entry in }\left[\mathbf{J}(\hat{\theta},\hat{\beta},\hat{\kappa})\right]^{-1}}/\hat{\theta}\right\} < \theta$$
$$\hat{\theta}\exp\left\{z_{\alpha/2}\sqrt{(1,1)\text{ entry in }\left[\mathbf{J}(\hat{\theta},\hat{\beta},\hat{\kappa})\right]^{-1}}/\hat{\theta}\right\}$$

The coverage rates for this confidence interval (with $\alpha = 0.95$) are between 90% and 95% for sample sizes as large as 20. This method can be applied to the parameters β and κ, yielding confidence intervals for the log of the parameter, rather than the parameter itself. For β and κ, such a transformation is not usually needed because the coverage rates for these parameters are much closer to the nominal level.

4.10.4 An Example

Example 4.22 Consider the data in Table 4.8, which give the failure times of an aircraft generator data. The failure times do seem to be spaced, but the spacings tend to get larger. Thus, we expect the estimate of β to be less than one (since there appears to be reliability improvement) and the estimate of κ to be greater than one (since the failures are spaced out, rather than clustered.) The MLEs of θ, β, and κ are

$$\hat{\theta} = 0.218$$
$$\hat{\beta} = 0.420$$
$$\hat{\kappa} = 4.800.$$

Confidence intervals (95%) for these parameters are

$$-0.48 < \theta < 0.91$$

$$0.32 < \beta < 0.53$$

$$1.34 < \kappa < 8.26.$$

If we find the asymptotic confidence interval for $\log\theta$ and then invert, we obtain

$$0.009 < \theta < 5.29.$$

An estimate of κ so far from one, together with a confidence interval that excludes one, indicates that the power law process is an inadequate model for this system. □

4.10 INFERENCE FOR THE PIECEWISE EXPONENTIAL MODEL

The piecewise exponential (PEXP) model of Sen and Bhattacharyya (1993) and Sen (1998) was introduced in Chapter 3. This model assumes that the times between failures $X_1, X_2, \ldots, X_n$ are independent but not identically distributed random variables. The assumption is that X_i is exponential with mean

$$E(X_i) = \frac{\delta}{\mu} i^{\delta-1}, \qquad i = 1, 2, \ldots .$$

The likelihood function for the failure truncated case is then

$$\begin{aligned} \ell(\delta, \mu) &= \prod_{i=1}^{n} \left(\frac{\delta}{\mu} i^{\delta-1} \right)^{-1} \exp\left(-x_i \left(\frac{\delta}{\mu} i^{\delta-1} \right)^{-1} \right) \\ &= \left(\frac{\mu}{\delta} \right)^n \left(\prod_{i=1}^{n} i \right)^{1-\delta} \exp\left(-\frac{\mu}{\delta} \sum_{i=1}^{n} i^{1-\delta} x_i \right). \end{aligned}$$

Sen and Bhattacharyya (1993) reparametrize their model as follows. Let

$$\beta = \begin{bmatrix} \beta_1 \\ \beta_2 \end{bmatrix} = \begin{bmatrix} \log(\delta/\mu) \\ \delta - 1 \end{bmatrix}$$

and

$$\mathbf{y}_i = \begin{bmatrix} 1 \\ \log i \end{bmatrix}.$$

With this formulation, we have

$$\begin{aligned} L(\delta, \mu) = \exp&\left(-n \log(\delta/\mu) - (\delta - 1) \sum_{i=1}^{n} \log i \right) \\ &\times \exp\left(-\sum_{i=1}^{n} \exp\left(-\log(\delta/\mu) - (\delta - 1) \log i \right) x_i \right). \end{aligned}$$

The log-likelihood function is therefore

$$\begin{aligned}\ell(\delta,\mu) = \log L(\delta,\mu) &= -n\log\frac{\delta}{\mu} - (\delta-1)\sum_{i=1}^{n}\log i \\ &\quad - \sum_{i=1}^{n}\exp\left(-\log(\delta/\mu) - (\delta-1)\log i\right)x_i \\ &= -\begin{bmatrix}\log(\delta/\mu) & \delta-1\end{bmatrix}\sum_{i=1}^{n}\begin{bmatrix}1\\ \log i\end{bmatrix} \\ &\quad - \sum_{i=1}^{n}\exp\left(-\begin{bmatrix}\log(\delta/\mu) & \delta-1\end{bmatrix}\sum_{i=1}^{n}\begin{bmatrix}1\\ \log i\end{bmatrix}x_i\right).\end{aligned}$$

In terms of β we have

$$\ell(\beta_1,\beta_2) = \log L(\beta_1,\beta_2) = -\beta'\sum_{i=1}^{n}\mathbf{y}_i - \sum_{i=1}^{n}\exp(-\beta'\mathbf{y}_i)\,x_i.$$

The derivatives with respect to β_1 and β_2 are

$$\begin{aligned}\frac{\partial\ell}{\partial\beta_1} &= -n + \sum_{i=1}^{n}x_i\exp(-\beta_1 - \beta_2\log i) \\ \frac{\partial\ell}{\partial\beta_2} &= -\sum_{i=1}^{n}\log i + \sum_{i=1}^{n}x_i\exp(-\beta_1 - \beta_2\log i)\log i\,.\end{aligned}$$

Solving the equations $\partial\ell/\partial\beta_1 = 0$ and $\partial\ell/\partial\beta_2 = 0$, using an iterative method such an the Newton-Raphson method, leads to the MLE's $\hat{\delta}$ and $\hat{\mu}$. Sen and Bhattacharyya (1993) suggest that because of the correspondence between the power law process and the PEXP model the following are reasonable estimates of δ and μ:

$$\delta^* = \frac{\sum_{i=1}^{n}\log(T_n/T_i)}{n} \tag{4.43}$$

$$\mu^* = \frac{n^{\delta^*}}{T_n}\,. \tag{4.44}$$

The values δ^* and μ^* are easily computed from the observed data and should make reasonable starting values for an iterative method.

Sen and Bhattacharyya also showed that the asymptotic distributions of the MLE's $\hat{\delta}$ and $\hat{\mu}$ satisfy

$$\begin{aligned}&\sqrt{n}\left(\hat{\delta}-\delta\right) \xrightarrow{d} N(0,1) \\ &\frac{\sqrt{n}}{\log n}(\hat{\mu}-\mu) \xrightarrow{d} N(0,1),\end{aligned}$$

where $\xrightarrow{d}$ indicates convergence in distribution (see Bain and Engelhardt 1992, Chapter 7) as $n \to \infty$. The former result leads to the confidence interval

$$\hat{\delta} - \frac{z_{\alpha/2}}{\sqrt{n}} < \delta < \hat{\delta} + \frac{z_{\alpha/2}}{\sqrt{n}} \tag{4.45}$$

for δ. Because convergence is slow in the latter case, they suggest using the asymptotic normal distribution for $\log \mu$ and then transforming the result to μ. Using this method, a confidence interval for μ is given by

$$\hat{\mu}\left[n^{-z_{\alpha/2}/\sqrt{n}}\right] < \mu < \hat{\mu}\left[n^{z_{\alpha/2}/\sqrt{n}}\right]. \tag{4.46}$$

Example 4.23 Consider the aircraft generator data from Table 4.8. An analysis of this data set assuming the power law process indicated that there was reliability improvement. The simple estimators, called the Weibull process analog estimators (WPAE's), from equations (4.43) and (4.44) are

$$\delta^* = \frac{\sum_{i=1}^{n} \log(T_n/T_i)}{n} = \frac{28.97}{14} = 2.069$$

and

$$\mu^* = \frac{n^{\delta^*}}{T_n} = \frac{14^{2.069}}{4596} = 0.00512.$$

The log likelihood function of β_1 and β_2, shown in Figures 4.30 and 4.31 contains a sharp ridge where the function is nearly constant. The MLE's are

$$\hat{\delta} = 2.589$$

and

$$\hat{\mu} = 0.227.$$

Confidence intervals for δ and μ computed from the results of equations (4.45) and (4.46) are

$$2.589 - 1.96/\sqrt{14} < \delta < 2.589 + 1.96/\sqrt{14}$$
$$2.065 < \delta < 3.113$$

and

$$0.227\left[14^{-1.96/\sqrt{14}}\right] < \mu < 0.227\left[14^{1.96/\sqrt{14}}\right]$$
$$0.057 < \mu < 0.904.$$

A confidence interval for δ that lies completely to the right of one is evidence of reliability improvement. □

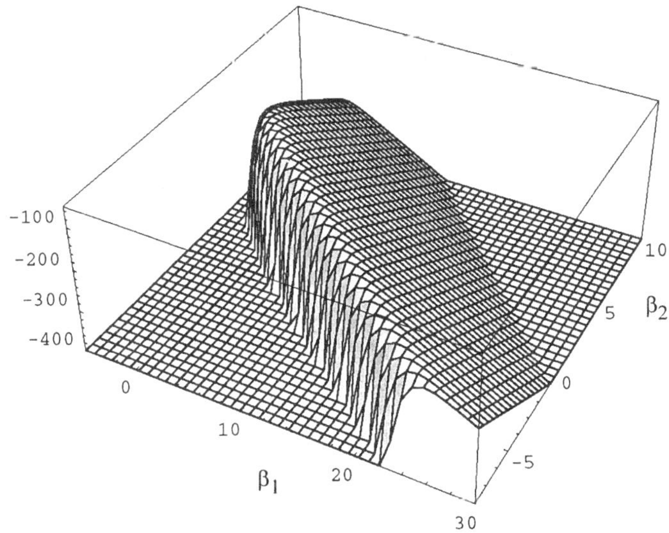

Fig. 4.30 Three-dimensional plot of log likelihood function for PEXP model for the data on the aircraft generator.

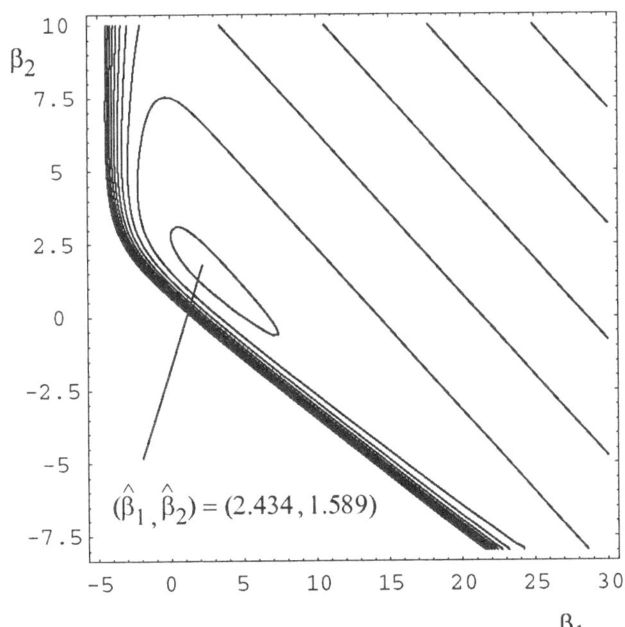

Fig. 4.31 Contour plot of log likelihood function for PEXP model for the data on the aircraft generator.

4.11 STANDARDS

4.11.1 MIL-HDBK-189

MIL-HDBK-189 "Reliability Growth Management" deals exclusively with reliability of repairable systems. The assumption seems to be that the reliability will improve during the design, development, and testing phases of a system, but that once released the system failures will follow an HPP with MTBF equal to that value characteristic of the system when testing stops. The handbook begins with these statements.

> 1. SCOPE
>
> 1.1 Purpose This handbook provides procuring activities and development contractors with an understanding of the concepts and principles of reliability growth, advantages of managing reliability growth, and guidelines and procedures to be used in managing reliability growth. ...
>
> 1.2 Application This handbook is intended for use on systems/equipment during their development phase by both contractor and government personnel.

Plots on pp. 37–38 indicate that reliability, as measured by the reciprocal of the intensity function $1/\lambda(t)$, called the MTBF (or instantaneous MTBF) is a step function that can be approximated by a smooth curve. A step function for the MTBF or $\lambda(t)$ is consistent with the piecewise exponential (PEXP) model of Sen and Bhattacharyya (1993). The power law process, or any NHPP for that matter, serves an approximation to this step function.

Section 5.3.5 suggests testing for trend using the test statistics

$$2n/\hat{\beta} \quad \text{(failure truncated case)}$$
$$2N/\hat{\beta} \quad \text{(time truncated case)},$$

where

$$\hat{\beta} = \frac{n}{\sum_{i=1}^{n-1} \ln(t_n/t_i)} \quad \text{(failure truncated case)}$$
$$\hat{\beta} = \frac{N}{\sum_{i=1}^{N} \ln(t/t_i)} \quad \text{(time truncated case)}$$

and t is the truncation time for the time truncated case. At this point in the handbook, the power law process has yet to be introduced, so a reader will likely not know where these formulas come from. Section 5.3.5.3 suggests a χ^2 goodness of fit test for trend. Suppose we partition the test time into k

disjoint subintervals and observe the number N_i of failures in the ith interval which has length T_i. The test statistic is then

$$X^2 = \sum_{i=1}^{k} \frac{(\text{observed \# of failures} \ - \ \text{expected \# of failures})^2}{\text{expected \# of failures}}$$

$$= \sum_{i=1}^{k} \frac{(N_i - NP_i)^2}{NP_i},$$

where $N = \sum_{i=1}^{k} N_i$ and $P_i = T_i/t$. The hypothesis of no trend is rejected for large values of X^2.

Section 5.3.6 introduces the power law process, called the AMSAA (Army Material Systems Analysis Activity) model. That handbook uses the parametrization $r(t) = \lambda\beta t^{\beta-1}$, which is, of course, equivalent to our parameterization with $\lambda = 1/\theta^\beta$. The function $r(t)$ is called the failure rate; it is the equivalent of what we have called the intensity function $\lambda(t)$. The MLE's of θ and β are given, but not derived. The test statistic for the Cramér-von Mises test (4.15) is given. This test is a goodness of fit test for the power law process. Section 5.3.6 refers the reader to Appendices B and C where the power law process (AMSAA model) is discussed in more detail.

Appendix B describes nine models for discrete reliability growth (DRG) and eight models for continuous reliability growth (CRG). The DRG models specify the probability of success R_n for the nth trial. These models range from simple models, such as

$$R_n = 1 - A\exp\left[-c(n-1)\right], \qquad n = 1, 2, \ldots,$$

a model given by Lloyd and Lipow (1962), to an ARIMA time series model (see Singpurwalla (1978) and Box and Jenkins (1976)). The CRG models include the homogeneous and nonhomogeneous Poisson processes, Duane's model, the power law process (AMSAA model), the NHPPs with intensity functions $\lambda(t) = \exp(\alpha + \gamma t)$, and $\lambda(t) = \exp\left(\alpha_0 + \alpha_1 t + \cdots + a_{10}t^{10}\right)$, and a number of models involving differential equations.

Appendix C "The AMSAA Reliability Growth Model" deals exclusively with the power law process. Properties of the power law process are described, but not derived, and a number of examples are given.

4.11.2 MIL-HDBK-781 and MIL-STD-781

The Department of Defense seems to be moving away from military standards and toward military handbooks. One example of this is the replacement of MIL-STD-781D with MIL-HDBK-781 in 1996. In this case, however, the replacement is a significant revision of the standard. The title of MIL-STD-781D was "Reliability Design Qualification and Production Acceptance Tests: Exponential Tests." Ascher (1987) criticized MIL-STD-781C on the grounds that if fails to clearly distinguish between repairable and nonrepairable systems. The title of the standard contains the words "exponential distribution" suggesting that it applies to nonrepairable systems. The mention of a particular distribution implies that it serves as a model for a lifetime, presumably a single lifetime. Ascher argues persuasively that the standard applies to repairable systems. He gives four reasons for such a conclusion. (1) The standard describes actions to be taken after a failure. Specifically, Section 5.6 says "After a failed and repaired equipment has been returned to operable condition, it shall be returned to test with appropriate entries in the data logs, and with minimum interruption to the other equipment continuing on test." The last part of this sentence suggests that the standard applies to multiple repairable systems. (2) A minimum operating time for each equipment is specified. An early failure of a nonrepairable system would render it incapable of meeting this requirement if the standard applied to nonrepairable systems. (3) The standard suggests in Section 4.5.2.3 that between 2 and 20 equipments be tested. The test plan described in Figure C-1 (actually a table) gives entries up to 41 failures. If 41 failures are to occur on 20 units, some units must have multiple failures. (4) Section 1.3 describes six "broad categories of equipment" to which the standard applies. These include "fixed ground equipment" and "equipment for jet aircraft." Systems this complex are usually repairable.

MIL-HDBK-781 "Handbook for Reliability Test Methods, Plans, and Environments for Engineering, Development Qualification, and Production" was issued on April 1, 1996, superseding MIL-STD-781D. The words "exponential distribution" have been dropped from the title, but the document does not state whether the methods apply to repairable or nonrepairable systems. The first two sections state

> 1. SCOPE
>
> 1.1 Purpose. This handbook provides test methods, test plans, and test environmental profiles which can be used in reliability testing during the development, qualification, and production of systems and equipment.
>
> 1.2 Application. This handbook explains techniques for use in reliability tests performed during integrated test programs. Procedures, plans, and environments which can be used in Reliability Development/Growth Tests (RD/GT), Reliability Qualification Tests (RQT), Production Reliability Acceptance Tests (PRAT), Environmental Stress Screening (ESS) methods, and Durability/Economic Life Test are provided.

As was the case for MIL-STD-781, the newer document MIL-HDBK-781 seems to apply exclusively to repairable systems, for mostly the same reasons Ascher cited for MIL-STD-781. References to Duane plots and MTBF clearly imply repairable systems. The equipment categories to which the handbook applies, listed in Section 1.4, include fixed ground equipment, mobile ground equipment, and naval surface craft. Complex systems like this are repairable.

The statistical parts of MIL-HDBK-781, which cover only about half of the entire standard, concentrate on two types of statistical methods:

1. Inference for the power law process (including Duane plots, point and interval estimates of the parameters of the power law process), and

2. Hypothesis testing and confidence for the MTBF θ (assuming that the failure process is an HPP with intensity $1/\theta$).

The Duane plots described in MIL-HDBK-781 put the quantity $N(t_i)/t_i$ on the vertical axis, rather than the reciprocal $t_i/N(t_i)$ as Duane had originally done. The power law process is called the AMSAA (Army Material Systems Analysis Activity) model, and it is parametrized with the intensity function having the form $\lambda(t) = \lambda\beta t^{\beta-1}$. The handbook suggests first a test for trend. For both failure truncation and time truncation, it then gives the MLEs of β and λ, along with an unbiased estimator for β. It also gives an estimator for the value of the intensity function at the time that testing terminates. Finally, the standard suggests applying the Cramér-von Mises goodness of fit test (see Sections 4.5.4 and 4.6.4).

Inference for the mean time between failure (MTBF) is based on the assumption that the failure process is an HPP, that is, the times between failures are iid exponential. Much of this part of the handbook deals with hypothesis tests of the form H_0: $\theta = \theta_0$ versus H_1: $\theta = \theta_1$. The quantity θ_0 is the upper test MTBF, defined as an "acceptable value of MTBF," and θ_1 is the lower test MTBF, defined as "that value which is unacceptable." The definition of the lower test MTBF adds "The standard test plans will reject, with high probability, equipment with a true MTBF that approaches θ_1."

4.11.3 ANSI/IEC/ASQ 61164

This document is both an American National Standard and a standard of the International Electrotechnical Commission (IEC). It was published in 1997 by the American Society for Quality (ASQ). The document is concisely written, consisting of only 29 pages, compared to 392 pages for MIL-HDBK-781.

The first section "Scope" states

> This International Standard gives models and numerical methods for reliability growth assessments based on failure data from a single system which were generated in a reliability improvement programme. These procedures deal with growth, estimation, confidence intervals for system reliability and goodness-of-fit tests.

Sections 2–4 give references, definitions, and symbols. Section 5 describes the power law process, which is called the power law model. Section 6 explains how the power law process can be used in reliability improvement programs. By far, the longest section is Section 7, which covers inference procedures (point and interval estimation and tests) for the power law process. Like the military handbooks described previously, ANSI/IEC/ASQ 61164 writes the intensity function as $z(t) = \lambda\beta t^{\beta-1}$.

The standard begins with growth tests and point estimation for the following cases:

1. Time data for every relevant failure (i.e. ungrouped failure data) — time truncation (called a type I test);

2. Time data for every relevant failure — failure truncation (called a type II test);

3. Grouped failure data (i.e. the data consist of the number of failures in nonoverlapping intervals)— time truncation;

4. Grouped failure data — failure truncation.

Laplace's test for trend is suggested; then point estimators (the MLEs of θ and β, along with unbiased estimators of β) are given. The next three subsections describe the Cramér-von Mises goodness-of-fit test, interval estimation for the MTBF (which is defined as the reciprocal of the intensity function, $1/z(T)$, where T is the time of the last failure in the failure truncated case or the truncation time in the time truncated case), and confidence intervals for the MTBF. The last subsection of Section 7 describes a projection technique, which is used to assess the effect of corrective modifications made after testing is terminated.

Annex A gives some numerical examples and Annex B gives further properties of the power law process.

4.12 OTHER INFERENCE PROCEDURES FOR REPAIRABLE SYSTEMS

In this chapter, we have presented some of the most commonly used procedures for analyzing data from repairable systems. A number of authors have studied statistical models and inference for repairable systems. In this section we mention briefly some other results.

Engelhardt and Bain (1978) developed prediction intervals for the time of the next, or the time of the mth subsequent, failure of a system modeled by a power law process. In a pair of papers, Engelhardt and Bain (1992) and Engelhardt, Williams, and Bain studied inference procedures for the power law process with left-censored and left-truncated data.

Goodness-of-fit tests have been studied by Park and Kim (1992), Park and Seoh (1994), and Baker (1996). Lee (1980a) developed a test for the adequacy of the power law process by embedding the power law process into a more general model.

Ebrahimi (1996) studied reliability for the case when the time of design changes is known. Bayesian methods of inference for the power law process have been studied by Higgins and Tsokos (1981), Calabria, Guida, and Pulcini (1990, 1992). Bain and Engelhardt (1982) presented sequential probability ratio tests for the parameter β in a power law process. Jang and Bai (1987) extended this method to other nonhomogeneous Poisson processes. Engelhardt (1995) presented a review paper, along with over 60 references) on the power law process. Discrete reliability growth models, discussed briefly in MIL-HDBK-189, are studied in more detail in Fries and Sen (1996). Doganaksoy and Nelson (1998) presented a method to compare two independent samples of failure data from repairable systems.

NHPP models have been applied to software reliability by a number of authors, including Goel and Okumoto (1979), Hossain and Dahiya (1993) and Lin and Chen (1993). Many software reliability models are more involved than simply applying an NHPP; Van Pul (1993) is a monograph devoted to software reliability models.

Whitaker and Samaniego (1989) estimated the parameters of the Brown-Proschan imperfect repair model. They pointed out that their results can be extended to the BBS age-dependent imperfect repair model. Presnell, Hollander, and Sethurman (1994) presented a method to test the minimal repair assumption in the BBS model. Agustin and Peña (1999) developed a goodness-of-fit test for the BBS model.

4.13 EXERCISES

1. For the following data sets, sketch a Duane plot and a plot of $N(t_n)$ against t_n. Estimate the parameters β and θ of the power law process, and obtain a confidence interval for the parameter β. Also, give point and interval estimates for the value of the intensity function at the time of the last failure. Assume failure truncation. These failure times have been read from the Duane plots in Duane (1964) and may be accurate only to about 5%. (The data for system #4 was analyzed earlier in this chapter.)

   (a) **System #1**

i	X_i	T_i
1	32	32
2	71	103
3	121	224
4	776	1000
5	420	1420
6	1692	3112
7	3392	6504

(b) **System #2**

i	X_i	T_i
1	12784	12784
2	85244	98028
3	64554	162582
4	223345	385927
5	940970	1326897

(c) **System #3**

i	X_i	T_i
1	6608	6608
2	6714	13322
3	4425	17747
4	7153	24900
5	5007	29907
6	8994	38901
7	11901	50802
8	14470	65295
9	17300	82595
10	15434	98029
11	28972	127001

(d) **System #5**

i	X_i	T_i
1	823	823
2	1209	2032
3	2548	4580
4	5341	9921
5	11396	21317
6	54348	75665

2. Newby (1991) attributed this data set to Balaban (1978), however Balaban's data set, which we saw in Chapter 1, was grouped into intervals. Compute the times between failure, X_i. What do you notice? Can you speculate on the cause for such regularities?

i	T_i	i	T_i
1	150	13	6054
2	360	14	6913
3	541	15	7782
4	775	16	9978
5	1009	17	10433
6	1244	18	10889
7	1478	19	11344
8	1712	20	13598
9	2197	21	14322
10	2682	22	15048
11	3166	23	15773
12	3651		

3. Newby (1991) also gave the following data set, which is the execution times (in 0.01 sec.) between software failure. Thoroughly analyze this data set.

i	X_i	T_i	i	X_i	T_i	i	X_i	T_i
1	479	479	31	277	19015	61	490	54456
2	266	745	32	1300	20315	62	503	55049
3	277	1022	33	821	21136	63	1769	56818
4	554	1576	34	213	21349	64	85	56903
5	1034	2610	35	1620	22969	65	2836	59739
6	949	3559	36	1601	24570	66	213	59952
7	693	4252	37	298	24868	67	1866	61818
8	597	4849	38	874	25742	68	490	62308
9	117	4966	39	618	26360	69	1487	63795
10	170	5136	40	2640	29000	70	4322	68117
11	117	5253	41	5	29005	71	1418	69535
12	1274	6527	42	149	29154	72	1023	70558
13	469	6996	43	1034	30188	73	5490	76048
14	1174	8170	44	2441	32629	74	1520	77568
15	696	8866	45	460	33089	75	3281	80849
16	1908	10774	46	565	33654	76	2716	83565
17	135	10909	47	1119	34773	77	2175	85740
18	277	11186	48	437	35210	78	3505	89245
19	596	11782	49	927	36137	79	725	89970
20	757	12539	50	4462	40599	80	1963	91933
21	437	12976	51	714	41313	81	3979	95912
22	2230	15206	52	181	41494	82	1090	97002
23	437	15643	53	1485	42979	83	245	97247
24	340	15983	54	757	43736	84	1194	98441
25	405	16388	55	3154	46890	85	994	99435
26	575	16963	56	2115	49005	86	3902	103337
27	277	17240	57	884	49889			
28	363	17603	58	2037	51926			
29	522	18125	59	1481	53407			
30	613	18738	60	559	53966			

4. These times between accidents at DuPont were presented by Lucas (1985). The times are measured in days. Thoroughly analyze this data set.

i	X_i	i	X_i	i	X_i	i	X_i	i	X_i	i	X_i
1	15	31	13	61	16	91	30	121	21	151	67
2	16	32	55	62	9	92	10	122	19	152	58
3	33	33	10	63	34	93	1	123	36	153	4
4	4	34	1	64	23	94	1	124	14	154	28
5	51	35	6	65	3	95	8	125	8	155	22
6	3	36	31	66	18	96	2	126	1	156	72
7	15	37	18	67	3	97	3	127	98	157	53
8	19	38	22	68	13	98	15	128	20	158	43
9	14	39	7	69	8	99	5	129	173	159	86
10	7	40	9	70	2	100	6	130	49	160	26
11	13	41	36	71	3	101	24	131	15	161	72
12	2	42	0	72	38	102	13	132	40	162	43
13	4	43	4	73	1	103	18	133	60	163	35
14	0	44	55	74	9	104	48	134	35	164	36
15	7	45	7	75	10	105	12	135	34	165	2
16	20	46	11	76	28	106	27	136	66	166	2
17	0	47	23	77	36	107	46	137	44	167	68
18	18	48	53	78	11	108	7	138	3	168	9
19	1	49	53	79	32	109	7	139	7	169	7
20	30	50	0	80	3	110	22	140	39	170	23
21	0	51	14	81	5	111	2	141	0	171	9
22	12	52	22	82	7	112	13	142	1	172	20
23	15	53	12	83	39	113	14	143	11	173	14
24	9	54	3	84	27	114	19	144	29	174	60
25	0	55	20	85	18	115	3	145	11	175	21
26	2	56	0	86	3	116	0	146	3	176	11
27	6	57	6	87	7	117	6	147	22	177	25
28	56	58	24	88	27	118	14	148	7		
29	26	59	4	89	2	119	8	149	0		
30	20	60	14	90	14	120	34	150	14		

5. Work out the details for the inference for the HPP when the prior for λ is the noninformative prior $p(\lambda) = 1/\lambda$.

6. Consider the posterior distribution of β for the power law process given in equation (4.39). Show that $2n\beta/\hat{\beta}$ has a chi-square distribution with $2(n-1)$ degrees of freedom. *Hint:*

$$\beta^{n-2}\left(\frac{P_n}{t_n^n}\right)^{\beta} = \beta^{n-2}\exp\left[-\beta\log\left(\frac{t_n^n}{P_n}\right)\right].$$

7. Derive the result in equation (4.40).

5

Analyzing Data from Multiple Repairable Systems

When data are available from more than one system, there are a number of models that could be applicable. The choice of a model depends on the assumptions that a user is willing to make. Are the systems identical? Is the HPP a reasonable model for each system? Are the systems dissimilar enough that the process parameters for each system should be estimated independently? Or are the systems similar enough that some "pooling" of data across systems should be applied? Users faced with the problem of analyzing data from many systems must address such questions.

This chapter covers several situations that may arise, beginning with the assumption that all systems can be modeled by the same HPP.

5.1 IDENTICAL HOMOGENEOUS POISSON PROCESSES

Suppose we have k identical repairable systems that we assume can be modeled by a homogeneous Poisson process. Since the systems are identical, the failure process of each system is an HPP with intensity $1/\theta$, and θ is the same for all systems. As in the previous chapter, we caution the reader about the implicit assumptions about aging that are made when assuming the HPP. In this section we discuss estimation and testing for the parameter θ.

5.1.1 Point Estimation for θ

Let t_{ij} denote the jth failure from the ith system. Suppose that n_i failures are observed for system i. We define T_i to be the time that data collection ceased for system i. Thus, if the ith system is failure truncated then $T_i = t_{i,n_i}$ and if the ith system is time truncated then $T_i > t_{i,n_i}$. In a given application, it may be possible that some systems are failure truncated and some are time truncated. We consider separately the systems that are failure and time truncated.

Suppose that the first a systems are failure truncated and the remaining $k-a$ systems are time truncated. The observed data is then of the form

$$0 < t_{11} < t_{12} < \cdots < t_{1n_1} = T_1$$

$$0 < t_{21} < t_{22} < \cdots < t_{2n_2} = T_2$$

$$\vdots$$

$$0 < t_{a1} < t_{a2} < \cdots < t_{an_a} = T_a$$

$$0 < t_{a+1,1} < t_{a+1,2} < \cdots < t_{a+1,n_{a+1}} < T_{a+1}$$

$$\vdots$$

$$0 < t_{k1} < t_{k2} < \cdots < t_{kn_k} < T_k.$$

Since the times between failures $T_{i,j+1} - T_{i,j}$ are iid exponential, the likelihood function for the failure truncated systems (i.e., the joint density of $T_{11}, T_{12}, \ldots, T_{1n_1}, T_{21}, \ldots, T_{a,n_a}$) is

$$\begin{aligned} L_1(\theta) &= \prod_{i=1}^{a} \prod_{j=1}^{n_i} \frac{1}{\theta} \exp\left[-\left(t_{i,j} - t_{i,j-1}\right)/\theta\right] \\ &= \theta^{-\sum_{i=1}^{a} n_i} \exp\left[-\sum_{i=1}^{a} t_{i,n_i}/\theta\right] \\ &= \theta^{-\sum_{i=1}^{a} n_i} \exp\left[-\sum_{i=1}^{a} T_i/\theta\right]. \end{aligned} \tag{5.1}$$

By convention, we have $t_{i,0} = 0$. The last $k-a$ systems are time truncated. The number of failures N_i in the interval $[0, T_i]$ has a Poisson distribution with mean

$$\int_0^{T_i} \frac{1}{\theta} dx = \frac{T_i}{\theta}.$$

The likelihood for the time truncated systems is therefore

$$\begin{aligned} L_2(\theta) &= \prod_{i=a+1}^{k} P(N_i = n_i) \\ &= \prod_{i=a+1}^{k} \frac{(T_i/\theta)^{n_i} \exp[-T_i/\theta]}{n_i!} \\ &= \left(\prod_{i=a+1}^{k} \frac{T_i^{n_i}}{n_i!} \right) \theta^{-\sum_{i=1}^{k} n_i} \exp\left[- \sum_{i=a+1}^{k} T_i/\theta \right]. \end{aligned}$$

The first expression in this product is independent of the parameter θ, so it can be excluded from the likelihood function. The likelihood function from all k systems is thus

$$\begin{aligned} L(\theta) &= L_1(\theta) L_2(\theta) \\ &\propto \left\{ \theta^{-\sum_{i=1}^{a} n_i} \exp\left[-\sum_{i=1}^{a} T_i/\theta \right] \right\} \left\{ \theta^{-\sum_{i=1}^{k} n_i} \exp\left[- \sum_{i=a+1}^{k} T_i/\theta \right] \right\} \\ &= \theta^{-\sum_{i=1}^{k} n_i} \exp\left[-\sum_{i=1}^{k} T_i/\theta \right]. \end{aligned} \tag{5.2}$$

The log-likelihood function is

$$\ell(\theta) = \log L(\theta) = -\left(\sum_{i=1}^{k} n_i \right) \log\theta - \sum_{i=1}^{k} T_i/\theta.$$

Differentiating with respect to zero and setting the result equal to zero yields

$$\frac{d\ell}{d\theta} = -\frac{\sum_{i=1}^{k} n_i}{\theta} + \frac{\sum_{i=1}^{k} T_i}{\theta^2} = 0.$$

Solving this equation for θ yields the MLE

$$\hat{\theta} = \frac{\sum_{i=1}^{k} T_i}{\sum_{i=1}^{k} n_i}. \tag{5.3}$$

5.1.2 Interval Estimation for θ

We will assume for now that all systems are failure truncated. If the jth system is failure truncated, then $2T_{j,n_j}/\theta$ has a chi-square distribution with

$2n_j$ degrees of freedom. Since the random variables $T_{1,n_1}, T_{2,n_2}, \ldots, T_{k,n_k}$ are independent and since the sum of independent chi-square random variables has a chi-square distribution having degrees of freedom equal to the sum of the degrees of freedom of the random variables in the summation, we may conclude that the statistic

$$\begin{aligned}\frac{2T_{1,n_1}}{\theta} + \frac{2T_{2,n_2}}{\theta} + \cdots + \frac{2T_{k,n_k}}{\theta} &= \frac{2}{\theta}\left(T_{1,n_1} + T_{2,n_2} + \cdots + T_{k,n_k}\right) \\ &= \frac{2}{\theta}\left(T_1 + T_2 + \cdots + T_k\right) \\ &= \frac{2}{\theta}\sum_{i=1}^{k} T_i\end{aligned}$$

has a chi-square distribution with

$$2\sum_{i=1}^{k} n_i$$

degrees of freedom. Thus

$$P\left(\chi^2_{1-\alpha/2}\left(2\sum_{i=1}^{k} n_i\right) < \frac{2}{\theta}\sum_{i=1}^{k} T_i < \chi^2_{\alpha/2}\left(2\sum_{i=1}^{k} n_i\right)\right) = 1 - \alpha.$$

This result leads to the confidence interval

$$\frac{2\sum_{i=1}^{k} T_{i,n_i}}{\chi^2_{\alpha/2}\left(2\sum_{i=1}^{k} n_i\right)} < \theta < \frac{2\sum_{i=1}^{k} T_{i,n_i}}{\chi^2_{1-\alpha/2}\left(2\sum_{i=1}^{k} n_i\right)}.$$

Example 5.1 For illustration, we consider a subset of Proschan's (1963) famous aircraft air-conditioning data set. The data we consider are from planes 1 through 4, and are shown in Table 5.1. Assuming that these four systems are identical and that none of them wear out, then it may be reasonable to assume that the times between failure are iid exponential. The MLE for θ is thus

$$\hat{\theta} = \frac{493 + 687 + 456 + 722}{6 + 6 + 6 + 6} = \frac{2358}{24} = 98.25.$$

The appropriate degrees of freedom is $2\sum_{i=1}^{4} n_i = 48$. From the chi-square table (Table A.1) we find that $\chi^2_{0.975}(48) = 30.754$ and $\chi^2_{0.025}(48) = 69.023$. Thus, the 95% confidence interval for θ is

$$\frac{2 \times 2358}{69.023} < \theta < \frac{2 \times 2358}{30.754},$$

Plane			
1	**2**	**3**	**4**
194	413	90	74
209	427	100	131
250	485	160	179
279	522	346	208
312	622	407	710
493	687	456	722

Table 5.1 A subset of the data on failure times of the airconditioning equipment on Boeing 720 aircraft.

which can be written as

$$68.33 < \theta < 153.35.$$

□

We now return to the case where some of all of the systems are time truncated. The MLE of θ is still given by 5.3, but the T_i's are interpreted differently for the failure truncated and time truncated cases. In the failure truncated case, $T_i = t_{i,n_i}$ whereas in the time truncated case T_i is the truncation time and $T_i > t_{i,n_i}$. To obtain an approximate confidence interval for θ when some of the systems are time truncated, we consider a likelihood ratio test of $H_0 : \theta = \theta_0$ versus $H_a : \theta \neq \theta_0$. A confidence interval for θ will then consist of all those values of θ_0 that do not lead to rejection of the null hypothesis. Since $\theta = \theta_0$ when the null hypothesis is true, the likelihood ratio statistic is

$$\begin{aligned}
\mathbf{LR}(\theta_0) &= \frac{\max_{H_0 \text{ true}} L(\theta)}{\max L(\theta)} \\
&= \frac{L(\theta_0)}{L(\hat{\theta})} \\
&= \frac{\theta_0^{-\sum_{i=1}^k n_i} \exp\left(\sum_{i=1}^k T_i/\theta_0\right)}{\hat{\theta}^{-\sum_{i=1}^k n_i} \exp\left(\sum_{i=1}^k T_i/\hat{\theta}\right)} \\
&= \left(\frac{\hat{\theta}}{\theta_0}\right)^{\sum_{i=1}^k n_i} \exp\left(\sum_{i=1}^k n_i - \sum_{i=1}^k T_i/\theta_0\right).
\end{aligned}$$

This statistic is distributed approximately as a chi-square with one degree of freedom. We would reject the null hypothesis $H_0 : \theta = \theta_0$ when $-2 \log \mathbf{LR}(\theta_0)$

is sufficiently large. Using the asymptotic chi-square distribution, the rule would be to reject the null hypothesis when $-2\log \mathbf{LR}(\theta_0) > \chi^2_{0.05}(1)$. A confidence interval for θ would consist of those values of θ_0 that do not lead to rejection of $H_0 : \theta = \theta_0$. In practice, this confidence interval can be obtained by plotting $-2\log \mathbf{LR}(\theta_0)$ against θ_0 and including in the confidence interval all those values of θ_0 for which $-2\log \mathbf{LR}(\theta_0) < \chi^2_{0.05}(1)$. The following example illustrates this procedure.

Example 5.2 We consider again the failure data on the aircraft air-conditioning equipment from Table 5.1. Suppose now that the data collection schemes for systems 3 and 4 were time truncated at times $T_3 = 460$ and $T_4 = 750$, respectively. Systems 1 and 2 are still failure truncated. The MLE of θ is now

$$\hat{\theta} = \frac{493 + 687 + 460 + 750}{6 + 6 + 6 + 6} = 99.58.$$

This estimate is slightly larger than before; this is to be expected since two of the systems were operated failure-free for an additional period of time. The likelihood ratio is

$$\Lambda(\theta_0) = \left(\frac{99.58}{\theta_0}\right)^{24} \exp\left(24 - 2390/\theta_0\right).$$

Hence,

$$\begin{aligned} -2\log \mathbf{LR}\,(\theta_0) &= -48\log 99.58 + 48\log\theta_0 - 48 + 2 \times 2390/\theta_0 \\ &= -268.846 + 48\log\theta_0 + 2 \times 2390/\theta_0. \end{aligned}$$

Figure 5.1 shows a plot of $-2\log \mathbf{LR}\,(\theta_0)$ against θ_0. The 95% confidence interval for θ includes all those values of θ_0 for which $-2\log \mathbf{LR}\,(\theta_0) < \chi^2_{0.05}(1) = 3.841$. The figure also shows the line $\lambda = 3.841$; the confidence interval for θ is thus

$$(68.4, 152.9).$$

This confidence interval is just slightly narrower than the confidence interval we calculated previously when we assumed that all four systems were failure truncated. We would expect the confidence interval to be slightly narrower in this case because we have a bit more information; we know that systems numbered 3 and 4 were operated without failure for an additional 4 and 28 hours, respectively. □

5.1.3 Hypothesis Testing for θ

The results from the previous section can be used to obtain hypothesis tests for θ. If all systems are failure truncated, then

$$\frac{2}{\theta}\sum_{i=1}^{k} T_i = \frac{2}{\theta}\sum_{i=1}^{k} t_{i,n_i}$$

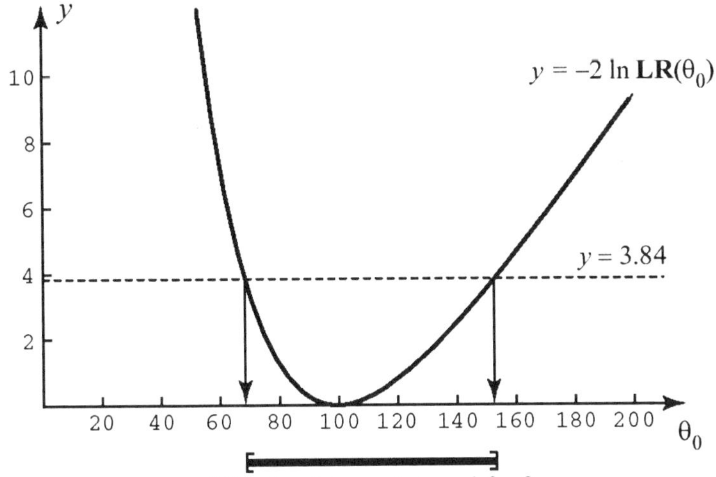

Fig. 5.1 $-2\ln \mathbf{LR}(\theta_0)$ for data on four systems.

has a chi-square distribution with $2\sum_{i=1}^{k} n_i$ degrees of freedom. To test $H_0 : \theta \geq \theta_0$ against $H_a : \theta < \theta_0$ the rule would be to reject the null hypothesis if

$$\frac{2}{\theta_0}\sum_{i=1}^{k} t_{i,n_i} > \chi^2_{\alpha}\left(2\sum_{i=1}^{k} n_i\right)$$

or equivalently

$$\sum_{i=1}^{k} t_{i,n_i} > \frac{\theta_0}{2}\chi^2_{\alpha}\left(2\sum_{i=1}^{k} n_i\right).$$

Notice the way we posed the null and alternative hypotheses. This formulation is testing whether the mean time between failures, θ, is sufficiently large ($\theta > \theta_0$) against the alternative that the mean is not sufficiently large ($\theta < \theta_0$).

The null and alternative hypotheses can be formulated in different ways and the rejection regions can be constructed in a similar manner. For example, for a two-sided test with $H_0 : \theta = \theta_0$ and $H_a : \theta \neq \theta_0$ we would reject the null hypothesis if

$$\sum_{i=1}^{k} t_{i,n_i} > \frac{\theta_0}{2}\chi^2_{\alpha/2}\left(2\sum_{i=1}^{k} n_i\right)$$

or

$$\sum_{i=1}^{k} t_{i,n_i} < \frac{\theta_0}{2}\chi^2_{1-\alpha/2}\left(2\sum_{i=1}^{k} n_i\right).$$

If one or more of the systems in time truncated, then we must resort to the asymptotic likelihood ratio test. We reject $H_0 : \theta = \theta_0$ in favor of the two-sided alternative $H_a : \theta \neq \theta_0$ if

$$-2 \log \mathbf{LR}(\theta_0) > \chi^2_\alpha(1)$$

or equivalently, if

$$-2\left(\sum_{i=1}^{k} n_i\right)\log\hat{\theta} + 2\left(\sum_{i=1}^{k} n_i\right)\log\theta_0 - 2\sum_{i=1}^{k} n_i + 2\sum_{i=1}^{k} T_i/\theta_0 > \chi^2_\alpha(1).$$

This can then be written as

$$-2\left(\sum_{i=1}^{k} n_i\right)\log\hat{\theta} + 2\left(\sum_{i=1}^{k} n_i\right)\log\theta_0 - 2\sum_{i=1}^{k} n_i + 2\frac{\hat{\theta}}{\theta_0}\sum_{i=1}^{k} n_i > \chi^2_\alpha(1).$$

This procedure must be adjusted when a one-sided test is desired. Suppose, for example, we wish to test $H_0 : \theta \geq \theta_0$ versus $H_a : \theta < \theta_0$. In order to have a size α test, we must double the desired α and reject only for small values of $\hat{\theta}$. The rule is to reject the null hypothesis when $\hat{\theta} < \theta$ and

$$-2 \log \mathbf{LR}(\theta_0) > \chi^2_{2\alpha}(1).$$

This last relationship is equivalent to

$$-2\left(\sum_{i=1}^{k} n_i\right)\log\hat{\theta} + 2\left(\sum_{i=1}^{k} n_i\right)\log\theta_0 - 2\sum_{i=1}^{k} n_i + 2\frac{\hat{\theta}}{\theta_0}\sum_{i=1}^{k} n_i > \chi^2_{2\alpha}(1).$$

Example 5.3 We illustrate this asymptotic test with the data from Table 5.1, modified so that systems 3 and 4 are time truncated at times 460 and 750, respectively. Suppose we wish to test the null hypothesis $H_0 : \theta \geq 200$ versus $H_a : \theta < 200$. An approximate size α test would reject the null hypothesis if

$$\begin{aligned}-2\log\mathbf{LR}(200) &= -2\times 24\log 99.58 + 2\times 24\log 200 - 2\times 24 + 2\times 2390/200\\ &= 9.373\end{aligned}$$

is larger than

$$\chi^2_{2\alpha}(1) = 2.706.$$

Thus we would reject the null hypothesis and conclude that there is evidence that the mean time between failures θ is less than 200. The P-value for the one-sided test can be obtained, but a bit of care must be taken. If this were a two-sided test, then the P-value would be the right tail area beyond 9.373 under the $\chi^2(1)$ distribution. This tail probability is 0.002202. Since we have a one-sided test, the P-value would be half of that value, namely, 0.00101. This small P-value is strong evidence that the mean is less than 200. □

5.2 NONIDENTICAL HOMOGENEOUS POISSON PROCESSES

It may be the case that the homogeneous Poisson process is a reasonable model for *each* system, but that the systems are not identical in the sense that their MTBFs are different. In this case we would normally estimate $\theta_1, \theta_2, \ldots, \theta_k$ separately for the k systems. The estimation problem can then be broken into k separate estimation problems, and the methods are described in the previous chapter.

It may be of interest to test whether all k systems do have the same θ.

5.2.1 Two Failure Truncated Systems

If there are $k = 2$ systems and both systems are failure truncated, then the hypothesis that the θ's are equal, $H_0 : \theta_1 = \theta_2$, can be tested as follows. The MLE's of θ_1 and θ_2 are

$$\hat{\theta}_i = \frac{T_{i,n_i}}{n_i}, \quad i = 1, 2.$$

Since the statistics

$$\frac{2T_{i,n_i}}{\theta_i} \sim \chi^2(2n_i)$$

are independent, we can conclude that the ratio

$$\frac{\dfrac{2T_{1,n_1}/\theta_1}{2n_1}}{\dfrac{2T_{2,n_2}/\theta_2}{2n_2}} = \frac{\hat{\theta}_1}{\hat{\theta}_2}\frac{\theta_2}{\theta_1}$$

has an F distribution with $2n_1$ numerator and $2n_2$ denominator degrees of freedom. When the null hypothesis is true, then

$$F = \frac{\hat{\theta}_1}{\hat{\theta}_2} \sim F(2n_1, 2n_2).$$

We would therefore reject the null hypothesis that $\theta_1 = \theta_2$ for $F < F_{1-\alpha/2}(2n_1, 2n_2)$ or $F > F_{\alpha/2}(2n_1, 2n_2)$.

If one or both of the systems is time truncated, then an approach based on likelihood ratios can be used.

Example 5.4 Let us consider the two systems whose failure times are shown in Table 5.2.

System 1	System 2
7	21
8	36
15	41
23	48
35	68
39	81
66	122
75	
111	
120	
126	
129	
152	

Table 5.2 Failure times for two systems.

For these systems, $n_1 = 13$ and $n_2 = 7$, and $t_{1,13} = 152$ and $t_{2,7} = 122$. The MLE's of the two means θ_1 and θ_2 are

$$\hat{\theta}_1 = \frac{152}{13} \approx 11.692$$

and

$$\hat{\theta}_2 = \frac{122}{7} \approx 17.429.$$

The test statistic is

$$F = \frac{\hat{\theta}_1}{\hat{\theta}_2} = \frac{152/13}{122/7} = 0.67087.$$

A size $\alpha = 0.05$ test would reject the null hypothesis $H_0 : \theta_1 = \theta_2$ when

$$F > F_{0.025}(26, 14) = 2.77$$

or

$$F < F_{0.975}(26, 14) = 1/F_{0.025}(14, 26) = 0.41.$$

Here, we obtained the critical values from Table A.2. In this case, since $0.41 < 0.67087 < 2.77$ we would not reject H_0. The P-value for this test is

$$P\text{-value} = 2\,P\left(F(26, 14) < 0.67087\right) = 2 \times 0.1835 = 0.367.$$

This conclusion may seem counterintuitive because the MTBF for system 2 is almost 50% higher than for system 1. The exponential distribution has a large standard deviation relative to its mean, so there is quite a bit of variation in estimates of the parameter θ. Because of the large variability in the estimates of θ_1 and θ_2 an F statistic as small as 0.67087 is not so unlikely. This example shows why it is necessary to conduct formal significance tests. □

5.2.2 k Systems

Suppose we want to test whether all k systems have the same parameter θ. We are thus testing $H_0 : \theta_1 = \theta_2 = \cdots = \theta_k$. For the failure truncated case, time truncated case, or the mixed case, we must base our test on the likelihood ratio and use the chi-square approximation of the test statistic. The likelihood ratio is

$$\mathbf{LR} = \frac{\max_{H_0 \text{ true}} L(\theta_1, \theta_2, \ldots, \theta_k)}{\max L(\theta_1, \theta_2, \ldots, \theta_k)}$$

$$= \frac{\max_\theta \theta^{-\sum_{i=1}^k n_i} \exp\left(-\sum_{i=1}^k T_i/\theta\right)}{\max_{\theta_1,\theta_2,\ldots,\theta_k} \prod_{i=1}^k \frac{1}{\theta_i^{n_i}} \exp(-T_i/\theta_i)}.$$

The function being maximized in the numerator is the likelihood function given in (5.1). This expression is maximized when θ is the MLE $\hat{\theta}$ given in (5.3). The denominator is just the product of k likelihoods, which are maximized when θ_i is equal to

$$\tilde{\theta}_i = \frac{T_i}{n_i};$$

this is the MLE of θ_i if we were to treat each system separately. The likelihood thus becomes

$$\mathbf{LR} = \frac{\hat{\theta}^{-\sum_{i=1}^k n_i} \exp\left[-\sum_{i=1}^k T_i/\hat{\theta}\right]}{\prod_{i=1}^k \left(1/\tilde{\theta}_i\right)^{n_i} \exp\left[-T_i/\tilde{\theta}_i\right]}$$

$$= \frac{\hat{\theta}^{-\sum_{i=1}^k n_i} \exp\left[-\sum_{i=1}^k n_i\right]}{\prod_{i=1}^k \left(1/\tilde{\theta}_i\right)^{n_i} \exp[-n_i]}$$

$$= \frac{\hat{\theta}^{-\sum_{i=1}^k n_i}}{\prod_{i=1}^k \tilde{\theta}_i^{-n_i}}$$

$$= \frac{\prod_{i=1}^k \tilde{\theta}_i^{n_i}}{\hat{\theta}^{\sum_{i=1}^k n_i}}.$$

The statistic

$$-2\log \mathbf{LR} = \left(2\sum_{i=1}^k n_i\right) \log\hat{\theta} - 2\sum_{i=1}^k n_i \log\tilde{\theta}_i$$

is distributed approximately as a chi-square with $k-1$ degrees of freedom. Large values of $-2\log\mathbf{LR}$ lead to rejection of the null hypothesis that the θ's are equal. For moderate to large values of n_i this approximation is good. For small values of n_i Lawless (1982) suggests the modified statistic

$$\frac{-2\log\mathbf{LR}}{1+\frac{1}{6(k-1)}\left[\sum_{i=1}^{k} n_i^{-1}-\left(\sum_{i=1}^{k} n_i\right)^{-1}\right]}, \tag{5.4}$$

which is approximately chi-square with $k-1$ degrees of freedom. This modification increases the accuracy of the chi-square approximation.

Example 5.5 We consider again the data on the air-conditioning systems in Table 5.1. Assuming that all four systems are failure truncated, we test whether the four systems have the same MTBF θ. The null hypothesis is $H_0 : \theta_1 = \theta_2 = \theta_3 = \theta_4$. From the previous discussion we have $\hat{\theta} = 98.25$. The separate estimates for $\theta_1, \theta_2, \theta_3,$ and θ_4 are

$$\begin{aligned}\tilde{\theta}_1 &= \frac{493}{6} = 82.17\\ \tilde{\theta}_2 &= \frac{687}{6} = 114.50\\ \tilde{\theta}_3 &= \frac{456}{6} = 76.00\\ \tilde{\theta}_4 &= \frac{722}{6} = 120.33.\end{aligned}$$

The likelihood ratio statistic is thus

$$\begin{aligned}-2\log\mathbf{LR} &= 2\times 24\log 98.95\\ &\quad -2\,(6\log 82.17+6\log 114.50+6\log 76.00+6\log 120.33)\\ &= 1.30.\end{aligned}$$

For a test with $\alpha = 0.05$, we would reject the null hypothesis when $-2\log\mathbf{LR} >$ 7.8147 and for a test with $\alpha = 0.10$, we would reject the null hypothesis when $-2\log\mathbf{LR} > 6.25139$. The P-value for this test is equal to the probability that a chi-square random variable with 3 degrees of freedom exceeds the test statistic of 1.30. This probability is equal to 0.7292. Such a large P-value does not give any evidence that the null hypothesis is false, so in this case we would conclude that the MTBFs are equal. Use of the modified test statistic 5.4 leads to the same conclusion. The modified chi-square statistic is

$$\frac{1.30}{1+[6\,(3-1)]^{-1}\left[\frac{1}{6}+\frac{1}{6}+\frac{1}{6}+\frac{1}{6}-(6+6+6+6)^{-1}\right]} = 1.236,$$

which yields a P-value of 0.7444. □

5.3 PARAMETRIC EMPIRICAL BAYES AND HIERARCHICAL BAYES MODELS FOR THE HOMOGENEOUS POISSON PROCESS

If the k systems are similar, but not identical, a parametric empirical Bayes (PEB) approach may be useful. With a PEB model, we assume that the system parameters (in this case the θ_i's) are drawn from some prior distribution $p(\theta)$. This assumption allows the θ's to vary from system to system, while still using the information that the systems are "similar." Specifically, we assume that $\theta_1, \theta_2, \ldots, \theta_k$ make up a random sample selected from the prior $p(\theta)$; that is, $\theta_1, \theta_2, \ldots, \theta_k$ are iid random variables from the distribution with pdf $p(\theta)$. This assumption is reasonable if the k systems were made from the same manufacturing process. When we apply control charts to maintain a manufacturing process, we often make the assumption that the process output is iid $N(\mu, \sigma^2)$. Our assumption here is similar except that the process output is the MTBF θ_i. An important difference between control charts and our current situation is that the output is directly observable in the control chart case, whereas the MTBF output is not directly observable. We observe the MTBF θ_i only indirectly by observing the failure data from system i.

If the prior distribution is known completely, then the approach is Bayesian, and the methods described in the last chapter would be applied; that is, for each system, we would use the prior along with the likelihood to obtain the posterior distribution for $\theta_1, \theta_2, \ldots, \theta_k$. Inferences for each θ_i can then be made based on the marginal posterior distribution.

If the prior distribution is not known completely, we can reasonably assume that the prior comes from some family of distributions, such as the gamma distribution. We can take one of two approaches. First, we may think of the parameters of the prior p as being fixed but unknown values. This is the PEB approach, which we describe in the Section 5.3.1. Second, the parameters of the prior p may be considered as random variables having their own prior distribution. This completely Bayesian approach, also called a hierarchical Bayes (HB) model, is discussed in Section 5.3.2. The parameters of the prior distribution are often called hyper-parameters. Figures 5.2 and 5.3 illustrate these two models. The HB model is fully Bayesian, in the sense that all unknown parameters have a prior distribution. The PEB model is not fully Bayesian because the hyper-parameters are considered fixed but unknown. Actually, the PEB approach is not Bayesian at all. This distinction shows that the real difference between the classical and Bayesian approaches is not the existence of priors (because the prior exists in both the PEB and HB approaches) but the way unknown parameters are treated. In the Bayesian paradigm the prior represents the degree of belief in the unknown parameter, whereas the prior for the PEB model exists in a physical sense.

When no parametric distribution form is assumed for θ we have the nonparametric empirical Bayes (NEB) model. Such a model is often called empirical Bayes. The empirical Bayes method was introduced by Robbins (1955).

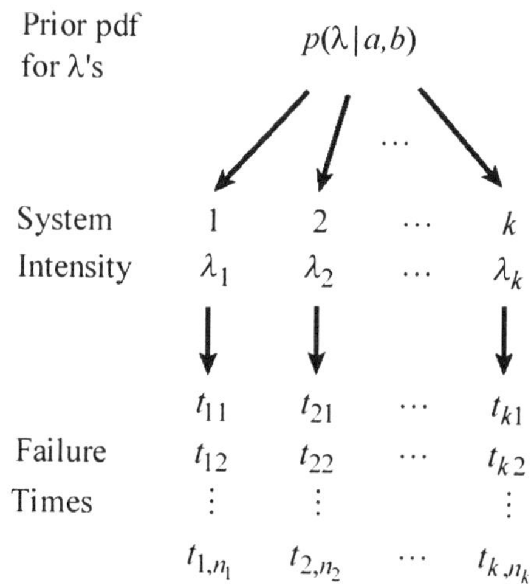

Fig. 5.2 Parametric empirical Bayes model for failure intensities $\lambda_1, \lambda_2, \dots, \lambda_k$.

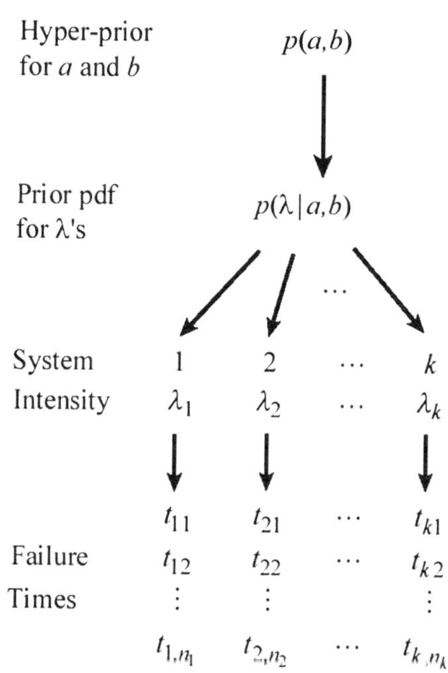

Fig. 5.3 Heierarchical Bayes model for failure intensities $\lambda_1, \lambda_2, \dots, \lambda_k$.

Both the PEB and HB models represent a compromise between the assumptions that

1. All k systems are identical and have the same θ; and

2. The systems are all different, in which case we treat each system separately using only data from system i to make inferences about θ_i.

5.3.1 Parametric Empirical Bayes Models

Let $\theta_1, \theta_2, \ldots, \theta_k$ denote the MTBFs for the k systems. The analysis for the PEB model is easier if we work with the intensities

$$\lambda_i = \frac{1}{\theta_i}, \quad i = 1, 2, \ldots, k$$

rather than with the θ_i's. Suppose that the prior distribution for $\lambda_1, \lambda_2, \ldots, \lambda_k$ is $p(\lambda)$. For now we will assume that the prior is a gamma distribution with scale parameter b and shape parameter a; that is,

$$p(\lambda \mid a, b) = \frac{b^a \lambda^{a-1}}{\Gamma(a)} \exp(-b\lambda), \quad \lambda > 0.$$

Here, the hyper-parameters a and b are considered fixed but unknown. Let $\mathbf{t}_i$ denote the vector of failure times for system i, let

$$\mathbf{T} = [\mathbf{t}_1, \mathbf{t}_2, \ldots, \mathbf{t}_k]$$

denote the two-dimensional array of failure times, and let $\lambda = (\lambda_1, \lambda_2, \ldots, \lambda_k)$ denote the vector of unknown intensities. Because the systems operate independently, the pdf of $\mathbf{T}$ given λ is

$$p(\mathbf{T} \mid \lambda) = \prod_{i=1}^{k} \lambda_i^{n_i} \exp(-\lambda_i t_{i,n_i}).$$

The pdf of λ given the hyper-parameters a and b is

$$p(\lambda \mid a, b) = \left(\frac{b^a}{\Gamma(a)}\right)^k \prod_{i=1}^{k} \lambda_i^{a-1} \exp(-b\lambda_i).$$

We should point out that the λ's are unobservable random variables. We can observe the intensity λ_i only indirectly by observing the sequence of failure times from system i. The above densities were derived so that the pdf of $\mathbf{T}$ given the hyper-parameters a and b could be obtained. This pdf is important because inference for a and b is based solely on the information contained in

T. The pdf is

$$
\begin{aligned}
p(\mathbf{T} \mid a, b) &= \int_0^\infty \cdots \int_0^\infty p\left(\mathbf{T}, \lambda \mid a, b\right) d\lambda_1 \cdots d\lambda_k \\
&= \int_0^\infty \cdots \int_0^\infty p\left(\mathbf{T} \mid \lambda\right) p(\lambda \mid a, b)\, d\lambda_1 \cdots d\lambda_k \\
&= \int_0^\infty \cdots \int_0^\infty \left[\prod_{i=1}^{k} \lambda_i^{n_i} \exp\left(-\lambda_i t_{i,n_i}\right)\right] \\
&\qquad \times \left[\left(\frac{b^a}{\Gamma(a)}\right)^k \prod_{i=1}^{k} \lambda_i^{a-1} \exp\left(-b\lambda_i\right)\right] d\lambda_1 \cdots d\lambda_k \\
&= \left(\frac{b^a}{\Gamma(a)}\right)^k \int_0^\infty \cdots \int_0^\infty \left\{\prod_{i=1}^{k} \lambda_i^{n_i+a-1} \exp\left[-\left(b + t_{i,n_i}\right)\lambda_i\right]\right\} \\
&\qquad d\lambda_1 \cdots d\lambda_k . \\
&= \left(\frac{b^a}{\Gamma(a)}\right)^k \prod_{i=1}^{k} \int_0^\infty \lambda_i^{n_i+a-1} \exp\left[-\left(b + t_{i,n_i}\right)\lambda_i\right] d\lambda_i \,.
\end{aligned}
$$

The ith expression in the product is a function of λ_i only. Thus these integrals can be evaluated one by one, yielding

$$
p(\mathbf{T} \mid a, b) = \left(\frac{b^a}{\Gamma(a)}\right)^k \prod_{i=1}^{k} \int_0^\infty \lambda_i^{n_i+a-1} \exp\left[-\left(b + t_{i,n_i}\right)\lambda_i\right] d\lambda_i .
$$

To evaluate this integral, let's make the change of variable

$$
x = \left(b + t_{i,n_i}\right)\lambda_i .
$$

With this change of variable, the limits on the integral remain the same, and we have $dx = \left(b + t_{i,n_i}\right) d\lambda_i$. The pdf of **T** given a and b then becomes

$$
\begin{aligned}
p(\mathbf{T} \mid a, b) &= \left(\frac{b^a}{\Gamma(a)}\right)^k \prod_{i=1}^{k} \int_0^\infty \frac{x^{n_i+a-1}}{\left(b + t_{i,n_i}\right)^{n_i+a-1}} \exp\left(-x\right) \frac{dx}{\left(b + t_{i,n_i}\right)} \\
&= \left(\frac{b^a}{\Gamma(\alpha)}\right)^k \prod_{i=1}^{k} \frac{1}{\left(a + t_{i,n_i}\right)^{n_i+a-1}} \int_0^\infty x^{n_i+a-1} \exp\left(-x\right) dx \\
&= \left(\frac{b^a}{\Gamma(a)}\right)^k \prod_{i=1}^{k} \frac{\Gamma\left(n_i + a\right)}{\left(b + t_{i,n_i}\right)^{n_i+a}} .
\end{aligned}
$$

This is then the likelihood function of (a, b). The log-likelihood function is thus

$$\ell(a, b) = ka \log b - k \log \Gamma(a) + \sum_{i=1}^{k} \log \Gamma(n_i + a) - \sum_{i=1}^{k} (n_i + a) \log (b + t_{i,n_i}). \tag{5.5}$$

The values of a and b that maximize this function are called the **marginal maximum likelihood estimators.**

To maximize (5.5) we take derivatives with respect to a and b and set them equal to zero. This yields the system of equations:

$$\frac{\partial \ell}{\partial a} = k \log b - k \frac{\Gamma'(a)}{\Gamma(a)} + \sum_{i=1}^{k} \frac{\Gamma'(a + n_i)}{\Gamma(a + n_i)} - \sum_{i=1}^{k} \log (b + t_{i,n_i}) = 0 \tag{5.6}$$

$$\frac{\partial \ell}{\partial b} = \frac{ka}{b} - \sum_{i=1}^{k} \frac{a + n_i}{b + t_{i,n_i}} = 0. \tag{5.7}$$

The function

$$\psi(z) = \frac{d}{dz} \log \Gamma(z) = \frac{\Gamma'(z)}{\Gamma(z)}$$

is called the digamma function, and its derivative, $\psi'(x)$ is called the trigamma function. Algorithms for evaluating the digamma and trigamma functions have been given by Bernardo (1976) and Schneider (1978). Many mathematical software packages, such as *Mathematica*, have built-in functions that evaluate the digamma and trigamma functions. In *Mathematica*, the function `PolyGamma[z]` gives the digamma function and `PolyGamma[n,z]` gives $\psi^{(n)}(z)$. Equation (5.7) can be written in the form

$$a \left[\frac{k}{b} - \sum_{i=1}^{k} (b + t_{i,n_i})^{-1} \right] = \sum_{i=1}^{k} \frac{n_i}{b + t_{i,n_i}},$$

which can be solved for a in terms of b to obtain

$$a = \frac{\sum_{i=1}^{k} n_i / (b + t_{i,n_i})}{k/b - \sum_{i=1}^{k} (b + t_{i,n_i})^{-1}}. \tag{5.8}$$

This expression for a can then be substituted into (5.6), yielding one nonlinear equation in the one unknown b. The solution can be approximated using Newton's method. The resulting equation is long and messy, but is straightforward to program. In terms of the digamma function ψ, the equation can be written as

$$k \log b - k\psi \left(\frac{\sum_{i=1}^{k} n_i/(b + t_{i,n_i})}{k/b - \sum_{i=1}^{k} (b + t_{i,n_i})^{-1}} \right)$$

$$+ \sum_{i=1}^{k} \psi \left(\frac{\sum_{i=1}^{k} n_i/(b + t_{i,n_i})}{k/b - \sum_{i=1}^{k} (b + t_{i,n_n})^{-1}} + n_i \right) - \sum_{i=1}^{k} \log (b + t_{i,n_i}) = 0. \tag{5.9}$$

The procedure for approximating the MLE's of a and b is then to (1) use Newton's method to solve (5.9) for b, and (2) use that value of b (call the solution $\hat{b}$) to compute a from (5.8). The resulting solution estimates $\hat{b}$ and $\hat{a}$ are the marginal maximum likelihood estimates (MMLEs).

Point estimates of $\lambda_1, \lambda_2, \ldots, \lambda_k$ can then be obtained by taking the posterior expectations. The posterior distribution of λ given that $\hat{a}$ and $\hat{b}$ are equal to the hyper-parameters is proportional to the prior times the likelihood. That is,

$$\begin{aligned} p(\lambda \mid \mathbf{T}, \hat{a}, \hat{b}) &= cp\left(\lambda \mid \hat{a}, \hat{b}\right) p(\mathbf{T} \mid \lambda) \\ &= c\left[\left(\frac{\hat{b}^{\hat{a}}}{\Gamma(\hat{a})}\right)^k \prod_{i=1}^{k} \lambda_i^{\hat{a}-1} \exp\left(-\hat{b}\lambda_i\right)\right]\left[\prod_{i=1}^{k} \lambda_i^{n_i} \exp\left(-\lambda_i t_{i,n_i}\right)\right] \\ &= c\prod_{i=1}^{k} \frac{\hat{b}^{\hat{a}} \lambda_i^{\hat{a}+n_i-1}}{\Gamma(\hat{a})} \exp\left[-\left(\hat{b} + t_{i,n_i}\right)\lambda_i\right]. \end{aligned} \tag{5.10}$$

This joint pdf is the product of independent gamma densities. Thus the marginal posterior distribution of λ_i is

$$p(\lambda_i \mid \mathbf{T}) = c_i \lambda_i^{\hat{a}+n_i-1} \exp\left[-\left(\hat{b} + t_{i,n_i}\right)\lambda_i\right], \quad \lambda_i > 0.$$

Here, $\hat{b}^{\hat{a}}$ and $\Gamma(\hat{a})$ have been absorbed into the constant c_i. In order for this posterior distribution to integrate to one,we must have

$$\begin{aligned} c_i &= \left\{\int_0^{\infty} \lambda_i^{\hat{a}+n_i-1} \exp\left[-\left(\hat{b} + t_{i,n_i}\right)\lambda_i\right] d\lambda_i\right\}^{-1} \\ &= \left\{\Gamma(\hat{a} + n_i)\left(\hat{b} + t_{i,n_i}\right)\right\}^{-1}. \end{aligned}$$

The posterior mean of λ_i is thus the mean of this gamma distribution. The point estimate for λ_i is then

$$\tilde{\lambda}_i = \frac{\hat{a} + n_i}{\hat{b} + t_{i,n_i}}.$$

Since, conditioned on $\hat{a}$ and $\hat{b}$, $2\left(\hat{b} + t_{i,n_i}\right)\lambda_i$ has a chi-square distribution with $2(\hat{a} + n_i)$ degrees of freedom, we can obtain approximate confidence intervals for λ_i. Many computer algorithms that compute the tabled values of the chi-square distribution will work even for noninteger degrees of freedom, like we have in this case.

The posterior mean for $\theta_i = 1/\lambda_i$ is then

$$\tilde{\theta}_i = \frac{\int_0^\infty \frac{1}{\lambda_i} \lambda_i^{\hat{a}+n_i-1} \exp\left[-\left(\hat{b} + t_{i,n_i}\right)\lambda_i\right] d\lambda_i}{\int_0^\infty \lambda_i^{\hat{a}+n_i-1} \exp\left[-\left(\hat{b} + t_{i,n_i}\right)\lambda_i\right] d\lambda_i}.$$

In both integrals, make the change of variable $y = \left(\hat{b} + t_{i,n_i}\right)\lambda_i$ so that $\lambda_i = y/\left(\hat{b} + t_{i,n_i}\right)$ and $dy = \left(\hat{b} + t_{i,n_i}\right) d\lambda_i$. With this change of variables we have

$$\begin{aligned}
\tilde{\theta}_i &= \frac{\int_0^\infty \left(\frac{y}{\hat{b} + t_{i,n_i}}\right)^{\hat{a}+n_i-2} \exp(-y) \frac{dy}{\hat{b} + t_{i,n_i}}}{\int_0^\infty \left(\frac{y}{\hat{b} + t_{i,n_i}}\right)^{\hat{a}+n_i-1} \exp(-y) \frac{dy}{\hat{b} + t_{i,n_i}}} \\
&= \left(\hat{b} + t_{i,n_i}\right) \frac{\int_0^\infty y^{\hat{a}+n_i-1-1} \exp(-y)\, dy}{\int_0^\infty y^{\hat{a}+n_i-1} \exp(-y)\, dy} \\
&= \left(\hat{b} + t_{i,n_i}\right) \frac{\Gamma(\hat{a} + n_i - 1)}{\Gamma(\hat{a} + n_i)} \\
&= \left(\hat{b} + t_{i,n_i}\right) \frac{\Gamma(\hat{a} + n_i - 1)}{(\hat{a} + n_i - 1)\Gamma(\hat{a} + n_i - 1)} \\
&= \frac{\hat{b} + t_{i,n_i}}{\hat{a} + n_i - 1}.
\end{aligned}$$

Notice that this is not equal to $1/\hat{\lambda}_i$ which is equal to

$$\frac{1}{\hat{\lambda}_i} = \frac{\hat{b} + t_{i,n_i}}{\hat{a} + n_i}.$$

Given observed data and given $\hat{a}$ and $\hat{b}$, the random variable $2\left(\hat{b} + t_{i,n_i}\right)\lambda_i$ has a chi-square distribution with $\nu = 2(\hat{a} + n_i)$ degrees of freedom. This result can be used to obtain posterior probability intervals for λ_i. Many computer algorithms that compute the tabled values of the chi-square distribution will work even for noninteger degrees of freedom like we have in the present case. *Minitab*, for example will handle noninteger degrees of freedom. Thus we have

$$P\left(\chi^2_{1-\gamma/2}(\nu) < 2\left(\hat{b} + t_{i,n_i}\right)\lambda_i < \chi^2_{\gamma/2}(\nu) \,\middle|\, \mathbf{T}\right) = 1 - \gamma$$

$$P\left(\frac{\chi^2_{1-\gamma/2}(\nu)}{\hat{b} + t_{i,n_i}} < \lambda_i < \frac{\chi^2_{\gamma/2}(\nu)}{\hat{b} + t_{i,n_i}} \,\middle|\, \mathbf{T}\right) = 1 - \gamma.$$

This gives

$$\left(\frac{\chi^2_{1-\gamma/2}(\nu)}{\hat{b} + t_{i,n_i}}, \frac{\chi^2_{\gamma/2}(\nu)}{\hat{b} + t_{i,n_i}}\right)$$

as a posterior probability interval for λ_i. An interval for θ_i would be

$$\left(\frac{\hat{b} + t_{i,n_i}}{\chi^2_{\gamma/2}(\nu)}, \frac{\hat{b} + t_{i,n_i}}{\chi^2_{1-\gamma/2}(\nu)}\right).$$

These intervals are sometimes called the *naive* PEB confidence intervals because they do not take into account the fact that the hyper-parameters b and a are estimated. Because of the uncertainty in the estimates for a and b, the naive PEB confidence intervals are narrower than they should be. [See Morris (1983a, 1983b) and Carlin and Gelfand (1990).]

We began this section by assuming that the prior distribution for λ was a gamma distribution with unknown parameters. The choice of the gamma was arbitrary, and other parametric forms could be chosen. For moderate or large sample sizes, the parametric form of the prior has little effect on the point and interval estimates of λ. Basu and Rigdon (1984) applied a gamma prior distribution, as we described here, and a log-normal prior distribution. The analysis of the log-normal prior is more complicated and the EM algorithm of Dempster, Laird, and Rubin (1977) was used to obtain the marginal maximum likelihood estimates of the prior parameters. The log-normal prior has the form

$$p(\lambda \mid \mu, \sigma^2) = (2\pi\lambda^2\sigma^2)^{-1/2} \exp\left[-\left(\log\lambda - \mu\right)^2 / 2\sigma^2\right], \quad \lambda > 0.$$

Gaver and O'Muircheartaigh (1987) derived similar results using gamma and log-student t prior distributions. As the next examples show, the choice of the parametric form for the prior has little effect on the point and interval estimates of λ_i.

Example 5.6 We consider the failure times of aircraft air-conditioning units that were given by Proschan (1963). Table 5.3 gives these data.

Plane	Failure Times							
1	194	209	250	279	312	493			
2	413	427	485	522	622	687	696	865	1312
	1496	1532	1733	1851	1885	1916	1934	1952	2019
	2076	2138	2145	2167	2201				
3	90	100	160	346	407	456	470	494	550
	570	649	733	777	836	965	983	1008	1164
	1474	1550	1576	1620	1643	1705	1835	2043	2113
	2214	2422							
4	74	131	179	208	710	722	792	813	842
	1228	1287	1314	1467	1403	1819			
5	55	375	431	535	755	994	1041	1287	1463
	1645	1678	1693	1797	1832				
6	23	284	371	378	498	512	574	621	846
	917	1163	1184	1226	1246	1251	1263	1383	1364
	1397	1411	1482	1493	1507	1518	1534	1624	1625
	1641	1693	1788						
7	97	148	159	163	304	322	464	532	609
	689	690	706	812	1018	1100	1154	1185	1401
	1447	1558	1597	1660	1678	1869	1887	2055	2079
8	50	94	196	268	290	329	332	347	544
	732	811	899	945	950	955	991	1013	1152
	1362	1459	1489	1512	1525	1539			
9	359	368	380	650	1253	1256	1360	1362	1800
10	50	304	309	592	627	639			
11	130	623							
12	487	505	605	612	710	715	800	891	934
	1164	1167	1297						
13	102	311	325	382	436	468	535	594	728
	880	907	921	1151	1217	1278	1312		

Table 5.3 Failure times for air-conditioning equipment on 13 aircraft.

The marginal maximum likelihood estimates of a and b are

$$\hat{a} = 18.41 \quad \text{and} \quad \hat{b} = 1733.10.$$

If the log-normal prior is assumed, then the estimates of the hyper-parameters are

$$\hat{\mu} = -4.57 \quad \text{and} \quad \hat{\sigma}^2 = 0.0552.$$

The computations involved in getting $\hat{\mu}$ and $\hat{\sigma}^2$ are rather complicated and are not discussed here. See Basu and Rigdon (1984) for a further discussion. Table 5.4 gives the MLE's and the PEB posterior means for a gamma prior and a lognormal prior. Table 5.5 gives 95% confidence intervals along with the 95% posterior probability intervals assuming gamma and lognormal priors. The units for the λ_i's are failures per thousand hours. □

i	n_i	t_{i,n_i}	MLE $\hat{\lambda}_i$	Gamma Prior $\tilde{\lambda}_i$	Lognormal Prior $\tilde{\lambda}_i$
1	6	493	12.17	10.97	10.92
2	23	2201	10.45	10.53	10.49
3	29	2422	11.97	11.41	11.33
4	15	1819	8.25	9.41	9.41
5	14	1832	7.64	9.09	9.12
6	30	1788	16.78	13.75	13.84
7	27	2079	12.99	11.93	11.89
8	24	1539	15.59	12.96	12.99
9	9	1800	5.00	7.76	7.96
10	6	639	9.39	10.29	10.27
11	2	623	3.21	8.66	8.81
12	12	1297	9.25	10.04	10.01
13	16	1312	12.20	11.30	11.24

Table 5.4 Point estimates for aircraft air-conditioning equipment.

i	95% Confidence Interval	Gamma Prior 95% Prob. Interval	Lognormal Prior 95% Prob. Interval
1	(4.46, 23.67)	(7.06, 15.72)	(6.58, 15.27)
2	(6.62, 15.07)	(7.57, 13.97)	(7.33, 13.65)
3	(8.02, 16.71)	(8.40, 15.12)	(8.11, 14.55)
4	(4.62, 12.91)	(6.49, 12.85)	(6.34, 12.49)
5	(4.18, 12.13)	(6.23, 12.48)	(6.13, 12.12)
6	(11.32, 23.29)	(10.15, 17.88)	(9.78, 17.90)
7	(8.58, 18.37)	(8.71, 15.64)	(8.40, 15.38)
8	(9.99, 22.42)	(9.36, 17.14)	(8.89, 17.09)
9	(2.29, 8.76)	(5.13, 10.92)	(5.26, 10.57)
10	(3.44, 18.26)	(6.62, 14.76)	(6.28, 14.25)
11	(0.39, 8.94)	(5.32, 12.81)	(5.31, 12.32)
12	(4.78, 15.17)	(6.79, 13.90)	(6.54, 13.47)
13	(6.97, 18.86)	(7.84, 15.38)	(7.44, 15.03)

Table 5.5 Intervals for λ_i for aircraft air-conditioning data.

We point out several features from Tables 5.4 and 5.5. First, the PEB point and interval estimates with gamma and log-normal priors are nearly the same, which means that the choice of the parametric form for the prior is not very important. Second, the PEB point estimates of the λ's are less disperse than the MLE. This contraction occurs because data from all k systems are used to estimate the λ for each individual system. The PEB point estimate for λ, assuming a gamma prior is

$$\tilde{\lambda}_i = \frac{\hat{a} + n_i}{\hat{b} + t_{i,n_i}}.$$

Data from system i, namely, n_i and t_{i,n_i}, directly enter into the estimate for $\hat{\lambda}_i$. Data from all other systems affect the estimates $\hat{a}$ and $\hat{b}$ and thus affect $\tilde{\lambda}_i$ indirectly. This indirect effect is desirable if the systems are similar but not identical. For a system like the air-conditioning unit in plane 11, which experienced only two failures, it makes sense to use information from the other twelve systems. Notice that the systems that had the fewest failures, also had the most contraction. For systems with many failures, the MLE's and the PEB posterior means are nearly the same.

Third, the 95% intervals are narrower for the PEB approach than for the classical approach. There are several explanations for this. Bayesian probability intervals and classical confidence intervals are interpreted differently, so a direct comparison is impossible. Another explanation is that more information is incorporated in the PEB interval than in the classical interval. Finally, we used the point estimates $\hat{\delta}$ and $\hat{\alpha}$ (or $\hat{\mu}$ and $\hat{\sigma}$ for the log-normal prior) as if these were the true parameters of the prior distribution. Of course, these are just estimates of the prior parameters and uncertainty in these estimates should be taken into account when estimating the λ's. The PEB probability intervals for which we use the hyper-parameter estimates as if they were the true hyper-parameters are called the naive PEB probability intervals. In the next section, we present a hierarchical Bayes approach that addresses this problem.

Example 5.7 As a second example, we consider a data set given by Davis (1952) on the number of entries between errors for five bookkeepers. Since the entire data set is rather voluminous we do not reproduce it here. A summary of the data, along with the classical and PEB point and interval estimates are given in Tables 5.6 and 5.7. (The units for the λ_i's are in errors per thousand entries.) For this data set the number of entries between errors must necessarily be an integer. The number of entries between errors is typically very large so the HPP may be a reasonable model. This example, however, is no more "discrete" than the failure times on aircraft air-conditioners, because the failure times were rounded to the nearest hour.

i	n_i	t_{i,n_i}	$\hat{\lambda}_i$	Gamma Prior $\tilde{\lambda}_i$	Lognormal Prior $\tilde{\lambda}_i$
1	31	17991	1.72	1.79	1.78
2	26	17533	1.48	1.57	1.57
3	26	18742	1.39	1.48	1.48
4	54	18273	2.96	2.92	2.90
5	81	15446	5.24	5.00	5.03

Table 5.6 MLE and PEB estimates of λ's for the bookkeeper data.

i	95% CI	Gamma Prior 95% PI	Lognormal Prior 95% PI
1	$(1.17, 2.38)$	$(1.25, 2.43)$	$(1.20, 2.36)$
2	$(0.97, 2.10)$	$(1.06, 2.18)$	$(1.06, 2.11)$
3	$(0.91, 1.97)$	$(1.00, 2.05)$	$(0.97, 1.99)$
4	$(2.22, 3.79)$	$(2.22, 3.72)$	$(2.15, 3.65)$
5	$(4.16, 6.45)$	$(3.99, 6.11)$	$(3.94, 6.12)$

Table 5.7 Confidence and probability intervals of λ's for the bookkeeper data.

The point estimates of the hyper-parameters are

$$\hat{a} = 4.074 \text{ and } \hat{b} = 1597.44$$

if the gamma prior is assumed, and

$$\hat{\mu} = -6.10 \text{ and } \hat{\sigma}^2 = 0.24$$

if the log-normal prior is assumed. Tables 5.6 and 5.7 show much the same features as Tables 5.4 and 5.5 for the previous example on the aircraft air-conditioning units. Since there a large number of errors (failures) per bookkeeper (system), and since k is so small, there is much less contraction in the PEB estimates than there was in the previous example. □

5.3.2 Hierarchical Bayes Models

As we mentioned in the previous subsection, the problem with the PEB approach is that we assume that the *estimates* of the prior parameters are the prior parameters themselves. The PEB approach does not account for uncertainty in the estimates of these hyper-parameters. Variation in these estimates would lead to more variation in the estimates of the λ's.

One approach to this problem is to employ a hierarchical Bayes (HB) model. This is a fully Bayesian approach in which we assign a hyper-prior distribution to the parameters for the prior of the λ's. The hierarchical nature of this model is shown in Figure 5.3. The hyper-prior distribution g expresses our uncertainty in the prior parameters a and b. (We will consider here only the gamma prior distribution.) The hyper-prior $p(a, b)$ gives rise to the λ's. Finally $\lambda_1, \lambda_2, \ldots, \lambda_k$ give rise to the failure times; that is, λ_i gives $t_{i,1} < t_{i,2} < \cdots < t_{i,n_i}$.

Point estimates for the λ's are the posterior means, which are obtained as follows. Since for any random variables X and Y

$$E\,(E\,(X \,|\, Y)) = E(X),$$

we have

$$\begin{aligned} E\,(\lambda_i \,|\, \mathbf{t}) &= E\,(E\,(\lambda_i \,|\, a, b, \mathbf{t}) \,|\, \mathbf{t}) \\ &= E\left(\frac{a + n_i}{b + t_{i,n_i}} \,|\, \mathbf{t}\right). \end{aligned}$$

This expectation is found by integrating the joint posterior distribution of a and b. The posterior distribution has pdf

$$\begin{aligned} p(a, b \,|\, \mathbf{t}) &= c\,p(a, b)\, p(\mathbf{t} \,|\, a, b) \\ &= c\,p(a, b) \prod_{i=1}^{k} p\,(\mathbf{t}_i \,|\, a, b) \\ &= c\,p(a, b) \prod_{i=1}^{k} \int_0^\infty p\,(\mathbf{t}_i, \lambda_i \,|\, a, b)\; d\lambda_i \\ &= c\,p(a, b) \prod_{i=1}^{k} \int_0^\infty \frac{b^a \lambda_i^{a-1}}{\Gamma(a)} \exp\left(-b\lambda_i\right)\, \lambda_i^{n_i} \exp\left(-t_{i,n_i}\lambda_i\right)\, d\lambda_i \\ &= c\,p(a, b) \prod_{i=1}^{k} \int_0^\infty \frac{b^a \lambda_i^{a+n_i-1}}{\Gamma(a)} \exp\left(-\left(b + t_{i,n_i}\right)\lambda_i\right)\, d\lambda_i\,, \end{aligned}$$

where c is a constant chosen to make the posterior pdf integrate to one. This integral can be evaluated by making a transformation of variables and using the properties of the gamma function. This yields

$$p(a, b \,|\, \mathbf{t}) = c\,p(a, b) \prod_{i=1}^{k} \frac{b^a}{\Gamma(a)} \frac{\Gamma\left(a + n_i\right)}{\left(b + t_{i,n_i}\right)^{a+n_i}}\,.$$

The posterior expectation of λ_i is therefore

$$E(\lambda_i \mid \mathbf{t}) = E\left(\frac{a+n_i}{b+t_{i,n_i}} \,\middle|\, \mathbf{t} \right)$$

$$= \frac{\displaystyle\int_0^\infty \int_0^\infty \frac{a+n_i}{b+t_{i,n_i}} p(b,a) \left(\frac{b^a}{\Gamma(a)} \right)^k \prod_{i=1}^k \frac{\Gamma(a+n_i)}{(b+t_{i,n_i})^{a+n_i}} \, da \, db}{\displaystyle\int_0^\infty \int_0^\infty p(b,a) \left(\frac{b^a}{\Gamma(a)} \right)^k \prod_{i=1}^k \frac{\Gamma(a+n_i)}{(b+t_{i,n_i})^{a+n_i}} \, da \, db}.$$

These integrals must be approximated using numerical integration. An improper prior, for example a prior such as $p(a,b) = c$, can lead to an improper posterior. We suggest using a prior distribution that is flat over a bounded set of plausible values for a and b.

Example 5.8 We applied this HB model to the data from Table 5.6 on the entries between errors for five bookkeepers. The posterior means, posterior distributions, and approximate 95% probability intervals are shown in Table 5.7. To make the computations simpler, we looked at the number of failures per thousand entries. The prior distribution was chosen to be

$$p(a,b) = \begin{cases} c, & 0.1 < a < 20, \; 0.1 < b < 10 \\ 0, & \text{otherwise.} \end{cases}$$

The HB point estimates are close to the MLE's and to the PEB estimates. The posterior probability intervals were computed assuming a normal posterior; that is,

$$\left(E(\lambda_i \mid \mathbf{t}) - 1.96\sqrt{V(\lambda_i \mid \mathbf{t})}, E(\lambda_i \mid \mathbf{t}) - 1.96\sqrt{V(\lambda_i \mid \mathbf{t})} \right).$$

The posterior variance is

$$\begin{aligned} V(\lambda_i \mid \mathbf{t}) &= E(\lambda_i^2 \mid \mathbf{t}) - [E(\lambda_i \mid \mathbf{t})]^2 \\ &= E(E(\lambda_i \mid a, b, \mathbf{t}) \mid \mathbf{t}) - [E(\lambda_i \mid \mathbf{t})]^2 \\ &= E\left(\frac{(n_i + a + 1)(n_i + a)}{(b + t_{i,n_i})^2} \,\middle|\, \mathbf{t} \right) - [E(\lambda_i \mid \mathbf{t})]^2. \end{aligned}$$

The results are shown in Table 5.8. □

i	n_i	t_{i,n_i}	**HB Pt Estimate**	**HB 95% Prob. Interval**
1	31	17,991	1.83	$(1.24, 2.42)$
2	26	17,553	1.62	$(1.05, 2.20)$
3	26	18,742	1.54	$(0.99, 2.08)$
4	54	18,273	2.89	$(2.15, 3.62)$
5	81	15,446	4.80	$(3.70, 5.89)$

Table 5.8 Hierarchical Bayes point and interval estimates.

5.4 POWER LAW PROCESS FOR IDENTICAL SYSTEMS

In this section we assume that all k systems are identical, and that the failure patterns are modeled by the same power law process, which has parameters θ and β. This assumption of identical systems is a strong one and should not be made without some justification. Simple graphical methods can often help to assess this assumption. Later in this section we will discuss methods for testing the equality of parameters from multiple systems.

Let t_{ij} denote the jth failure on the ith system. Suppose that system i is observed until time T_i, where $T_i = t_{i,n_i}$ if system i was failure truncated and $T_i > t_{i,n_i}$ if system i was time truncated. Let n_i be the number of observed failures for system i, and $N = n_1 + n_2 + \cdots + n_k$. Since the failures on separate systems are independent, the likelihood function is

$$\begin{aligned} L(\theta,\beta) &= \prod_{i=1}^{k} \left\{ \left[\prod_{j=1}^{n_i} \lambda(t_{i,j}) \right] \exp\left[-\Lambda(T_i)\right] \right\} \\ &= \prod_{i=1}^{k} \left\{ \left[\prod_{j=1}^{n_i} \frac{\beta}{\theta} \left(\frac{t_{i,j}}{\theta}\right)^{\beta-1} \right] \exp\left[-(T_i/\theta)^{\beta}\right] \right\} \\ &= \beta^{\sum_{i=1}^{k} n_i}\, \theta^{-\beta \sum_{i=1}^{k} n_i} \left(\prod_{i=1}^{k} \prod_{j=1}^{n_i} t_{ij} \right)^{\beta-1} \exp\left[-\sum_{i=1}^{k} (T_i/\theta)^{\beta} \right]. \end{aligned}$$

The log-likelihood $\ell(\theta,\beta) = \log L(\theta,\beta)$ is readily obtained. Differentiating $\ell(\theta,\beta)$ with respect to θ and with respect to β yields (after some simplification)

$$\theta = \left(\frac{\sum_{i=1}^{k} T_i^{\beta}}{N} \right)^{1/\beta} \tag{5.11}$$

and

$$\beta = \frac{N}{\theta^{-\beta} \sum_{i=1}^{k} T_i^{\beta} \log T_i - \sum_{i=1}^{k} \sum_{j=1}^{n_i} \log t_{ij}}. \tag{5.12}$$

If all k systems are time truncated at the same time T, then (5.11) becomes

$$\theta = \left(\frac{kT^{\beta}}{N} \right)^{1/\beta} = \frac{k^{1/\beta} T}{N^{1/\beta}}.$$

When this is substituted into (5.12) we obtain

$$\begin{aligned}
\beta &= \frac{N}{\left(\dfrac{k^{1/\beta}T}{N^{1/\beta}}\right)^{-\beta} \sum_{i=1}^{k} T^{\beta} \log T - \sum_{i=1}^{k}\sum_{j=1}^{n_i} \log t_{ij}} \\
&= \frac{N}{\dfrac{k^{-1}T^{-\beta}}{N^{-1}} T^{\beta} k \log T - \sum_{i=1}^{k}\sum_{j=1}^{n_i} \log t_{ij}} \\
&= \frac{N}{N \log T - \sum_{i=1}^{k}\sum_{j=1}^{n_i} \log t_{ij}} \\
&= \frac{N}{\sum_{i=1}^{k}\sum_{j=1}^{n_i} \log T - \sum_{i=1}^{k}\sum_{j=1}^{n_i} \log t_{ij}} \\
&= \frac{N}{\sum_{i=1}^{k}\sum_{j=1}^{n_i} \log (T/t_{ij})} .
\end{aligned}$$

In all other cases, (5.11) and (5.12) cannot be solved explicitly for θ and β, although substitution of (5.11) into (5.12) yields one nonlinear equation in the one unknown β. This equation can then be solved by some iterative algorithm such as Newton's method or fixed-point iteration $[\beta_{\nu+1} = g(\beta_{\nu})]$.

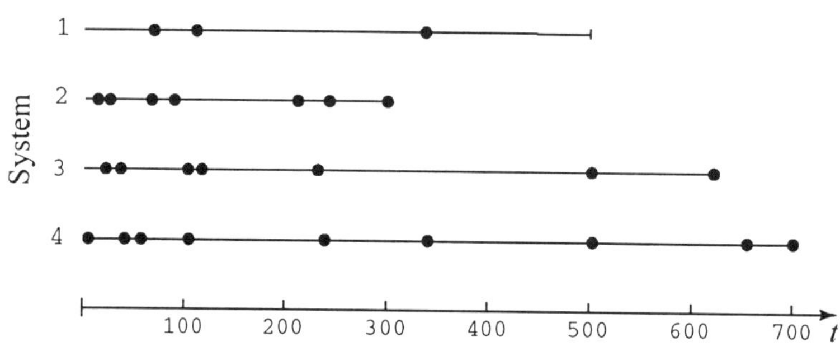

Fig. 5.4 Time dot plots for four systems. System 1 is time truncated, while the others are failure truncated.

Example 5.9 For example, consider the failure times given in Table 5.9. System 1 is time truncated, and the other three are failure truncated. The time dot plots, shown in Figure 5.4, indicates that there is reliability improvement.

i	n_i	Failure Times $t_{ij},\ j = 1, 2, \ldots, n_i$							Termination of Test T_i
1	3	69	112	338					500
2	7	14	26	67	90	213	244	301	301
3	7	22	37	104	118	233	502	622	622
4	9	5	41	57	105	240	341	503	701
		655	701						

Table 5.9 Failure times of four systems.

Substitution of (5.11) into (5.12) yields one nonlinear equation in the one unknown β

$$\beta = \frac{N}{\left(\dfrac{N}{\sum_{i=1}^{k} T_i^{\beta}}\right) \sum_{i=1}^{k} T_i^{\beta} \log T_i - \sum_{i=1}^{k} \sum_{j=1}^{n_i} \log t_{ij}}. \tag{5.13}$$

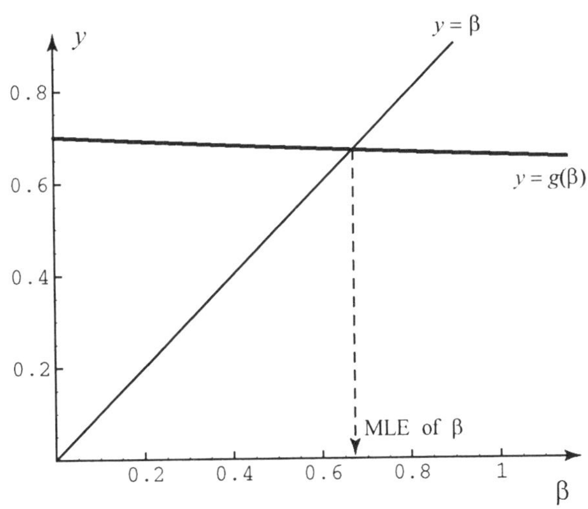

Fig. 5.5 MLE as the solution to $\beta = g(\beta)$.

Figure 5.5 shows a plot of the right side of (5.13) along with the line $y = \beta$; the point of intersection is the solution of (5.13) and therefore it is the MLE of β. Five iterations of fixed-point iteration yield the values shown in Table 5.10; the MLE of β is therefore $\hat{\beta} = 0.670$.

ν	β_ν
0	0.6
1	0.672611
2	0.669744
3	0.669855
4	0.669851
5	0.669851

Table 5.10 Results of fixed point iteration.

The MLE of θ is then obtained by substituting this value for β into (5.11), yielding

$$\hat{\theta} = \left(\frac{\sum_{i=1}^{4} T_i^{0.670}}{26} \right)^{1/0.670} = 31.99.$$

Figure 5.6 shows the estimated intensity function $\lambda(t) = (\hat{\beta}/\hat{\theta})(t/\hat{\theta})^{\hat{\beta}-1}$. □

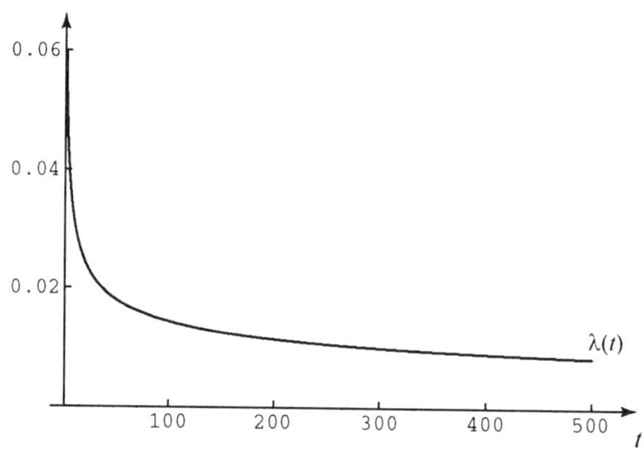

Fig. 5.6 Estimate of common intensity function for four systems.

Conditioning on the time of the nth failure (failure truncated case) or on the number of observed failures (time truncated case) yields a conditional likelihood function. We can use this conditional likelihood to find the **conditional MLEs**. The conditional MLEs for β and θ are (Crow, 1974)

$$\tilde{\beta} = \frac{M}{\sum_{i=1}^{k} \sum_{j=1}^{n_i} \log(T_i/t_{ij})}$$

and

$$\tilde{\theta} = \left(\frac{\sum_{i=1}^{k} T_i^{\tilde{\beta}}}{M} \right)^{1/\tilde{\beta}},$$

where

$$m_i = \begin{cases} n_i - 1, & \text{if data are failure truncated} \\ n_i & \text{if data are time truncated} \end{cases}$$

and

$$M = \sum_{i=1}^{k} m_i.$$

The estimator $\tilde{\beta}$ has the property that $2n\beta/\tilde{\beta} \sim \chi^2(2M)$. This result gives

$$P\left(\chi^2_{1-\alpha/2}(2M) \leq \frac{2M\beta}{\tilde{\beta}} \leq \chi^2_{\alpha/2}(2M) \right) = 1 - \alpha$$

$$P\left(\frac{\chi^2_{1-\alpha/2}(2M)\tilde{\beta}}{2M} < \beta < \frac{\chi^2_{\alpha/2}(2M)\tilde{\beta}}{2M} \right) = 1 - \alpha.$$

A $100\times(1-\alpha)\%$ confidence interval for β is therefore

$$\frac{\chi^2_{1-\alpha/2}(2M)\tilde{\beta}}{2M} < \beta < \frac{\chi^2_{\alpha/2}(2M)\tilde{\beta}}{2M}.$$

Example 5.10 Nelson (1995) presented the data set shown in Table 5.11 that gave the replacement time of valve seats on 41 diesel engines. (See also Lawless and Nadeau, 1995). Although each engine contained 16 valves, we will ignore this information and model the replacement times as a point process on each system. Further, we assume that the systems are identical and therefore the failure processes are all power law processes with a common θ and β. Figure 5.7 shows a time dot plot for the failure times the engine. (Here "failure" is defined as the replacement of a valve seat.)

The MLEs of β and θ, obtained by using an iterative method to solve (5.12) for β and substituting the result into (5.11), are

$$\hat{\beta} = 1.4204$$

$$\hat{\theta} = 548.86.$$

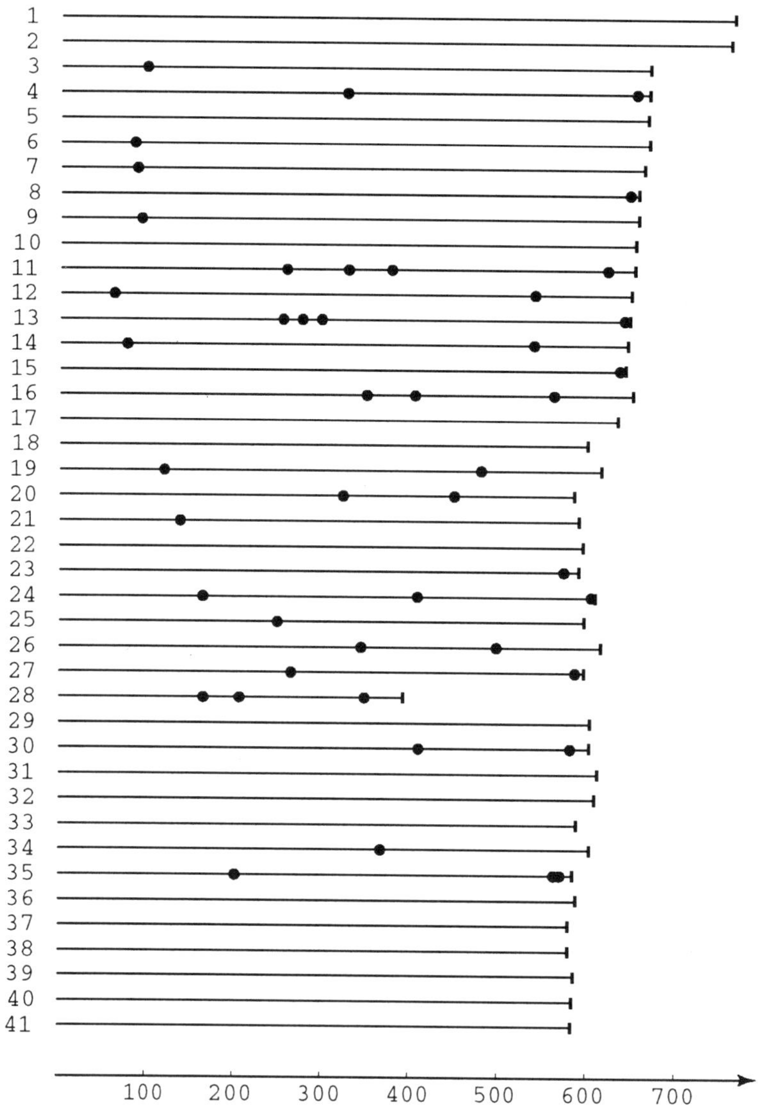

Fig. 5.7 Time dot plot for 41 systems with valve seat replacements. All systems were time truncated.

The estimated intensity function is

$$\hat{\lambda}(t) = \frac{\hat{\beta}}{\hat{\theta}}\left(\frac{t}{\hat{\theta}}\right)^{\hat{\beta}} = 0.0001825\, t^{0.4204}.$$

The conditional MLE of β is given by

$$\tilde{\beta} = \frac{48}{\sum_{i=1}^{41}\sum_{j=1}^{n_i}\log(T_i/t_{ij})} \approx 1.4513.$$

A 95% confidence interval for β is thus

$$\frac{\chi^2_{0.975}(2 \times 48)1.4513}{2 \times 48} < \beta < \frac{\chi^2_{0.025}(2 \times 48)1.4513}{2 \times 48}$$

$$\frac{70.783 \times 1.4513}{2 \times 48} < \beta < \frac{125.000 \times 1.4513}{2 \times 48}$$

$$1.070 < \beta < 1.890\,,$$

giving evidence that the intensity of valve seat replacements is increasing with time. □

5.5 TESTING FOR THE EQUALITY OF THE GROWTH PARAMETERS IN THE POWER LAW PROCESS

In this section we address the problem of testing whether the growth parameters for the k systems are the same,

$$H_0 : \beta_1 = \beta_2 = \cdots = \beta_k$$

against the alternative that at least two of the β's are different. For a general k, this requires a chi-square approximation, but for the case of $k = 2$, an exact procedure is available. We begin with the special case of $k = 2$.

System i	Failure Times $t_{i,1}, \ldots, t_{i,n_i}$				Truncation Time T_i
1					761
2					759
3	98				667
4	326	653	653		667
5					665
6	84				667
7	87				663
8	646				653
9	92				653
10					651
11	258	328	377	621	650
12	61	539			648
13	254	276	298	640	644
14	76	538			642
15	635				641
16	349	404	561		649
17					631
18					596
19	120	479			614
20	323	449			582
21	139	139			589
22					593
23	573				589
24	165	408	604		606
25	249				594
26	344	497			613
27	265	586			595
28	166	206	348		389
29					601
30	410	581			601
31					611
32					608
33					587
34	367				603
35	202	563	570		585
36					587
37					578
38					578
39					586
40					585
41					582

Table 5.11 Valve-seat replacement data.

5.5.1 Testing Equality of β's for Two Systems

With two systems, we can form separate estimators $\hat{\beta}_1$ and $\hat{\beta}_2$ of the growth parameters β_1 and β_2. The two estimators have the properties that

$$\frac{2n_1\beta_1}{\hat{\beta}_1} \sim \chi^2(2M_1) \quad \text{and} \quad \frac{2n_2\beta_2}{\hat{\beta}_2} \sim \chi^2(2M_2),$$

where $m_i = n_i - 1$ if system i was failure truncated and $m_i = n_i$ if system i was time truncated. Furthermore, the random variables $2n_1\beta_1/\hat{\beta}_1$ and $2n_2\beta_2/\hat{\beta}_2$ are independent. Thus, the statistic

$$F = \frac{\dfrac{2n_1\beta_1}{\hat{\beta}_1} \Big/ 2m_1}{\dfrac{2n_2\beta_2}{\hat{\beta}_2} \Big/ 2m_2}$$

has an F distribution with $2m_1$ and $2m_2$ degrees of freedom. Under the null hypothesis that $\beta_1 = \beta_2$, we have

$$F = \frac{\dfrac{n_1}{\hat{\beta}_1} \Big/ m_1}{\dfrac{n_2}{\hat{\beta}_2} \Big/ m_2} = \frac{\dfrac{n_1}{m_1\hat{\beta}_1}}{\dfrac{n_2}{m_2\hat{\beta}_2}} .$$

If we define

$$\tilde{\beta}_i = \frac{m_i}{\sum_{j=1}^{n_i} \log\left(T_i/t_{ij}\right)} \tag{5.14}$$

then we can write

$$F = \frac{\tilde{\beta}_2}{\tilde{\beta}_1} .$$

If both systems are failure truncated, this expression becomes

$$F = \frac{\dfrac{n_1}{(n_1-1)\hat{\beta}_1}}{\dfrac{n_2}{(n_2-1)\hat{\beta}_2}} = \frac{n_1(n_2-1)}{n_2(n_1-1)} \frac{\tilde{\beta}_2}{\tilde{\beta}_1}$$

and if both systems are time truncated, the F statistic simplifies to

$$F = \frac{1/\hat{\beta}_1}{1/\hat{\beta}_2} = \frac{\hat{\beta}_2}{\hat{\beta}_1} .$$

A size α test of H_0: $\beta_1 = \beta_2$ against the two-sided alternative H_a: $\beta_1 \neq \beta_2$ is to reject H_0 if

$$F < F_{1-\alpha/2}(2M_1, 2M_2) \quad \text{or} \quad F > F_{\alpha/2}(2M_1, 2M_2),$$

where $F_\gamma(\nu_1, \nu_2)$ is that value leaving an area of γ to the right under the F distribution with ν_1 and ν_2 degrees of freedom. The test of H_0: $\beta_1 \leq \beta_2$ against the two-sided alternative H_a: $\beta_1 > \beta_2$ leads to rejection of H_0 when $F < F_{1-\alpha}(2m_1, 2m_2)$. Similarly, the null hypothesis H_0: $\beta_1 \geq \beta_2$ is rejected in favor of the alternative H_a: $\beta_1 < \beta_2$ when $F > F_\alpha(2m_1, 2m_2)$.

System 1 $t_{1,j}$	**System 2** $t_{2,j}$
4.3	0.1
4.4	5.6
10.2	18.6
23.5	19.5
23.8	24.2
26.4	26.7
74.0	45.1
77.1	45.8
92.1	75.7
197.2	79.7
	98.6
	120.1
	161.8
	180.6
	190.8
$T_1 = 200$	$T_2 = 200$

Table 5.12 Failure times for two systems.

Example 5.11 Crow (1974) simulated failure times for three systems, using failure truncation at time 200. Table 5.12 and Figure 5.8 show the failure data from systems 1 and 2. Both systems seem to show reliability improvement. This conclusion is supported by the estimates for the growth parameters β_1 and β_2 :

$$\hat{\beta}_1 = \frac{10}{\sum_{i=1}^{10} \log \dfrac{200}{t_{1i}}} \approx 0.5085$$

and

$$\hat{\beta}_2 = \frac{15}{\sum_{i=1}^{15} \log \dfrac{200}{t_{2i}}} \approx 0.5674.$$

The test statistic for testing the null hypothesis H_0: $\beta_1 = \beta_2$ against the

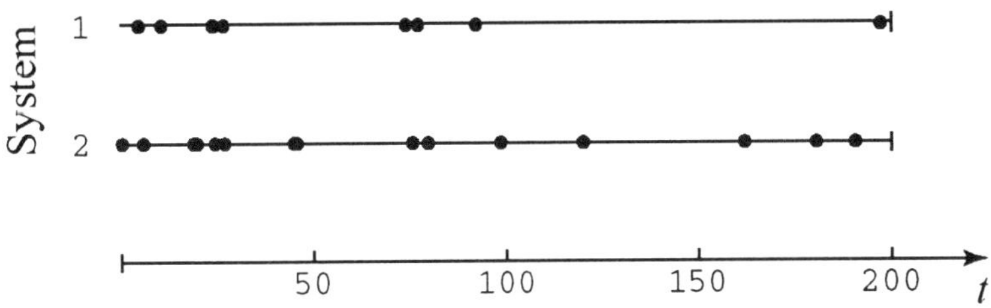

Fig. 5.8 Time dot plot for data in Table 5.12.

two-sided alternative H_a: $\beta_1 \neq \beta_2$ is

$$F = \frac{\frac{10}{10\hat{\beta}_1}}{\frac{15}{15\hat{\beta}_2}} = \frac{0.5674}{0.5085} \approx 1.12.$$

From Table A.2 we find that $F_{0.025}(20, 30) = 2.20$ and

$$F_{0.975}(20, 30) = 1/F_{0.025}(30, 20) = 1/2.35 \approx 0.426.$$

Since $F = 1.12$ is between 0.426 and 2.20, we fail to reject the null hypothesis. Thus, there is no evidence that these two systems have different values of the growth parameter β. □

5.5.2 Testing Equality of β's for k Systems

The likelihood ratio test for testing the null hypothesis

$$H_0 : \beta_1 = \beta_2 = \cdots = \beta_k$$

against the alternative that at least two of the β's are different leads to

$$\mathbf{LR} = M \log \beta^* - \sum_{i=1}^{k} m_i \log \tilde{\beta}_i\,,$$

where $\tilde{\beta}_i$ is the conditional MLE for the ith system from equation (5.14) and β^* is the weighted harmonic mean of the $\tilde{\beta}_i$'s

$$\beta^* = \frac{M}{\sum_{i=1}^{k} \frac{m_i}{\tilde{\beta}_i}}\,.$$

Using an approximation due to Bartlett (1937) we have

$$-\frac{2 \times \mathbf{LR}}{a} \mathrel{\dot\sim} \chi^2(k-1),$$

where

$$a = 1 + \frac{1}{6(k-1)} \left(\sum_{i=1}^{k} \frac{1}{m_i} - \frac{1}{M} \right).$$

The approximate size α test of H_0 is then rejected when $-2\mathbf{LR}/a > \chi^2_\alpha(k-1)$.

Example 5.12 Consider Crow's data on $k = 3$ systems, each time truncated at time $T = 200$. The CMLEs (also the MLEs since all systems were time truncated) of the growth parameters are

$$\tilde{\beta}_1 = 0.5085$$
$$\tilde{\beta}_2 = 0.5674$$
$$\tilde{\beta}_3 = 0.8870.$$

The constant a is

$$a = 1 + \frac{1}{6(3-1)} \left[\frac{1}{10} + \frac{1}{15} + \frac{1}{11} - \frac{1}{36} \right] \approx 1.019$$

and β^* is

$$\beta^* = \frac{36}{\dfrac{10}{0.5086} + \dfrac{15}{0.5675} + \dfrac{11}{0.8873}} \approx 0.6155.$$

The likelihood ratio statistic is therefore

$$\mathbf{LR} = 36 \log 0.6155 - (10 \log 0.5086 + 15 \log 0.5675 + 11 \log 0.8873) \approx -0.898.$$

The test statistics is then

$$-\frac{2 \times \mathbf{LR}}{a} \approx -\frac{2\,(-0.898)}{1.019} \approx 1.76.$$

From Table A.1 we find that $\chi^2_{0.05}(2) = 5.991$. We therefore fail to reject the null hypothesis and conclude that there is no evidence that the β's differ among the three systems. □

Lee (1980b) studied the problem of testing for equal rate functions, that is, common λ's ($\lambda = \theta^\beta$) as well as β's. He suggested first testing for equality of β's and if this hypothesis is accepted, testing the equality of the λ's.

5.6 POWER LAW PROCESS FOR NONIDENTICAL SYSTEMS

If copies of a system are considered different, with different β's and θ's, then a separate analysis of each system as suggested in Chapter 4 is appropriate. It may, however be reasonable to assume that all systems have the same growth parameter β, but different scale parameters, $\theta_1, \theta_2, \ldots, \theta_k$. Under such an assumption, closed form expressions for the MLE's of the parameters can be derived. The likelihood function is

$$L(\theta, \beta) = \prod_{i=1}^{k} \left\{ \left[\prod_{j=1}^{n_i} \frac{\beta}{\theta_i} \left(\frac{t_{i,j}}{\theta_i} \right)^{\beta-1} \right] \exp\left[-(T_i/\theta_i)^{\beta} \right] \right\}$$

$$= \beta^{\sum_{i=1}^{k} n_i} \left(\prod_{i=1}^{k} \theta_i^{-n_i \beta} \right) \left(\prod_{i=1}^{k} \prod_{j=1}^{n_i} t_{ij} \right)^{\beta-1} \exp\left[-\sum_{i=1}^{k} (T_i/\theta_i)^{\beta} \right].$$

The log-likelihood function is then

$$\ell(\theta, \beta) = N \log \beta - \beta \sum_{i=1}^{k} n_i \log \theta_i + (\beta - 1) \sum_{i=1}^{k} \sum_{j=1}^{n_i} \log t_{ij} - \sum_{i=1}^{k} \left(\frac{T_i}{\theta_i} \right)^{\beta},$$

from which we obtain the likelihood equations

$$\frac{\partial \ell}{\partial \theta_i} = -\frac{\beta n_i}{\theta_i} + \frac{\beta}{\theta_i} \left(\frac{T_i}{\theta_i} \right)^{\beta} = 0, \qquad i = 1, 2, \ldots, k$$

$$\frac{\partial \ell}{\partial \beta} = \frac{N}{\beta} - \sum_{i=1}^{k} n_i \log \theta_i + \sum_{i=1}^{k} \sum_{j=1}^{n_i} \log t_{ij} - \sum_{i=1}^{k} \left(\frac{T_i}{\theta_i} \right)^{\beta} \log \frac{T_i}{\theta_i} = 0. \tag{5.15}$$

The first equation leads to

$$\hat{\theta}_i = \frac{T_i}{n_i^{1/\hat{\beta}}}. \tag{5.16}$$

This is equivalent to

$$n = \left(\frac{T_i}{\hat{\theta}_i} \right)^{\hat{\beta}},$$

a result needed in the derivation of $\hat{\beta}$ (see Problem 2). The equation $\partial \ell / \partial \beta = 0$ leads to the MLE of β:

$$\hat{\beta} = \frac{N}{\displaystyle\sum_{i=1}^{k} \sum_{j=1}^{n_i} \log(T_i/t_{ij})}.$$

Once $\hat{\beta}$ is obtained, we can substitute its value into (5.16) to obtain $\hat{\theta}$.

The conditional MLE of β is

$$\tilde{\beta} = \frac{M}{\sum_{i=1}^{k}\sum_{j=1}^{n_i} \log(T_i/t_{ij})},$$

whether or not the θ_i's are equal. The estimator $\tilde{\beta}$ has the property that $2n\beta/\tilde{\beta} \sim \chi^2(2M)$ where M is as defined in Section 5.6. This result leads to

$$P\left(\frac{\chi^2_{1-\alpha/2}(2M)\tilde{\beta}}{2M} < \beta < \frac{\chi^2_{\alpha/2}(2M)\tilde{\beta}}{2M}\right) = 1 - \alpha.$$

A $100\times(1-\alpha)\%$ confidence interval for β is therefore

$$\frac{\chi^2_{1-\alpha/2}(2M)\tilde{\beta}}{2M} < \beta < \frac{\chi^2_{\alpha/2}(2M)\tilde{\beta}}{2M}.$$

Example 5.13 For the four systems shown in Table 5.9, the MLE of β is

$$\hat{\beta} = \frac{26}{\sum_{i=1}^{4}\sum_{j=1}^{n_i} \log(T_i/t_{ij})} \approx 0.68703$$

and the MLEs for $\theta_1, \theta_2, \theta_3$, and θ_4 are

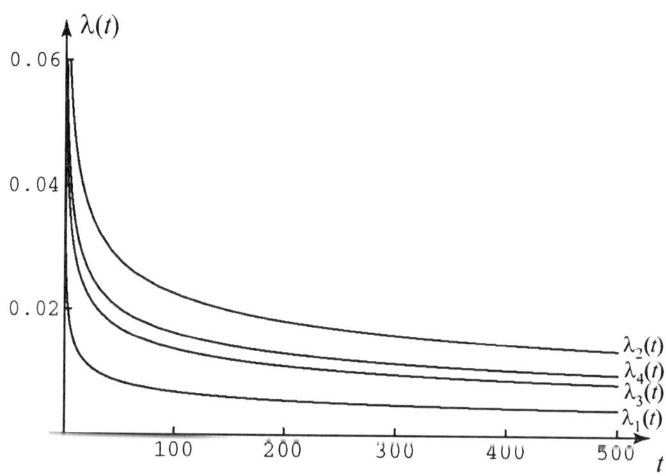

Fig. 5.9 Intensity functions for four systems, assuming a common β.

$$\hat{\theta}_1 = \frac{500}{3^{1/0.68703}} \approx 101.04$$

$$\hat{\theta}_2 = \frac{301}{7^{1/0.68703}} \approx 17.72$$

$$\hat{\theta}_3 = \frac{622}{7^{1/0.68703}} \approx 36.62$$

$$\hat{\theta}_4 = \frac{701}{9^{1/0.68703}} \approx 28.63\ .$$

With so few systems, the separate estimates of the θ's are subject to large variability. Figure 5.9 shows the four intensity functions in a single plot. The first system seems to be the best (smallest intensity) and system 2 seems to be the worst (largest intensity). A 95% confidence interval for β is then

$$\frac{\chi^2_{0.975}(2 \times 26)\tilde{\beta}}{2 \times 26} < \beta < \frac{\chi^2_{0.025}(2 \times 26)\tilde{\beta}}{2 \times 26}$$

$$\frac{33.968 \times 0.68703}{52} < \beta < \frac{73.810 \times 0.68703}{52}$$

$$0.449 < \beta < 0.975.$$

Since this confidence interval for β excludes the value $\beta = 1$, we have evidence of reliability improvement. □

5.7 PARAMETRIC EMPIRICAL BAYES MODELS FOR THE PLP

Parametric empirical Bayes procedures are easier to work with if the intensity function is parametrized as

$$\lambda(t) = \lambda\beta\, t^{\beta-1}, \quad t > 0.$$

Throughout most of this book, we have used the parametrization

$$\lambda(t) = \frac{\beta}{\theta}\left(\frac{t}{\theta}\right)^{\beta-1}, \quad t > 0.$$

Both parametrizations describe the power law process and the relationship between λ and θ is $\lambda = \theta^{-\beta}$.

We now suppose that the k systems have the same growth parameter β, but the parameters $\lambda_1, \lambda_2, \ldots, \lambda_k$ are drawn from a gamma prior distribution with pdf

$$p(\lambda \mid a, b) = \frac{b^a \lambda^{b-1}}{\Gamma(a)} \exp(-b\lambda), \quad \lambda > 0.$$

Here, the hyper-parameters a and b are considered fixed but unknown. Following the notation in Section 5.3, let $\mathbf{t}_i$ denote the vector of failure times for system i, and let

$$\mathbf{T} = [\mathbf{t}_1, \mathbf{t}_2, \ldots, \mathbf{t}_k]$$

denote the two-dimensional array of failure times. Since systems operate independently, the pdf of $\mathbf{T}$ given λ is

$$p(\mathbf{T} \,|\, \lambda, \beta) = \beta^N \left(\prod_{i=1}^{k} \lambda_i^{n_i} \right) \left(\prod_{i=1}^{k} \prod_{j=1}^{n_i} t_{ij} \right)^{\beta-1} \exp\left[-\sum_{i=1}^{k} \lambda_i T_i^{\beta} \right].$$

The pdf of $\mathbf{T}$ given a, b, and β is then

$$\begin{aligned}
p(\mathbf{T} \,|\, a, b, \beta) &= \int_0^\infty \cdots \int_0^\infty p(\mathbf{T}, \lambda \,|\, a, b, \beta)\, d\lambda_1 \cdots d\lambda_k \\
&= \int_0^\infty \cdots \int_0^\infty p(\mathbf{T} \,|\, \lambda, \beta)\, p(\lambda \,|\, a, b)\, d\lambda_1 \cdots d\lambda_k \\
&= \beta^N \frac{b^{ak}}{[\Gamma(a)]^k} \left(\prod_{i=1}^{k} \prod_{j=1}^{n_i} t_{ij} \right)^{\beta-1} \\
&\quad \times \int_0^\infty \cdots \int_0^\infty \left\{ \prod_{i=1}^{k} \lambda_i^{n_i+a-1} \exp\left[-\left(T_i^{\beta} + b \right) \lambda_i \right] \right\} d\lambda_1 \cdots d\lambda_k. \\
&= \beta^N \frac{b^{ak}}{[\Gamma(a)]^k} \left(\prod_{i=1}^{k} \prod_{j=1}^{n_i} t_{ij} \right)^{\beta-1} \\
&\quad \times \prod_{i=1}^{k} \int_0^\infty \lambda_i^{n_i+a-1} \exp\left[-\left(T_i^{\beta} + b \right) \lambda_i \right] d\lambda_i .
\end{aligned}$$

With the change of variable

$$\begin{aligned}
x &= (T_i + b)\, \lambda_i, \\
dx &= (T_i + b)\; d\lambda_i,
\end{aligned}$$

the interval of integration remains $(0, \infty)$. The pdf of $\mathbf{T}$ given a, b, and β is then

$$\begin{aligned} p(\mathbf{T} \mid a, b, \beta) &= \beta^N \frac{b^{ak}}{[\Gamma(a)]^k} \left(\prod_{i=1}^{k} \prod_{j=1}^{n_i} t_{ij} \right)^{\beta-1} \\ &\quad \times \prod_{i=1}^{k} \int_0^{\infty} \frac{x^{n_i+a-1}}{(T_i+b)^{n_i+\alpha-1}} \exp(-x) \frac{dx}{T_i+b} \\ &= \beta^N \frac{b^{ak}}{[\Gamma(a)]^k} \left(\prod_{i=1}^{k} \prod_{j=1}^{n_i} t_{ij} \right)^{\beta-1} \prod_{i=1}^{k} \frac{\Gamma(n_i+a)}{\left(T_i^{\beta}+b\right)^{n_i+a}} . \end{aligned}$$

The log-likelihood function of (a, b, β) is thus

$$\begin{aligned} \ell(a, b, \beta) &= N \log \beta + ak \log b - k \log \Gamma(a) \\ &\quad + \sum_{i=1}^{k} \sum_{j=1}^{n_i} \log t_{ij} + \sum_{i=1}^{k} \log \Gamma(n_i+a) - \sum_{i=1}^{k} (n_i+a) \log \left(T_i^{\beta}+b\right) . \end{aligned}$$

In general, the likelihood equations do not admit a closed form solution and an iterative procedure must be employed to approximate the MLEs of a, b, and β. However, for the special case when all k systems are time truncated at the same time T, we have

$$\begin{aligned} \ell(a, b, \beta) &= N \log \beta + ak \log b - k \log \Gamma(a) \\ &\quad + \sum_{i=1}^{k} \sum_{j=1}^{n_i} \log t_{ij} + \sum_{i=1}^{k} \log \Gamma(n_i+a) - (N+ka) \log \left(T^{\beta}+b\right) . \end{aligned}$$

Engelhardt and Bain (1987) showed that in this special case, the MLE of β is

$$\hat{\beta} = \frac{N}{\sum_{i=1}^{k} \sum_{j=1}^{n_i} \log(T/t_{ij})} \tag{5.17}$$

and the MLE $\hat{a}$ of the parameter a satisfies

$$\sum_{i=1}^{k} \sum_{j=1}^{n_i} \frac{1}{\hat{a}+j-1} = k \log\left(1+\frac{N}{k\hat{a}}\right) . \tag{5.18}$$

The solution $\hat{\alpha}$ to this equation can be approximated by a numerical method. Once $\hat{a}$ and $\hat{\beta}$ have been determined, the MLE of b can be found explicitly from

$$\hat{b} = \frac{k\hat{a}T^{\hat{\beta}}}{N} . \tag{5.19}$$

Engelhardt and Bain (1987) showed that (5.18) has a unique solution whenever

$$\frac{1}{k}\sum_{i=1}^{k}(n_i - \overline{n})^2 > \overline{n}, \tag{5.20}$$

where $\overline{n} = \sum_{i=1}^{k} n_i/k$. If the inequality in (5.20) fails to hold, then the likelihood function has a ridge that lies near the line b/a = constant and the likelihood function increases along this line as $a \to \infty$ and $b \to \infty$. (See the second example below.)

Once point estimates of a, b, and β are found, we can use them to estimate $\lambda_1, \lambda_2, \ldots, \lambda_k$. If we take these estimates as the true values of the parameters (a, b, and β), we find, using a derivation much like that found in (5.10), that the posterior distribution of λ_i is $\text{GAM}(n_i + \hat{a}, \hat{b} + t_i^{\hat{\beta}})$; hence a point estimate for λ_i is

$$\hat{\lambda}_i = \frac{\hat{a} + n_i}{\hat{b} + t_i^{\hat{\beta}}}. \tag{5.21}$$

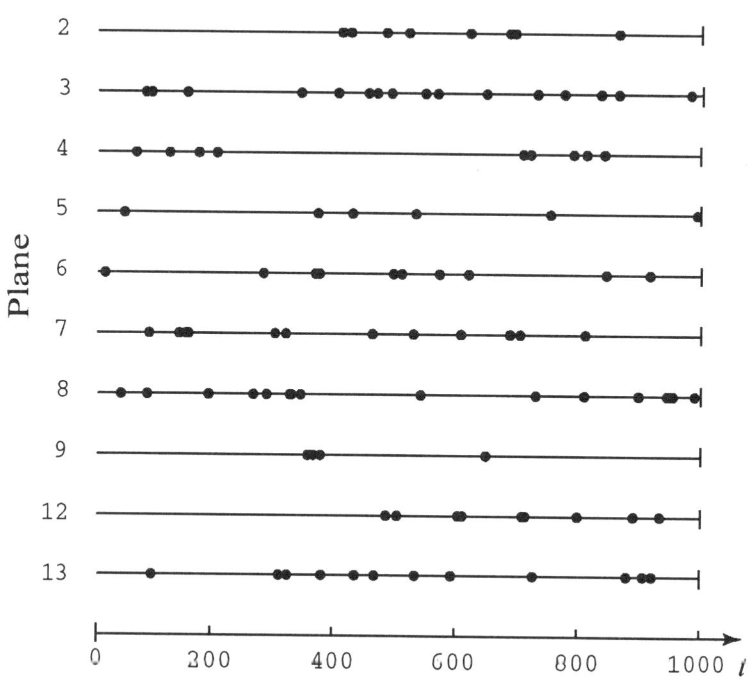

Fig. 5.10 Time dot plots for planes 2–9, 12, and 13, assuming time truncation at time $T = 1000$.

Example 5.14 Engelhardt and Bain (1987) considered a subset of the Proschan data given in Table 5.3. They considered all those planes whose air conditioning units lasted over 1000 hours (planes 2 through 9, 12 and 13), and assumed that each of these $k = 10$ systems was time truncated at time $T = 1000$. Time dot plots for these systems are shown in Figure 5.10. The MLE of β, obtained from (5.17) is

$$\hat{\beta} = \frac{103}{\sum_{i=1}^{k} \sum_{j=1}^{n_i} \log(1000/t_{ij})} = 1.20277.$$

Since

$$\frac{1}{k}\sum_{i=1}^{k}(n_i - \overline{n})^2 = 14.21 > 10.3 = \overline{n},$$

we know that (5.18) has a solution. Using an iterative method, we find that the solution is $\hat{a} = 25.0603$. Substituting this result into (5.19) yields

$$\hat{b} = \frac{k\hat{a}T^{\hat{\beta}}}{N} = \frac{10 \times 25.0603 \times 1000^{1.20277}}{103} \approx 9873.2.$$

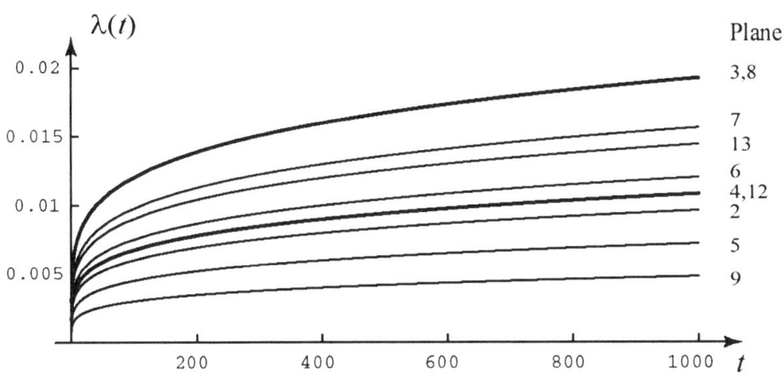

Fig. 5.11 MLEs of $\lambda_i(t)$ for aircraft air-conditioning equipment.

Estimates of the λ's can then be found from (5.21). Table 5.13 shows the MLEs and the PEB estimates. Figures 5.11 and 5.12 show the estimated intensity functions using both types of estimates of λ_i. These figures and the table show that the PEB estimates are shrunk toward the middle. □

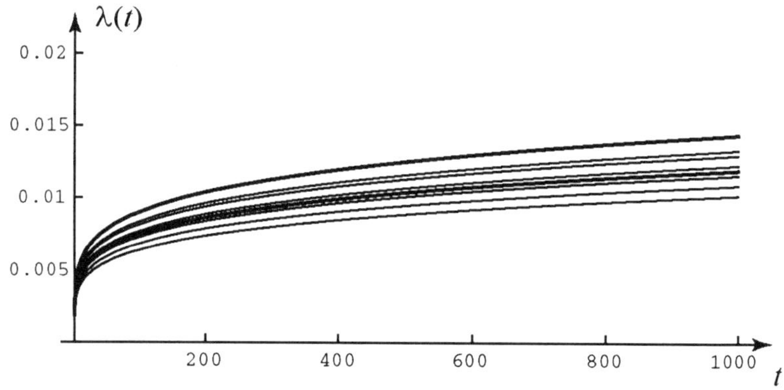

Fig. 5.12 Estimates of $\lambda_i(t)$ using marginal MLEs $\hat{a}$, $\hat{b}$, and $\hat{\beta}$.

Plane	MLE of λ	PEB Estimate of λ
2	0.00197	0.00237
3	0.00394	0.00295
4	0.00222	0.00244
5	0.00148	0.00223
6	0.00246	0.00252
7	0.00320	0.00273
8	0.00394	0.00295
9	0.00098	0.00209
12	0.00222	0.00244
13	0.00296	0.00266

Table 5.13 MLE and PEB estimates of λ's.

Example 5.15 Consider the four systems from Table 5.9. For this set of data, which we studied in Examples 5.9 and 5.13, we have

$$\frac{1}{k}\sum_{i=1}^{k}(n_i - \overline{n})^2 = 4.75 < 6.5 = \overline{n}.$$

Since this violates the inequality in (5.20), we cannot expect to have finite marginal maximum likelihood estimates for a and b. Figure 5.13 illustrates the problem; the likelihood function has a nearly flat ridge, roughly along the line $b = 10a$. This indicates that values (a, b) along this line fit the data equally well. Along this line, $a/b = 0.1$ and a/b is the prior mean of the λ's.

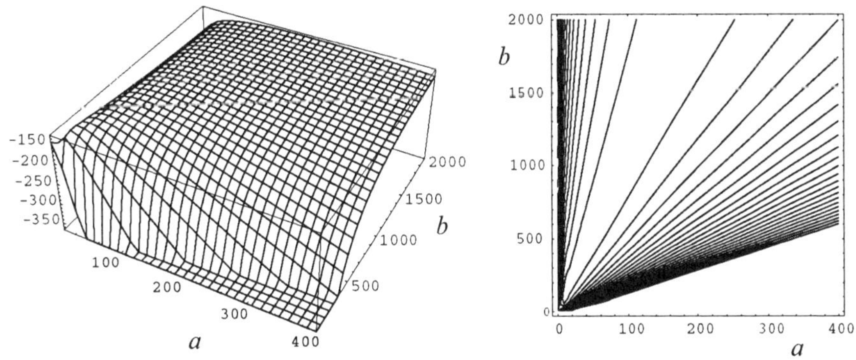

Fig. 5.13 Three-dimensional and contour plots of the likelihood function for a fixed value of β.

This is consistent with the point estimates for the λ's,

$$\hat{\lambda}_1 = 0.0420$$
$$\hat{\lambda}_2 = 0.1387$$
$$\hat{\lambda}_3 = 0.0843$$
$$\hat{\lambda}_4 = 0.0998,$$

which are all roughly 0.1. □

5.8 EXERCISES

1. Starting with the equation in (5.15) derive the result

$$\hat{\beta} = \frac{N}{\sum_{i=1}^{k} \sum_{j=1}^{n_i} \log(T_i/t_{ij})}$$

   for the case where the k systems have the same β but possibly different θ's.

2. Derive the MLE for β under the assumption of a common β but possibly different θ's.

3. Derive an unbiased estimator for β under (a) completely identical systems; (b) systems with the same β but possibly different θ's.

4. Show that the posterior distribution of λ_i given estimates $\hat{a}$, $\hat{b}$, and $\hat{\beta}$ for the case where systems have a common growth parameter β but

scale parameters λ_i, $i = 1, 2, \ldots, k$, that vary according to a $\text{GAM}(a, b)$ distribution, is $\text{GAM}(n_i + \hat{a}, \hat{b} + t_i^{\hat{\beta}})$.

Appendix A
Tables

Table A.1
Critical values for the chi-square distribution

ν	0.999	0.995	0.99	0.975	0.95	0.9	0.8	0.5
1	0.0000	0.0000	0.0002	0.001	0.004	0.016	0.064	0.455
2	0.002	0.010	0.020	0.051	0.103	0.211	0.446	1.386
3	0.024	0.072	0.115	0.216	0.352	0.584	1.005	2.366
4	0.091	0.207	0.297	0.484	0.711	1.064	1.649	3.357
5	0.210	0.412	0.554	0.831	1.145	1.610	2.343	4.351
6	0.381	0.676	0.872	1.237	1.635	2.204	3.070	5.348
7	0.599	0.989	1.239	1.690	2.167	2.833	3.822	6.346
8	0.857	1.344	1.647	2.180	2.733	3.490	4.594	7.344
9	1.152	1.735	2.088	2.700	3.325	4.168	5.380	8.343
10	1.479	2.156	2.558	3.247	3.940	4.865	6.179	9.342
11	1.834	2.603	3.053	3.816	4.575	5.578	6.989	10.341
12	2.214	3.074	3.571	4.404	5.226	6.304	7.807	11.340
13	2.617	3.565	4.107	5.009	5.892	7.041	8.634	12.340
14	3.041	4.075	4.660	5.629	6.571	7.790	9.467	13.339
15	3.483	4.601	5.229	6.262	7.261	8.547	10.307	14.339
16	3.942	5.142	5.812	6.908	7.962	9.312	11.152	15.338
17	4.416	5.697	6.408	7.564	8.672	10.085	12.002	16.338
18	4.905	6.265	7.015	8.231	9.390	10.865	12.857	17.338
19	5.407	6.844	7.633	8.907	10.117	11.651	13.716	18.338
20	5.921	7.434	8.260	9.591	10.851	12.443	14.578	19.337
21	6.447	8.034	8.897	10.283	11.591	13.240	15.445	20.337
22	6.983	8.643	9.542	10.982	12.338	14.041	16.314	21.337
23	7.529	9.260	10.196	11.689	13.091	14.848	17.187	22.337
24	8.085	9.886	10.856	12.401	13.848	15.659	18.062	23.337
25	8.649	10.520	11.524	13.120	14.611	16.473	18.940	24.337
26	9.222	11.160	12.198	13.844	15.379	17.292	19.820	25.336
27	9.803	11.808	12.878	14.573	16.151	18.114	20.703	26.336
28	10.391	12.461	13.565	15.308	16.928	18.939	21.588	27.336
29	10.986	13.121	14.256	16.047	17.708	19.768	22.475	28.336
30	11.588	13.787	14.953	16.791	18.493	20.599	23.364	29.336
31	12.196	14.458	15.655	17.539	19.281	21.434	24.255	30.336
32	12.810	15.134	16.362	18.291	20.072	22.271	25.148	31.336
33	13.431	15.815	17.073	19.047	20.867	23.110	26.042	32.336
34	14.057	16.501	17.789	19.806	21.664	23.952	26.938	33.336
35	14.688	17.192	18.509	20.569	22.465	24.797	27.836	34.336

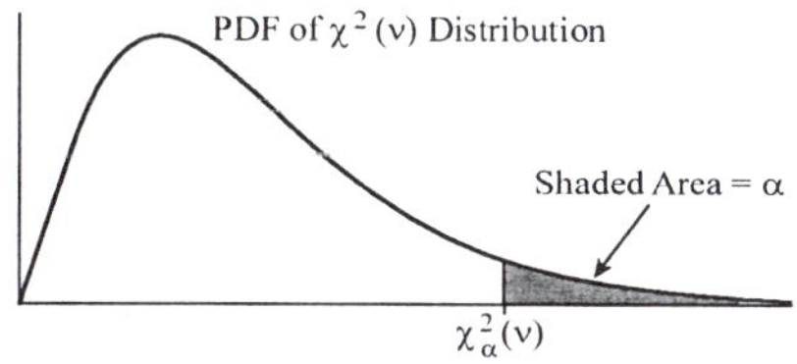

ν	**0.2**	**0.1**	**0.05**	**0.025**	**0.01**	**0.005**	**0.001**
1	1.642	2.706	3.841	5.024	6.635	7.879	10.827
2	3.219	4.605	5.991	7.378	9.210	10.597	13.815
3	4.642	6.251	7.815	9.348	11.345	12.838	16.266
4	5.989	7.779	9.488	11.143	13.277	14.860	18.466
5	7.289	9.236	11.070	12.832	15.086	16.750	20.515
6	8.558	10.645	12.592	14.449	16.812	18.548	22.457
7	9.803	12.017	14.067	16.013	18.475	20.278	24.321
8	11.030	13.362	15.507	17.535	20.090	21.955	26.124
9	12.242	14.684	16.919	19.023	21.666	23.589	27.877
10	13.442	15.987	18.307	20.483	23.209	25.188	29.588
11	14.631	17.275	19.675	21.920	24.725	26.757	31.264
12	15.812	18.549	21.026	23.337	26.217	28.300	32.909
13	16.985	19.812	22.362	24.736	27.688	29.819	34.527
14	18.151	21.064	23.685	26.119	29.141	31.319	36.124
15	19.311	22.307	24.996	27.488	30.578	32.801	37.698
16	20.465	23.542	26.296	28.845	32.000	34.267	39.252
17	21.615	24.769	27.587	30.191	33.409	35.718	40.791
18	22.760	25.989	28.869	31.526	34.805	37.156	42.312
19	23.900	27.204	30.144	32.852	36.191	38.582	43.819
20	25.038	28.412	31.410	34.170	37.566	39.997	45.314
21	26.171	29.615	32.671	35.479	38.932	41.401	46.796
22	27.301	30.813	33.924	36.781	40.289	42.796	48.268
23	28.429	32.007	35.172	38.076	41.638	44.181	49.728
24	29.553	33.196	36.415	39.364	42.980	45.558	51.179
25	30.675	34.382	37.652	40.646	44.314	46.928	52.619
26	31.795	35.563	38.885	41.923	45.642	48.290	54.051
27	32.912	36.741	40.113	43.195	46.963	49.645	55.475
28	34.027	37.916	41.337	44.461	48.278	50.994	56.892
29	35.139	39.087	42.557	45.722	49.588	52.335	58.301
30	36.250	40.256	43.773	46.979	50.892	53.672	59.702
31	37.359	41.422	44.985	48.232	52.191	55.002	61.098
32	38.466	42.585	46.194	49.480	53.486	56.328	62.487
33	39.572	43.745	47.400	50.725	54.775	57.648	63.869
34	40.676	44.903	48.602	51.966	56.061	58.964	65.247
35	41.778	46.059	49.802	53.203	57.342	60.275	66.619

Table A.1 (Continued)
Critical values for the chi-square distribution

ν	0.999	0.995	0.99	0.975	0.95	0.9	0.8	0.5
36	15.324	17.887	19.233	21.336	23.269	25.643	28.735	35.336
37	15.965	18.586	19.960	22.106	24.075	26.492	29.635	36.336
38	16.611	19.289	20.691	22.878	24.884	27.343	30.537	37.335
39	17.261	19.996	21.426	23.654	25.695	28.196	31.441	38.335
40	17.917	20.707	22.164	24.433	26.509	29.051	32.345	39.335
41	18.576	21.421	22.906	25.215	27.326	29.907	33.251	40.335
42	19.238	22.138	23.650	25.999	28.144	30.765	34.157	41.335
43	19.905	22.860	24.398	26.785	28.965	31.625	35.065	42.335
44	20.576	23.584	25.148	27.575	29.787	32.487	35.974	43.335
45	21.251	24.311	25.901	28.366	30.612	33.350	36.884	44.335
46	21.929	25.041	26.657	29.160	31.439	34.215	37.795	45.335
47	22.610	25.775	27.416	29.956	32.268	35.081	38.708	46.335
48	23.294	26.511	28.177	30.754	33.098	35.949	39.621	47.335
49	23.983	27.249	28.941	31.555	33.930	36.818	40.534	48.335
50	24.674	27.991	29.707	32.357	34.764	37.689	41.449	49.335
51	25.368	28.735	30.475	33.162	35.600	38.560	42.365	50.335
52	26.065	29.481	31.246	33.968	36.437	39.433	43.281	51.335
53	26.765	30.230	32.019	34.776	37.276	40.308	44.199	52.335
54	27.467	30.981	32.793	35.586	38.116	41.183	45.117	53.335
55	28.173	31.735	33.571	36.398	38.958	42.060	46.036	54.335
56	28.881	32.491	34.350	37.212	39.801	42.937	46.955	55.335
57	29.592	33.248	35.131	38.027	40.646	43.816	47.876	56.335
58	30.305	34.008	35.914	38.844	41.492	44.696	48.797	57.335
59	31.021	34.770	36.698	39.662	42.339	45.577	49.718	58.335
60	31.738	35.534	37.485	40.482	43.188	46.459	50.641	59.335
61	32.458	36.300	38.273	41.303	44.038	47.342	51.564	60.335
62	33.181	37.068	39.063	42.126	44.889	48.226	52.487	61.335
63	33.905	37.838	39.855	42.950	45.741	49.111	53.412	62.335
64	34.632	38.610	40.649	43.776	46.595	49.996	54.336	63.335
65	35.362	39.383	41.444	44.603	47.450	50.883	55.262	64.335
66	36.092	40.158	42.240	45.431	48.305	51.770	56.188	65.335
67	36.826	40.935	43.038	46.261	49.162	52.659	57.115	66.335
68	37.561	41.714	43.838	47.092	50.020	53.548	58.042	67.335
69	38.298	42.493	44.639	47.924	50.879	54.438	58.970	68.334
70	39.036	43.275	45.442	48.758	51.739	55.329	59.898	69.334

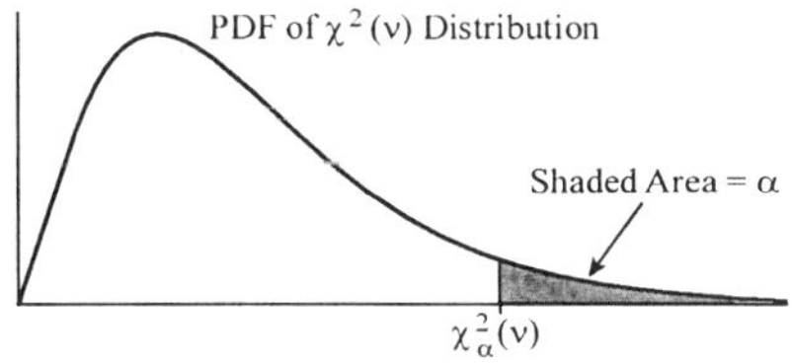

ν	0.2	0.1	0.05	0.025	0.01	0.005	0.001
36	42.879	47.212	50.998	54.437	58.619	61.581	67.985
37	43.978	48.363	52.192	55.668	59.893	62.883	69.348
38	45.076	49.513	53.384	56.895	61.162	64.181	70.704
39	46.173	50.660	54.572	58.120	62.428	65.475	72.055
40	47.269	51.805	55.758	59.342	63.691	66.766	73.403
41	48.363	52.949	56.942	60.561	64.950	68.053	74.744
42	49.456	54.090	58.124	61.777	66.206	69.336	76.084
43	50.548	55.230	59.304	62.990	67.459	70.616	77.418
44	51.639	56.369	60.481	64.201	68.710	71.892	78.749
45	52.729	57.505	61.656	65.410	69.957	73.166	80.078
46	53.818	58.641	62.830	66.616	71.201	74.437	81.400
47	54.906	59.774	64.001	67.821	72.443	75.704	82.720
48	55.993	60.907	65.171	69.023	73.683	76.969	84.037
49	57.079	62.038	66.339	70.222	74.919	78.231	85.350
50	58.164	63.167	67.505	71.420	76.154	79.490	86.660
51	59.248	64.295	68.669	72.616	77.386	80.746	87.967
52	60.332	65.422	69.832	73.810	78.616	82.001	89.272
53	61.414	66.548	70.993	75.002	79.843	83.253	90.573
54	62.496	67.673	72.153	76.192	81.069	84.502	91.871
55	63.577	68.796	73.311	77.380	82.292	85.749	93.167
56	64.658	69.919	74.468	78.567	83.514	86.994	94.46
57	65.737	71.040	75.624	79.752	84.733	88.237	95.75
58	66.816	72.160	76.778	80.936	85.950	89.477	97.04
59	67.894	73.279	77.930	82.117	87.166	90.72	98.32
60	68.972	74.397	79.082	83.298	88.38	91.95	99.61
61	70.049	75.514	80.232	84.476	89.59	93.19	100.89
62	71.125	76.630	81.381	85.654	90.80	94.42	102.17
63	72.201	77.745	82.529	86.83	92.01	95.65	103.44
64	73.276	78.860	83.675	88.00	93.22	96.88	104.72
65	74.351	79.973	84.82	89.18	94.42	98.10	105.99
66	75.424	81.085	85.96	90.35	95.63	99.33	107.26
67	76.498	82.20	87.11	91.52	96.83	100.55	108.53
68	77.571	83.31	88.25	92.69	98.03	101.78	109.79
69	78.643	84.42	89.39	93.86	99.23	103.00	111.06
70	79.71	85.53	90.53	95.02	100.43	104.21	112.32

Table A.1 (Continued)
Critical values for the chi-square distribution

ν	0.999	0.995	0.99	0.975	0.95	0.9	0.8	0.5
71	39.776	44.058	46.246	49.592	52.600	56.221	60.827	70.334
72	40.520	44.843	47.051	50.428	53.462	57.113	61.756	71.334
73	41.263	45.629	47.858	51.265	54.325	58.006	62.686	72.334
74	42.009	46.417	48.666	52.103	55.189	58.900	63.616	73.334
75	42.757	47.206	49.475	52.942	56.054	59.795	64.547	74.334
76	43.507	47.996	50.286	53.782	56.920	60.690	65.478	75.334
77	44.257	48.788	51.097	54.623	57.786	61.586	66.409	76.334
78	45.011	49.581	51.910	55.466	58.654	62.483	67.341	77.334
79	45.764	50.376	52.725	56.309	59.522	63.380	68.274	78.334
80	46.520	51.172	53.540	57.153	60.391	64.278	69.207	79.334
81	47.276	51.969	54.357	57.998	61.262	65.176	70.140	80.334
82	48.036	52.767	55.174	58.845	62.132	66.076	71.074	81.334
83	48.795	53.567	55.993	59.692	63.004	66.976	72.008	82.334
84	49.558	54.368	56.813	60.540	63.876	67.876	72.943	83.334
85	50.320	55.170	57.634	61.389	64.749	68.777	73.878	84.334
86	51.084	55.973	58.456	62.239	65.623	69.679	74.813	85.334
87	51.850	56.777	59.279	63.089	66.498	70.581	75.749	86.334
88	52.617	57.582	60.103	63.941	67.373	71.484	76.685	87.334
89	53.385	58.389	60.928	64.793	68.249	72.387	77.622	88.334
90	54.156	59.196	61.754	65.647	69.126	73.291	78.558	89.334
91	54.925	60.005	62.581	66.501	70.003	74.196	79.496	90.334
92	55.698	60.815	63.409	67.356	70.882	75.100	80.433	91.334
93	56.471	61.625	64.238	68.211	71.760	76.006	81.371	92.334
94	57.246	62.437	65.068	69.068	72.640	76.912	82.309	93.334
95	58.022	63.250	65.898	69.925	73.520	77.818	83.248	94.334
96	58.800	64.063	66.730	70.783	74.401	78.725	84.187	95.334
97	59.577	64.878	67.562	71.642	75.282	79.633	85.126	96.334
98	60.356	65.693	68.396	72.501	76.164	80.541	86.065	97.334
99	61.136	66.510	69.230	73.361	77.046	81.449	87.005	98.334
100	61.918	67.328	70.065	74.222	77.929	82.358	87.945	99.334

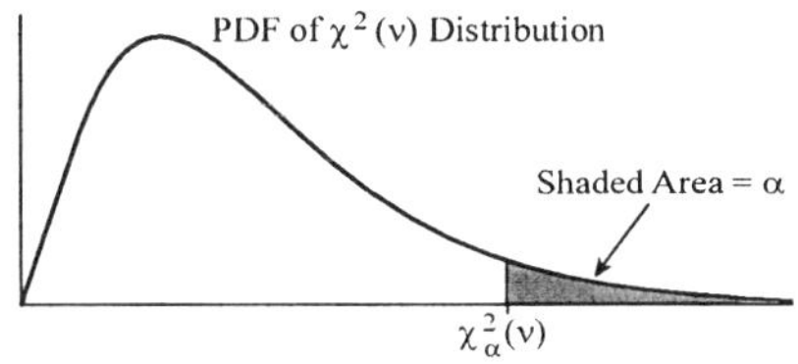

ν	0.2	0.1	0.05	0.025	0.01	0.005	0.001
71	80.79	86.64	91.67	96.19	101.62	105.43	113.58
72	81.86	87.74	92.81	97.35	102.82	106.65	114.83
73	82.93	88.85	93.95	98.52	104.01	107.86	116.09
74	84.00	89.96	95.08	99.68	105.20	109.07	117.35
75	85.07	91.06	96.22	100.84	106.39	110.29	118.60
76	86.13	92.17	97.35	102.00	107.58	111.50	119.85
77	87.20	93.27	98.48	103.16	108.77	112.70	121.10
78	88.27	94.37	99.62	104.32	109.96	113.91	122.35
79	89.34	95.48	100.75	105.47	111.14	115.12	123.59
80	90.41	96.58	101.88	106.63	112.33	116.32	124.84
81	91.47	97.68	103.01	107.78	113.51	117.52	126.08
82	92.54	98.78	104.14	108.94	114.69	118.73	127.32
83	93.60	99.88	105.27	110.09	115.88	119.93	128.56
84	94.67	100.98	106.39	111.24	117.06	121.13	129.80
85	95.73	102.08	107.52	112.39	118.24	122.32	131.04
86	96.80	103.18	108.65	113.54	119.41	123.52	132.28
87	97.86	104.28	109.77	114.69	120.59	124.72	133.51
88	98.93	105.37	110.90	115.84	121.77	125.91	134.75
89	99.99	106.47	112.02	116.99	122.94	127.11	135.98
90	101.05	107.57	113.15	118.14	124.12	128.30	137.21
91	102.12	108.66	114.27	119.28	125.29	129.49	138.44
92	103.18	109.76	115.39	120.43	126.46	130.68	139.67
93	104.24	110.85	116.51	121.57	127.63	131.87	140.89
94	105.30	111.94	117.63	122.72	128.80	133.06	142.12
95	106.36	113.04	118.75	123.86	129.97	134.25	143.34
96	107.43	114.13	119.87	125.00	131.14	135.43	144.57
97	108.49	115.22	120.99	126.14	132.31	136.62	145.79
98	109.55	116.32	122.11	127.28	133.48	137.80	147.01
99	110.61	117.41	123.23	128.42	134.64	138.99	148.23
100	111.67	118.50	124.34	129.56	135.81	140.17	149.45

Table A.2
Critical values for the F distribution

$\alpha = 0.05$

Denominator degrees of freedom ν_2	Numerator degrees of freedon ν_1: 2	4	6	8	10	12	14	16	18	20	22
2	19.00	19.25	19.33	19.37	19.40	19.41	19.42	19.43	19.44	19.45	19.45
4	6.94	6.39	6.16	6.04	5.96	5.91	5.87	5.84	5.82	5.80	5.79
6	5.14	4.53	4.28	4.15	4.06	4.00	3.96	3.92	3.90	3.87	3.86
8	4.46	3.84	3.58	3.44	3.35	3.28	3.24	3.20	3.17	3.15	3.13
10	4.10	3.48	3.22	3.07	2.98	2.91	2.86	2.83	2.80	2.77	2.75
12	3.89	3.26	3.00	2.85	2.75	2.69	2.64	2.60	2.57	2.54	2.52
14	3.74	3.11	2.85	2.70	2.60	2.53	2.48	2.44	2.41	2.39	2.37
16	3.63	3.01	2.74	2.59	2.49	2.42	2.37	2.33	2.30	2.28	2.25
18	3.55	2.93	2.66	2.51	2.41	2.34	2.29	2.25	2.22	2.19	2.17
20	3.49	2.87	2.60	2.45	2.35	2.28	2.22	2.18	2.15	2.12	2.10
22	3.44	2.82	2.55	2.40	2.30	2.23	2.17	2.13	2.10	2.07	2.05
24	3.40	2.78	2.51	2.36	2.25	2.18	2.13	2.09	2.05	2.03	2.00
26	3.37	2.74	2.47	2.32	2.22	2.15	2.09	2.05	2.02	1.99	1.97
28	3.34	2.71	2.45	2.29	2.19	2.12	2.06	2.02	1.99	1.96	1.93
30	3.32	2.69	2.42	2.27	2.16	2.09	2.04	1.99	1.96	1.93	1.91
32	3.29	2.67	2.40	2.24	2.14	2.07	2.01	1.97	1.94	1.91	1.88
34	3.28	2.65	2.38	2.23	2.12	2.05	1.99	1.95	1.92	1.89	1.86
36	3.26	2.63	2.36	2.21	2.11	2.03	1.98	1.93	1.90	1.87	1.85
38	3.24	2.62	2.35	2.19	2.09	2.02	1.96	1.92	1.88	1.85	1.83
40	3.23	2.61	2.34	2.18	2.08	2.00	1.95	1.90	1.87	1.84	1.81
42	3.22	2.59	2.32	2.17	2.06	1.99	1.94	1.89	1.86	1.83	1.80
44	3.21	2.58	2.31	2.16	2.05	1.98	1.92	1.88	1.84	1.81	1.79
46	3.20	2.57	2.30	2.15	2.04	1.97	1.91	1.87	1.83	1.80	1.78
48	3.19	2.57	2.29	2.14	2.03	1.96	1.90	1.86	1.82	1.79	1.77
50	3.18	2.56	2.29	2.13	2.03	1.95	1.89	1.85	1.81	1.78	1.76
60	3.15	2.53	2.25	2.10	1.99	1.92	1.86	1.82	1.78	1.75	1.72
70	3.13	2.50	2.23	2.07	1.97	1.89	1.84	1.79	1.75	1.72	1.70
80	3.11	2.49	2.21	2.06	1.95	1.88	1.82	1.77	1.73	1.70	1.68
90	3.10	2.47	2.20	2.04	1.94	1.86	1.80	1.76	1.72	1.69	1.66
100	3.09	2.46	2.19	2.03	1.93	1.85	1.79	1.75	1.71	1.68	1.65
120	3.07	2.45	2.18	2.02	1.91	1.83	1.78	1.73	1.69	1.66	1.63

$\alpha = 0.05$

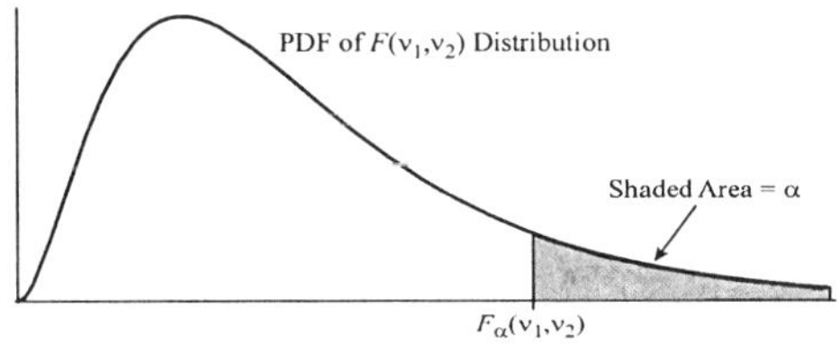

Numerator degrees of freedon ν_1											
24	**26**	**28**	**30**	**32**	**34**	**40**	**50**	**60**	**80**	**100**	**Denominator degrees of freedom ν_2**
19.45	19.46	19.46	19.46	19.46	19.47	19.47	19.48	19.48	19.48	19.49	**2**
5.77	5.76	5.75	5.75	5.74	5.73	5.72	5.70	5.69	5.67	5.66	**4**
3.84	3.83	3.82	3.81	3.80	3.79	3.77	3.75	3.74	3.72	3.71	**6**
3.12	3.10	3.09	3.08	3.07	3.06	3.04	3.02	3.01	2.99	2.97	**8**
2.74	2.72	2.71	2.70	2.69	2.68	2.66	2.64	2.62	2.60	2.59	**10**
2.51	2.49	2.48	2.47	2.46	2.45	2.43	2.40	2.38	2.36	2.35	**12**
2.35	2.33	2.32	2.31	2.30	2.29	2.27	2.24	2.22	2.20	2.19	**14**
2.24	2.22	2.21	2.19	2.18	2.17	2.15	2.12	2.11	2.08	2.07	**16**
2.15	2.13	2.12	2.11	2.10	2.09	2.06	2.04	2.02	1.99	1.98	**18**
2.08	2.07	2.05	2.04	2.03	2.02	1.99	1.97	1.95	1.92	1.91	**20**
2.03	2.01	2.00	1.98	1.97	1.96	1.94	1.91	1.89	1.86	1.85	**22**
1.98	1.97	1.95	1.94	1.93	1.92	1.89	1.86	1.84	1.82	1.80	**24**
1.95	1.93	1.91	1.90	1.89	1.88	1.85	1.82	1.80	1.78	1.76	**26**
1.91	1.90	1.88	1.87	1.86	1.85	1.82	1.79	1.77	1.74	1.73	**28**
1.89	1.87	1.85	1.84	1.83	1.82	1.79	1.76	1.74	1.71	1.70	**30**
1.86	1.85	1.83	1.82	1.80	1.79	1.77	1.74	1.71	1.69	1.67	**32**
1.84	1.82	1.81	1.80	1.78	1.77	1.75	1.71	1.69	1.66	1.65	**34**
1.82	1.81	1.79	1.78	1.76	1.75	1.73	1.69	1.67	1.64	1.62	**36**
1.81	1.79	1.77	1.76	1.75	1.74	1.71	1.68	1.65	1.62	1.61	**38**
1.79	1.77	1.76	1.74	1.73	1.72	1.69	1.66	1.64	1.61	1.59	**40**
1.78	1.76	1.75	1.73	1.72	1.71	1.68	1.65	1.62	1.59	1.57	**42**
1.77	1.75	1.73	1.72	1.71	1.69	1.67	1.63	1.61	1.58	1.56	**44**
1.76	1.74	1.72	1.71	1.69	1.68	1.65	1.62	1.60	1.57	1.55	**46**
1.75	1.73	1.71	1.70	1.68	1.67	1.64	1.61	1.59	1.56	1.54	**48**
1.74	1.72	1.70	1.69	1.67	1.66	1.63	1.60	1.58	1.54	1.52	**50**
1.70	1.68	1.66	1.65	1.64	1.62	1.59	1.56	1.53	1.50	1.48	**60**
1.67	1.65	1.64	1.62	1.61	1.60	1.57	1.53	1.50	1.47	1.45	**70**
1.65	1.63	1.62	1.60	1.59	1.58	1.54	1.51	1.48	1.45	1.43	**80**
1.64	1.62	1.60	1.59	1.57	1.56	1.53	1.49	1.46	1.43	1.41	**90**
1.63	1.61	1.59	1.57	1.56	1.55	1.52	1.48	1.45	1.41	1.39	**100**
1.61	1.59	1.57	1.55	1.54	1.53	1.50	1.46	1.43	1.39	1.37	**120**

Table A.2 (Continued)
Critical values for the *F* distribution

$\alpha = 0.025$

Denominator degrees of freedom ν_2	Numerator degrees of freedon ν_1										
	2	**4**	**6**	**8**	**10**	**12**	**14**	**16**	**18**	**20**	**22**
2	39.00	39.25	39.33	39.37	39.40	39.41	39.43	39.44	39.44	39.45	39.45
4	10.65	9.60	9.20	8.98	8.84	8.75	8.68	8.63	8.59	8.56	8.53
6	7.26	6.23	5.82	5.60	5.46	5.37	5.30	5.24	5.20	5.17	5.14
8	6.06	5.05	4.65	4.43	4.30	4.20	4.13	4.08	4.03	4.00	3.97
10	5.46	4.47	4.07	3.85	3.72	3.62	3.55	3.50	3.45	3.42	3.39
12	5.10	4.12	3.73	3.51	3.37	3.28	3.21	3.15	3.11	3.07	3.04
14	4.86	3.89	3.50	3.29	3.15	3.05	2.98	2.92	2.88	2.84	2.81
16	4.69	3.73	3.34	3.12	2.99	2.89	2.82	2.76	2.72	2.68	2.65
18	4.56	3.61	3.22	3.01	2.87	2.77	2.70	2.64	2.60	2.56	2.53
20	4.46	3.51	3.13	2.91	2.77	2.68	2.60	2.55	2.50	2.46	2.43
22	4.38	3.44	3.05	2.84	2.70	2.60	2.53	2.47	2.43	2.39	2.36
24	4.32	3.38	2.99	2.78	2.64	2.54	2.47	2.41	2.36	2.33	2.30
26	4.27	3.33	2.94	2.73	2.59	2.49	2.42	2.36	2.31	2.28	2.24
28	4.22	3.29	2.90	2.69	2.55	2.45	2.37	2.32	2.27	2.23	2.20
30	4.18	3.25	2.87	2.65	2.51	2.41	2.34	2.28	2.23	2.20	2.16
32	4.15	3.22	2.84	2.62	2.48	2.38	2.31	2.25	2.20	2.16	2.13
34	4.12	3.19	2.81	2.59	2.45	2.35	2.28	2.22	2.17	2.13	2.10
36	4.09	3.17	2.78	2.57	2.43	2.33	2.25	2.20	2.15	2.11	2.08
38	4.07	3.15	2.76	2.55	2.41	2.31	2.23	2.17	2.13	2.09	2.05
40	4.05	3.13	2.74	2.53	2.39	2.29	2.21	2.15	2.11	2.07	2.03
42	4.03	3.11	2.73	2.51	2.37	2.27	2.20	2.14	2.09	2.05	2.02
44	4.02	3.09	2.71	2.50	2.36	2.26	2.18	2.12	2.07	2.03	2.00
46	4.00	3.08	2.70	2.48	2.34	2.24	2.17	2.11	2.06	2.02	1.99
48	3.99	3.07	2.69	2.47	2.33	2.23	2.15	2.09	2.05	2.01	1.97
50	3.97	3.05	2.67	2.46	2.32	2.22	2.14	2.08	2.03	1.99	1.96
60	3.93	3.01	2.63	2.41	2.27	2.17	2.09	2.03	1.98	1.94	1.91
70	3.89	2.97	2.59	2.38	2.24	2.14	2.06	2.00	1.95	1.91	1.88
80	3.86	2.95	2.57	2.35	2.21	2.11	2.03	1.97	1.92	1.88	1.85
90	3.84	2.93	2.55	2.34	2.19	2.09	2.02	1.95	1.91	1.86	1.83
100	3.83	2.92	2.54	2.32	2.18	2.08	2.00	1.94	1.89	1.85	1.81
120	3.80	2.89	2.52	2.30	2.16	2.05	1.98	1.92	1.87	1.82	1.79

$\alpha = 0.025$

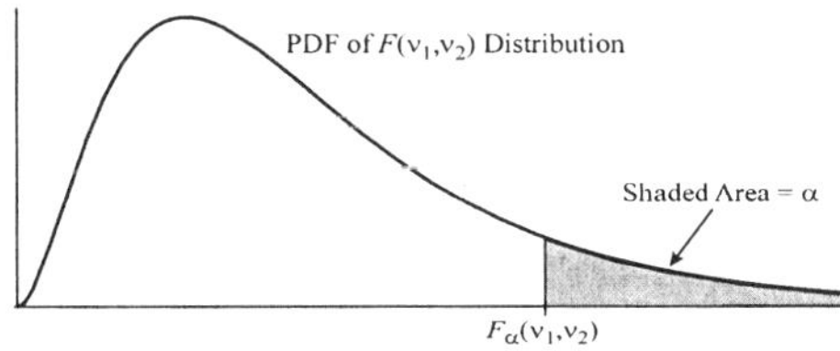

Numerator degrees of freedon ν_1											
24	**26**	**28**	**30**	**32**	**34**	**40**	**50**	**60**	**80**	**100**	Denominator degrees of freedom ν_2
39.46	39.46	39.46	39.46	39.47	39.47	39.47	39.48	39.48	39.49	39.49	**2**
8.51	8.49	8.48	8.46	8.45	8.44	8.41	8.38	8.36	8.33	8.32	**4**
5.12	5.10	5.08	5.07	5.05	5.04	5.01	4.98	4.96	4.93	4.92	**6**
3.95	3.93	3.91	3.89	3.88	3.87	3.84	3.81	3.78	3.76	3.74	**8**
3.37	3.34	3.33	3.31	3.30	3.29	3.26	3.22	3.20	3.17	3.15	**10**
3.02	3.00	2.98	2.96	2.95	2.94	2.91	2.87	2.85	2.82	2.80	**12**
2.79	2.77	2.75	2.73	2.72	2.71	2.67	2.64	2.61	2.58	2.56	**14**
2.63	2.60	2.58	2.57	2.55	2.54	2.51	2.47	2.45	2.42	2.40	**16**
2.50	2.48	2.46	2.44	2.43	2.42	2.38	2.35	2.32	2.29	2.27	**18**
2.41	2.39	2.37	2.35	2.33	2.32	2.29	2.25	2.22	2.19	2.17	**20**
2.33	2.31	2.29	2.27	2.26	2.24	2.21	2.17	2.14	2.11	2.09	**22**
2.27	2.25	2.23	2.21	2.19	2.18	2.15	2.11	2.08	2.05	2.02	**24**
2.22	2.19	2.17	2.16	2.14	2.13	2.09	2.05	2.03	1.99	1.97	**26**
2.17	2.15	2.13	2.11	2.10	2.08	2.05	2.01	1.98	1.94	1.92	**28**
2.14	2.11	2.09	2.07	2.06	2.04	2.01	1.97	1.94	1.90	1.88	**30**
2.10	2.08	2.06	2.04	2.02	2.01	1.98	1.93	1.91	1.87	1.85	**32**
2.07	2.05	2.03	2.01	2.00	1.98	1.95	1.90	1.88	1.84	1.82	**34**
2.05	2.03	2.00	1.99	1.97	1.96	1.92	1.88	1.85	1.81	1.79	**36**
2.03	2.00	1.98	1.96	1.95	1.93	1.90	1.85	1.82	1.79	1.76	**38**
2.01	1.98	1.96	1.94	1.93	1.91	1.88	1.83	1.80	1.76	1.74	**40**
1.99	1.96	1.94	1.92	1.91	1.89	1.86	1.81	1.78	1.74	1.72	**42**
1.97	1.95	1.93	1.91	1.89	1.88	1.84	1.80	1.77	1.73	1.70	**44**
1.96	1.93	1.91	1.89	1.88	1.86	1.82	1.78	1.75	1.71	1.69	**46**
1.94	1.92	1.90	1.88	1.86	1.85	1.81	1.77	1.73	1.69	1.67	**48**
1.93	1.91	1.89	1.87	1.85	1.83	1.80	1.75	1.72	1.68	1.66	**50**
1.88	1.86	1.83	1.82	1.80	1.78	1.74	1.70	1.67	1.63	1.60	**60**
1.85	1.82	1.80	1.78	1.76	1.75	1.71	1.66	1.63	1.59	1.56	**70**
1.82	1.79	1.77	1.75	1.73	1.72	1.68	1.63	1.60	1.55	1.53	**80**
1.80	1.77	1.75	1.73	1.71	1.70	1.66	1.61	1.58	1.53	1.50	**90**
1.78	1.76	1.74	1.71	1.70	1.68	1.64	1.59	1.56	1.51	1.48	**100**
1.76	1.73	1.71	1.69	1.67	1.66	1.61	1.56	1.53	1.48	1.45	**120**

Table A.2 (Continued)
Critical values for the *F* distribution

$\alpha = 0.01$

Denominator degrees of freedom ν_2	Numerator degrees of freedon ν_1										
	2	**4**	**6**	**8**	**10**	**12**	**14**	**16**	**18**	**20**	**22**
2	99.00	99.25	99.33	99.38	99.40	99.42	99.43	99.44	99.44	99.45	99.46
4	18.00	15.98	15.21	14.80	14.55	14.37	14.25	14.15	14.08	14.02	13.97
6	10.92	9.15	8.47	8.10	7.87	7.72	7.60	7.52	7.45	7.40	7.35
8	8.65	7.01	6.37	6.03	5.81	5.67	5.56	5.48	5.41	5.36	5.32
10	7.56	5.99	5.39	5.06	4.85	4.71	4.60	4.52	4.46	4.41	4.36
12	6.93	5.41	4.82	4.50	4.30	4.16	4.05	3.97	3.91	3.86	3.82
14	6.51	5.04	4.46	4.14	3.94	3.80	3.70	3.62	3.56	3.51	3.46
16	6.23	4.77	4.20	3.89	3.69	3.55	3.45	3.37	3.31	3.26	3.22
18	6.01	4.58	4.01	3.71	3.51	3.37	3.27	3.19	3.13	3.08	3.03
20	5.85	4.43	3.87	3.56	3.37	3.23	3.13	3.05	2.99	2.94	2.90
22	5.72	4.31	3.76	3.45	3.26	3.12	3.02	2.94	2.88	2.83	2.78
24	5.61	4.22	3.67	3.36	3.17	3.03	2.93	2.85	2.79	2.74	2.70
26	5.53	4.14	3.59	3.29	3.09	2.96	2.86	2.78	2.72	2.66	2.62
28	5.45	4.07	3.53	3.23	3.03	2.90	2.79	2.72	2.65	2.60	2.56
30	5.39	4.02	3.47	3.17	2.98	2.84	2.74	2.66	2.60	2.55	2.51
32	5.34	3.97	3.43	3.13	2.93	2.80	2.70	2.62	2.55	2.50	2.46
34	5.29	3.93	3.39	3.09	2.89	2.76	2.66	2.58	2.51	2.46	2.42
36	5.25	3.89	3.35	3.05	2.86	2.72	2.62	2.54	2.48	2.43	2.38
38	5.21	3.86	3.32	3.02	2.83	2.69	2.59	2.51	2.45	2.40	2.35
40	5.18	3.83	3.29	2.99	2.80	2.66	2.56	2.48	2.42	2.37	2.33
42	5.15	3.80	3.27	2.97	2.78	2.64	2.54	2.46	2.40	2.34	2.30
44	5.12	3.78	3.24	2.95	2.75	2.62	2.52	2.44	2.37	2.32	2.28
46	5.10	3.76	3.22	2.93	2.73	2.60	2.50	2.42	2.35	2.30	2.26
48	5.08	3.74	3.20	2.91	2.71	2.58	2.48	2.40	2.33	2.28	2.24
50	5.06	3.72	3.19	2.89	2.70	2.56	2.46	2.38	2.32	2.27	2.22
60	4.98	3.65	3.12	2.82	2.63	2.50	2.39	2.31	2.25	2.20	2.15
70	4.92	3.60	3.07	2.78	2.59	2.45	2.35	2.27	2.20	2.15	2.11
80	4.88	3.56	3.04	2.74	2.55	2.42	2.31	2.23	2.17	2.12	2.07
90	4.85	3.53	3.01	2.72	2.52	2.39	2.29	2.21	2.14	2.09	2.04
100	4.82	3.51	2.99	2.69	2.50	2.37	2.27	2.19	2.12	2.07	2.02
120	4.79	3.48	2.96	2.66	2.47	2.34	2.23	2.15	2.09	2.03	1.99

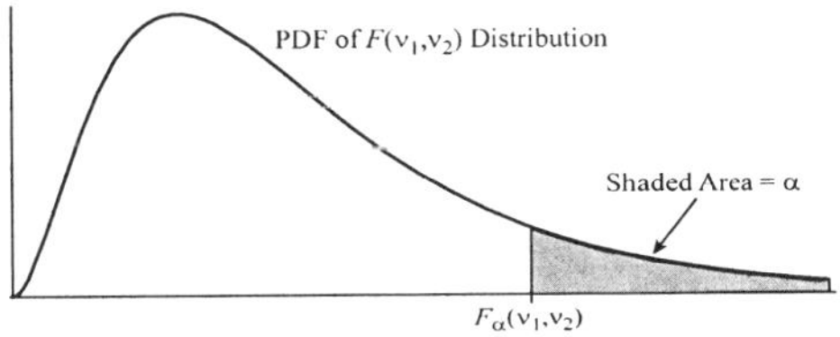

$\alpha = 0.01$

Numerator degrees of freedon ν_1											Denominator degrees of freedom ν_2
24	**26**	**28**	**30**	**32**	**34**	**40**	**50**	**60**	**80**	**100**	
99.46	99.46	99.46	99.47	99.47	99.47	99.48	99.48	99.48	99.48	99.49	**2**
13.93	13.89	13.86	13.84	13.81	13.79	13.75	13.69	13.65	13.61	13.58	**4**
7.31	7.28	7.25	7.23	7.21	7.19	7.14	7.09	7.06	7.01	6.99	**6**
5.28	5.25	5.22	5.20	5.18	5.16	5.12	5.07	5.03	4.99	4.96	**8**
4.33	4.30	4.27	4.25	4.23	4.21	4.17	4.12	4.08	4.04	4.01	**10**
3.78	3.75	3.72	3.70	3.68	3.66	3.62	3.57	3.54	3.49	3.47	**12**
3.43	3.40	3.37	3.35	3.33	3.31	3.27	3.22	3.18	3.14	3.11	**14**
3.18	3.15	3.12	3.10	3.08	3.06	3.02	2.97	2.93	2.89	2.86	**16**
3.00	2.97	2.94	2.92	2.90	2.88	2.84	2.78	2.75	2.70	2.68	**18**
2.86	2.83	2.80	2.78	2.76	2.74	2.69	2.64	2.61	2.56	2.54	**20**
2.75	2.72	2.69	2.67	2.65	2.63	2.58	2.53	2.50	2.45	2.42	**22**
2.66	2.63	2.60	2.58	2.56	2.54	2.49	2.44	2.40	2.36	2.33	**24**
2.58	2.55	2.53	2.50	2.48	2.46	2.42	2.36	2.33	2.28	2.25	**26**
2.52	2.49	2.46	2.44	2.42	2.40	2.35	2.30	2.26	2.22	2.19	**28**
2.47	2.44	2.41	2.39	2.36	2.35	2.30	2.25	2.21	2.16	2.13	**30**
2.42	2.39	2.36	2.34	2.32	2.30	2.25	2.20	2.16	2.11	2.08	**32**
2.38	2.35	2.32	2.30	2.28	2.26	2.21	2.16	2.12	2.07	2.04	**34**
2.35	2.32	2.29	2.26	2.24	2.22	2.18	2.12	2.08	2.03	2.00	**36**
2.32	2.28	2.26	2.23	2.21	2.19	2.14	2.09	2.05	2.00	1.97	**38**
2.29	2.26	2.23	2.20	2.18	2.16	2.11	2.06	2.02	1.97	1.94	**40**
2.26	2.23	2.20	2.18	2.16	2.14	2.09	2.03	1.99	1.94	1.91	**42**
2.24	2.21	2.18	2.15	2.13	2.11	2.07	2.01	1.97	1.92	1.89	**44**
2.22	2.19	2.16	2.13	2.11	2.09	2.04	1.99	1.95	1.90	1.86	**46**
2.20	2.17	2.14	2.12	2.09	2.07	2.02	1.97	1.93	1.88	1.84	**48**
2.18	2.15	2.12	2.10	2.08	2.06	2.01	1.95	1.91	1.86	1.82	**50**
2.12	2.08	2.05	2.03	2.01	1.99	1.94	1.88	1.84	1.78	1.75	**60**
2.07	2.03	2.01	1.98	1.96	1.94	1.89	1.83	1.78	1.73	1.70	**70**
2.03	2.00	1.97	1.94	1.92	1.90	1.85	1.79	1.75	1.69	1.65	**80**
2.00	1.97	1.94	1.92	1.89	1.87	1.82	1.76	1.72	1.66	1.62	**90**
1.98	1.95	1.92	1.89	1.87	1.85	1.80	1.74	1.69	1.63	1.60	**100**
1.95	1.92	1.89	1.86	1.84	1.82	1.76	1.70	1.66	1.60	1.56	**120**

Table A.3

Confidence limits for the mean of a Poisson distribution given an observation of c events.

	Confidence Level							
	0.99		0.95		0.9		0.8	
c	λ_L	λ_U	λ_L	λ_U	λ_L	λ_U	λ_L	λ_U
0	0.000	5.298	0.000	3.689	0.000	2.996	0.000	2.303
1	0.005	7.430	0.025	5.572	0.051	4.744	0.105	3.890
2	0.103	9.274	0.242	7.225	0.355	6.296	0.532	5.322
3	0.338	10.977	0.619	8.767	0.818	7.754	1.102	6.681
4	0.672	12.594	1.090	10.242	1.366	9.154	1.745	7.994
5	1.078	14.150	1.623	11.668	1.970	10.513	2.433	9.275
6	1.537	15.660	2.202	13.059	2.613	11.842	3.152	10.532
7	2.037	17.134	2.814	14.423	3.285	13.148	3.895	11.771
8	2.571	18.578	3.454	15.763	3.981	14.435	4.656	12.995
9	3.132	19.998	4.115	17.085	4.695	15.705	5.432	14.206
10	3.717	21.398	4.795	18.390	5.425	16.962	6.221	15.407
11	4.321	22.779	5.491	19.682	6.169	18.208	7.021	16.598
12	4.943	24.145	6.201	20.962	6.924	19.443	7.829	17.782
13	5.580	25.497	6.922	22.230	7.690	20.669	8.646	18.958
14	6.231	26.836	7.654	23.490	8.464	21.886	9.470	20.128
15	6.893	28.164	8.395	24.740	9.246	23.097	10.300	21.292
16	7.567	29.482	9.145	25.983	10.036	24.301	11.135	22.452
17	8.251	30.791	9.903	27.219	10.832	25.499	11.976	23.606
18	8.943	32.091	10.668	28.448	11.634	26.692	12.822	24.756
19	9.644	33.383	11.439	29.671	12.442	27.879	13.671	25.903
20	10.353	34.668	12.217	30.888	13.255	29.062	14.525	27.045
21	11.069	35.946	12.999	32.101	14.072	30.240	15.383	28.184
22	11.792	37.218	13.787	33.308	14.894	31.415	16.244	29.320
23	12.521	38.484	14.580	34.511	15.719	32.585	17.108	30.453
24	13.255	39.745	15.377	35.710	16.549	33.752	17.975	31.584
25	13.995	41.000	16.179	36.905	17.382	34.916	18.844	32.711
26	14.741	42.251	16.984	38.096	18.219	36.077	19.717	33.836
27	15.491	43.497	17.793	39.284	19.058	37.234	20.592	34.959
28	16.245	44.738	18.606	40.468	19.901	38.389	21.469	36.080
29	17.004	45.976	19.422	41.649	20.746	39.541	22.348	37.198
30	17.767	47.209	20.241	42.827	21.594	40.690	23.229	38.315
31	18.534	48.439	21.063	44.002	22.445	41.838	24.113	39.430
32	19.305	49.665	21.888	45.174	23.297	42.982	24.998	40.543
33	20.079	50.888	22.716	46.344	24.153	44.125	25.885	41.654
34	20.857	52.107	23.546	47.512	25.010	45.266	26.774	42.764
35	21.638	53.324	24.379	48.676	25.870	46.404	27.664	43.872

Table A.3 (Continued)
Confidence limits for the mean of a Poisson distribution given an observation of c events.

	Confidence Level							
	0.99		0.95		0.9		0.8	
c	λ_L	λ_U	λ_L	λ_U	λ_L	λ_U	λ_L	λ_U
36	22.422	54.537	25.214	49.839	26.731	47.541	28.556	44.978
37	23.208	55.748	26.051	51.000	27.595	48.675	29.450	46.083
38	23.998	56.955	26.891	52.158	28.460	49.808	30.345	47.187
39	24.791	58.160	27.733	53.314	29.327	50.940	31.241	48.289
40	25.586	59.363	28.577	54.469	30.196	52.069	32.139	49.390
42	27.184	61.761	30.270	56.772	31.938	54.324	33.938	51.589
44	28.791	64.149	31.970	59.068	33.687	56.573	35.742	53.783
46	30.407	66.529	33.678	61.358	35.441	58.816	37.550	55.972
48	32.032	68.901	35.391	63.641	37.200	61.054	39.363	58.158
50	33.664	71.266	37.111	65.919	38.965	63.287	41.179	60.339
52	35.303	73.623	38.836	68.191	40.734	65.516	42.999	62.518
54	36.949	75.974	40.566	70.458	42.507	67.740	44.823	64.693
56	38.602	78.319	42.302	72.721	44.285	69.960	46.649	66.864
58	40.261	80.657	44.042	74.978	46.067	72.177	48.479	69.033
60	41.926	82.990	45.786	77.232	47.852	74.390	50.312	71.199
62	43.596	85.317	47.535	79.481	49.641	76.599	52.147	73.362
64	45.271	87.639	49.288	81.727	51.434	78.805	53.985	75.523
66	46.952	89.956	51.044	83.968	53.229	81.008	55.826	77.681
68	48.638	92.269	52.805	86.206	55.028	83.208	57.669	79.836
70	50.327	94.577	54.568	88.441	56.830	85.405	59.515	81.990
72	52.022	96.880	56.336	90.672	58.634	87.599	61.362	84.141
74	53.720	99.180	58.106	92.900	60.441	89.790	63.212	86.291
76	55.423	101.476	59.879	95.125	62.251	91.979	65.064	88.438
78	57.129	103.768	61.656	97.347	64.063	94.166	66.917	90.583
80	58.840	106.056	63.435	99.567	65.878	96.350	68.773	92.727
82	60.553	108.340	65.217	101.784	67.695	98.532	70.630	94.869
84	62.271	110.621	67.002	103.998	69.514	100.712	72.489	97.009
86	63.991	112.899	68.789	106.209	71.336	102.889	74.350	99.147
88	65.715	115.174	70.579	108.418	73.159	105.065	76.212	101.284
90	67.442	117.445	72.371	110.625	74.984	107.239	78.076	103.419
92	69.172	119.714	74.165	112.830	76.812	109.410	79.942	105.553
94	70.905	121.980	75.962	115.032	78.641	111.580	81.809	107.686
96	72.641	124.243	77.760	117.232	80.472	113.748	83.677	109.817
98	74.379	126.503	79.561	119.431	82.305	115.915	85.547	111.946
100	76.120	128.760	81.364	121.627	84.139	118.079	87.418	114.075

Table A.4
Factors for obtaining a confidence interval for the intensity at the time of the last failure for a failure truncated power law process.

	Confidence Level						
n	**0.8**	**0.85**	**0.9**	**0.95**	**0.98**	**0.99**	**0.995**
2	0.02964	0.02157	0.01384	0.00624	0.00269	0.00165	0.00107
2	1.2399	1.4633	1.8012	2.4394	3.3971	4.2056	5.0852
3	0.1120	0.0918	0.0702	0.0455	0.0266	0.0180	0.0123
3	1.4620	1.6590	1.9468	2.4665	3.2066	3.8061	4.4393
4	0.1877	0.1608	0.1307	0.0939	0.0627	0.0469	0.0355
4	1.5150	1.6864	1.9326	2.3670	2.9687	3.4454	3.9408
5	0.2500	0.2194	0.1844	0.1399	0.1001	0.0789	0.0628
5	1.5226	1.6745	1.8902	2.2651	2.7752	3.1733	3.5825
6	0.3011	0.2684	0.2305	0.1811	0.1353	0.1102	0.0906
6	1.5152	1.6520	1.8448	2.1764	2.6216	2.9653	3.3157
7	0.3437	0.3098	0.2701	0.2176	0.1677	0.1396	0.1173
7	1.5025	1.6274	1.8025	2.1010	2.4979	2.8017	3.1095
8	0.3796	0.3452	0.3045	0.2499	0.1971	0.1669	0.1425
8	1.4880	1.6034	1.7643	2.0369	2.3963	2.6695	2.9449
9	0.4105	0.3758	0.3345	0.2786	0.2238	0.1919	0.1659
9	1.4735	1.5810	1.7303	1.9818	2.3112	2.5602	2.8102
10	0.4373	0.4026	0.3610	0.3043	0.2480	0.2150	0.1878
10	1.4594	1.5603	1.6999	1.9340	2.2389	2.4682	2.6976
11	0.4609	0.4263	0.3847	0.3275	0.2702	0.2362	0.2081
11	1.4461	1.5413	1.6726	1.8921	2.1765	2.3895	2.6019
12	0.4818	0.4474	0.4059	0.3484	0.2904	0.2558	0.2270
12	1.4336	1.5239	1.6482	1.8551	2.1221	2.3214	2.5195
13	0.5005	0.4664	0.4250	0.3675	0.3091	0.2740	0.2445
13	1.4219	1.5079	1.6260	1.8221	2.0741	2.2615	2.4475
14	0.5174	0.4836	0.4424	0.3850	0.3263	0.2909	0.2610
14	1.4110	1.4932	1.6059	1.7924	2.0314	2.2087	2.3841
15	0.5327	0.4992	0.4583	0.4011	0.3423	0.3066	0.2764
15	1.4008	1.4796	1.5875	1.7656	1.9931	2.1614	2.3276
16	0.5466	0.5135	0.4729	0.4160	0.3572	0.3213	0.2908
16	1.3913	1.4671	1.5706	1.7412	1.9585	2.1189	2.2770
17	0.5594	0.5266	0.4864	0.4298	0.3710	0.3350	0.3044
17	1.3823	1.4554	1.5551	1.7189	1.9271	2.0804	2.2313
18	0.5712	0.5388	0.4989	0.4426	0.3840	0.3480	0.3172
18	1.3740	1.4445	1.5407	1.6985	1.8985	2.0455	2.1899

Table A.4 (Continued)
Factors for obtaining a confidence interval for the intensity at the time of the last failure for a failure truncated power law process.

	Confidence Level						
n	**0.8**	**0.85**	**0.9**	**0.95**	**0.98**	**0.99**	**0.995**
19	0.5821	0.5500	0.5106	0.4546	0.3962	0.3601	0.3292
19	1.3661	1.4344	1.5273	1.6796	1.8722	2.0135	2.1521
20	0.5923	0.5605	0.5214	0.4659	0.4077	0.3716	0.3407
20	1.3587	1.4249	1.5149	1.6621	1.8480	1.9841	2.1175
21	0.6017	0.5703	0.5316	0.4764	0.4185	0.3825	0.3515
21	1.3517	1.4160	1.5033	1.6459	1.8255	1.9569	2.0855
22	0.6105	0.5795	0.5411	0.4864	0.4287	0.3928	0.3618
22	1.3451	1.4076	1.4924	1.6307	1.8047	1.9317	2.0559
23	0.6188	0.5881	0.5501	0.4957	0.4383	0.4026	0.3716
23	1.3389	1.3997	1.4822	1.6165	1.7853	1.9083	2.0284
24	0.6266	0.5962	0.5585	0.5046	0.4475	0.4119	0.3810
24	1.3329	1.3922	1.4725	1.6033	1.7671	1.8864	2.0028
25	0.6339	0.6038	0.5665	0.5130	0.4563	0.4207	0.3899
25	1.3273	1.3852	1.4635	1.5908	1.7501	1.8660	1.9789
26	0.6408	0.6110	0.5741	0.5210	0.4646	0.4292	0.3984
26	1.3219	1.3784	1.4549	1.5790	1.7342	1.8469	1.9566
27	0.6474	0.6179	0.5813	0.5286	0.4725	0.4372	0.4066
27	1.3168	1.3721	1.4468	1.5679	1.7191	1.8289	1.9357
28	0.6536	0.6244	0.5881	0.5358	0.4801	0.4450	0.4144
28	1.3119	1.3660	1.4390	1.5574	1.7050	1.8120	1.9160
29	0.6595	0.6306	0.5946	0.5428	0.4873	0.4524	0.4219
29	1.3073	1.3602	1.4317	1.5474	1.6916	1.7960	1.8974
30	0.6651	0.6364	0.6008	0.5494	0.4943	0.4595	0.4291
30	1.3028	1.3547	1.4247	1.5379	1.6788	1.7806	1.8795
31	0.6704	0.6421	0.6067	0.5557	0.5009	0.4663	0.4360
31	1.2986	1.3494	1.4181	1.5289	1.6668	1.7664	1.8632
32	0.6755	0.6474	0.6124	0.5617	0.5073	0.4728	0.4427
32	1.2945	1.3444	1.4117	1.5203	1.6552	1.7526	1.8470
33	0.6804	0.6525	0.6178	0.5675	0.5134	0.4791	0.4491
33	1.2905	1.3396	1.4056	1.5121	1.6444	1.7397	1.8322
34	0.6850	0.6575	0.6230	0.5731	0.5193	0.4852	0.4553
34	1.2868	1.3349	1.3998	1.5043	1.6339	1.7274	1.8179
35	0.6895	0.6622	0.6280	0.5785	0.5250	0.4911	0.4613
35	1.2831	1.3305	1.3942	1.4968	1.6240	1.7155	1.8042

Table A.4 (Continued)

Factors for obtaining a confidence interval for the intensity at the time of the last failure for a failure truncated power law process.

	Confidence Level						
n	**0.8**	**0.85**	**0.9**	**0.95**	**0.98**	**0.99**	**0.995**
36	0.6938	0.6667	0.6328	0.5836	0.5305	0.4967	0.4671
36	1.2796	1.3262	1.3888	1.4896	1.6144	1.7042	1.7911
37	0.6979	0.6710	0.6374	0.5886	0.5358	0.5022	0.4727
37	1.2762	1.3221	1.3837	1.4828	1.6054	1.6935	1.7788
38	0.7018	0.6752	0.6419	0.5934	0.5409	0.5075	0.4781
38	1.2730	1.3181	1.3787	1.4762	1.5966	1.6832	1.7669
39	0.7056	0.6792	0.6461	0.5980	0.5459	0.5126	0.4833
39	1.2699	1.3143	1.3739	1.4698	1.5882	1.6733	1.7554
40	0.7093	0.6831	0.6503	0.6025	0.5506	0.5175	0.4884
40	1.2668	1.3106	1.3693	1.4637	1.5802	1.6638	1.7446
41	0.7128	0.6869	0.6543	0.6068	0.5553	0.5223	0.4933
41	1.2639	1.3070	1.3649	1.4578	1.5724	1.6546	1.7340
42	0.7162	0.6905	0.6581	0.6110	0.5597	0.5270	0.4981
42	1.2610	1.3036	1.3606	1.4522	1.5650	1.6459	1.7239
43	0.7195	0.6940	0.6619	0.6150	0.5641	0.5315	0.5027
43	1.2583	1.3002	1.3565	1.4467	1.5578	1.6374	1.7141
44	0.7227	0.6973	0.6655	0.6189	0.5683	0.5359	0.5072
44	1.2556	1.2970	1.3525	1.4414	1.5509	1.6293	1.7048
45	0.7258	0.7006	0.6690	0.6227	0.5724	0.5401	0.5116
45	1.2530	1.2939	1.3486	1.4363	1.5442	1.6214	1.6958
46	0.7288	0.7038	0.6723	0.6264	0.5764	0.5442	0.5158
46	1.2505	1.2909	1.3449	1.4314	1.5377	1.6138	1.6870
47	0.7316	0.7068	0.6756	0.6300	0.5802	0.5482	0.5200
47	1.2481	1.2879	1.3412	1.4266	1.5315	1.6065	1.6786
48	0.7344	0.7098	0.6788	0.6335	0.5840	0.5522	0.5240
48	1.2458	1.2851	1.3377	1.4220	1.5255	1.5994	1.6706
49	0.7372	0.7127	0.6819	0.6368	0.5876	0.5560	0.5279
49	1.2434	1.2823	1.3343	1.4175	1.5196	1.5925	1.6626
50	0.7398	0.7155	0.6849	0.6401	0.5912	0.5597	0.5317
50	1.2412	1.2796	1.3310	1.4132	1.5139	1.5859	1.6550
51	0.7423	0.7182	0.6878	0.6433	0.5946	0.5633	0.5355
51	1.2391	1.2770	1.3278	1.4090	1.5085	1.5796	1.6480
52	0.7448	0.7208	0.6907	0.6464	0.5980	0.5668	0.5391
52	1.2369	1.2745	1.3247	1.4049	1.5032	1.5733	1.6407

Table A.4 (Continued)

Factors for obtaining a confidence interval for the intensity at the time of the last failure for a failure truncated power law process.

	Confidence Level						
n	0.8	0.85	0.9	0.95	0.98	0.99	0.995
53	0.7472	0.7234	0.6934	0.6495	0.6013	0.5702	0.5427
53	1.2349	1.2720	1.3216	1.4009	1.4980	1.5672	1.6338
54	0.7496	0.7259	0.6961	0.6524	0.6045	0.5736	0.5461
54	1.2329	1.2696	1.3187	1.3971	1.4930	1.5615	1.6272
55	0.7518	0.7284	0.6987	0.6553	0.6076	0.5768	0.5495
55	1.2309	1.2673	1.3158	1.3933	1.4882	1.5558	1.6207
56	0.7541	0.7307	0.7013	0.6581	0.6106	0.5800	0.5528
56	1.2290	1.2650	1.3130	1.3897	1.4834	1.5502	1.6144
57	0.7562	0.7330	0.7038	0.6608	0.6136	0.5831	0.5560
57	1.2271	1.2627	1.3103	1.3861	1.4787	1.5447	1.6079
58	0.7583	0.7353	0.7062	0.6635	0.6165	0.5862	0.5592
58	1.2253	1.2606	1.3076	1.3827	1.4743	1.5396	1.6022
59	0.7604	0.7375	0.7086	0.6661	0.6194	0.5892	0.5623
59	1.2236	1.2584	1.3050	1.3793	1.4699	1.5344	1.5962
60	0.7624	0.7396	0.7109	0.6686	0.6221	0.5921	0.5653
60	1.2218	1.2564	1.3025	1.3760	1.4658	1.5297	1.5910
61	0.7643	0.7417	0.7132	0.6711	0.6249	0.5949	0.5683
61	1.2201	1.2544	1.3000	1.3728	1.4616	1.5248	1.5853
62	0.7662	0.7438	0.7154	0.6735	0.6275	0.5977	0.5712
62	1.2185	1.2524	1.2976	1.3697	1.4576	1.5201	1.5799
63	0.7681	0.7458	0.7175	0.6759	0.6301	0.6004	0.5740
63	1.2169	1.2505	1.2953	1.3666	1.4536	1.5154	1.5746
64	0.7699	0.7477	0.7196	0.6782	0.6327	0.6031	0.5768
64	1.2153	1.2486	1.2930	1.3636	1.4498	1.5110	1.5696
65	0.7717	0.7496	0.7217	0.6805	0.6351	0.6057	0.5795
65	1.2138	1.2468	1.2908	1.3608	1.4461	1.5068	1.5649
66	0.7734	0.7515	0.7237	0.6827	0.6376	0.6083	0.5822
66	1.2122	1.2449	1.2885	1.3579	1.4424	1.5024	1.5598
67	0.7751	0.7533	0.7257	0.6849	0.6400	0.6108	0.5848
67	1.2108	1.2432	1.2864	1.3551	1.4389	1.4983	1.5553
68	0.7768	0.7551	0.7276	0.6870	0.6423	0.6133	0.5873
68	1.2093	1.2415	1.2843	1.3524	1.4354	1.4942	1.5506
69	0.7784	0.7568	0.7295	0.6891	0.6446	0.6157	0.5899
69	1.2079	1.2398	1.2823	1.3498	1.4320	1.4903	1.5462

Table A.4 (Continued)

Factors for obtaining a confidence interval for the intensity at the time of the last failure for a failure truncated power law process.

	Confidence Level						
n	0.8	0.85	0.9	0.95	0.98	0.99	0.995
70	0.7800	0.7585	0.7313	0.6912	0.6468	0.6180	0.5923
70	1.2065	1.2381	1.2802	1.3472	1.4286	1.4863	1.5415
71	0.7816	0.7602	0.7331	0.6932	0.6491	0.6204	0.5947
71	1.2051	1.2365	1.2783	1.3447	1.4254	1.4827	1.5375
72	0.7831	0.7618	0.7349	0.6952	0.6512	0.6227	0.5971
72	1.2038	1.2349	1.2763	1.3422	1.4222	1.4789	1.5331
73	0.7846	0.7634	0.7367	0.6971	0.6533	0.6249	0.5995
73	1.2025	1.2334	1.2744	1.3397	1.4191	1.4753	1.5290
74	0.7860	0.7650	0.7384	0.6990	0.6554	0.6271	0.6018
74	1.2012	1.2318	1.2726	1.3374	1.4161	1.4718	1.5252
75	0.7875	0.7665	0.7400	0.7008	0.6575	0.6292	0.6040
75	1.1999	1.2303	1.2708	1.3350	1.4130	1.4683	1.5210
76	0.7889	0.7681	0.7417	0.7027	0.6595	0.6314	0.6062
76	1.1987	1.2289	1.2690	1.3327	1.4101	1.4648	1.5171
77	0.7903	0.7695	0.7433	0.7044	0.6614	0.6334	0.6084
77	1.1975	1.2274	1.2673	1.3305	1.4072	1.4615	1.5133
78	0.7916	0.7710	0.7449	0.7062	0.6634	0.6355	0.6105
78	1.1963	1.2260	1.2655	1.3283	1.4044	1.4583	1.5097
79	0.7929	0.7724	0.7464	0.7079	0.6653	0.6375	0.6126
79	1.1951	1.2246	1.2639	1.3261	1.4017	1.4552	1.5062
80	0.7942	0.7738	0.7479	0.7096	0.6672	0.6395	0.6147
80	1.1940	1.2233	1.2623	1.3241	1.3991	1.4522	1.5030
81	0.7955	0.7752	0.7494	0.7113	0.6690	0.6414	0.6167
81	1.1929	1.2219	1.2606	1.3220	1.3964	1.4491	1.4995
82	0.7968	0.7765	0.7509	0.7129	0.6708	0.6433	0.6187
82	1.1917	1.2206	1.2590	1.3199	1.3938	1.4460	1.4959
83	0.7980	0.7779	0.7523	0.7145	0.6726	0.6452	0.6207
83	1.1906	1.2193	1.2575	1.3179	1.3912	1.4429	1.4923
84	0.7992	0.7792	0.7538	0.7161	0.6743	0.6471	0.6226
84	1.1896	1.2181	1.2560	1.3160	1.3888	1.4403	1.4895
85	0.8004	0.7805	0.7552	0.7177	0.6760	0.6489	0.6245
85	1.1885	1.2168	1.2544	1.3140	1.3862	1.4372	1.4858
86	0.8016	0.7817	0.7565	0.7192	0.6777	0.6507	0.6264
86	1.1875	1.2156	1.2529	1.3121	1.3838	1.4344	1.4827

Table A.4 (Continued)

Factors for obtaining a confidence interval for the intensity at the time of the last failure for a failure truncated power law process.

	Confidence Level						
n	**0.8**	**0.85**	**0.9**	**0.95**	**0.98**	**0.99**	**0.995**
87	0.8027	0.7830	0.7579	0.7207	0.6794	0.6524	0.6283
87	1.1865	1.2144	1.2515	1.3103	1.3815	1.4318	1.4798
88	0.8038	0.7842	0.7592	0.7222	0.6810	0.6542	0.6301
88	1.1855	1.2132	1.2501	1.3085	1.3792	1.4291	1.4767
89	0.8050	0.7854	0.7605	0.7236	0.6827	0.6559	0.6319
89	1.1845	1.2120	1.2487	1.3067	1.3769	1.4265	1.4738
90	0.8060	0.7865	0.7618	0.7251	0.6842	0.6576	0.6336
90	1.1835	1.2109	1.2473	1.3049	1.3746	1.4238	1.4706
91	0.8071	0.7877	0.7631	0.7265	0.6858	0.6592	0.6354
91	1.1825	1.2098	1.2459	1.3032	1.3725	1.4214	1.4680
92	0.8082	0.7888	0.7643	0.7279	0.6874	0.6609	0.6371
92	1.1816	1.2087	1.2446	1.3015	1.3703	1.4190	1.4653
93	0.8092	0.7900	0.7655	0.7292	0.6889	0.6625	0.6388
93	1.1807	1.2076	1.2433	1.2998	1.3682	1.4166	1.4627
94	0.8102	0.7911	0.7667	0.7306	0.6904	0.6641	0.6404
94	1.1798	1.2065	1.2420	1.2982	1.3661	1.4142	1.4600
95	0.8112	0.7922	0.7679	0.7319	0.6918	0.6656	0.6421
95	1.1789	1.2054	1.2407	1.2965	1.3640	1.4117	1.4571
96	0.8122	0.7932	0.7691	0.7332	0.6933	0.6672	0.6437
96	1.1780	1.2044	1.2394	1.2949	1.3619	1.4091	1.4539
97	0.8132	0.7943	0.7702	0.7345	0.6947	0.6687	0.6453
97	1.1771	1.2034	1.2382	1.2933	1.3600	1.4070	1.4517
98	0.8142	0.7953	0.7714	0.7358	0.6961	0.6702	0.6469
98	1.1762	1.2023	1.2370	1.2917	1.3579	1.4045	1.4487
99	0.8151	0.7963	0.7725	0.7370	0.6975	0.6717	0.6484
99	1.1754	1.2014	1.2358	1.2903	1.3562	1.4026	1.4469
100	0.8160	0.7973	0.7736	0.7383	0.6989	0.6731	0.6499
100	1.1746	1.2004	1.2346	1.2888	1.3542	1.4003	1.4443

Table A.5
Factors for obtaining a confidence interval for the intensity at the time of the last failure for a time truncated power law process.

	Confidence Level						
n	**0.8**	**0.85**	**0.9**	**0.95**	**0.98**	**0.99**	**0.995**
2	0.05361	0.03949	0.02587	0.01271	0.00503	0.00251	0.00125
2	3.8249	4.3084	5.0109	6.2702	8.0487	9.4802	10.9855
3	0.1581	0.1316	0.1027	0.0387	0.0415	0.0287	0.0200
3	3.0073	3.3322	3.7973	4.6141	5.7395	6.6273	7.5470
4	0.2357	0.2040	0.1682	0.1236	0.0847	0.0646	0.0497
4	2.5966	2.8463	3.2002	3.8132	4.6441	5.2905	5.9535
5	0.2953	0.2609	0.2214	0.1706	0.1243	0.0994	0.0802
5	2.3464	2.5521	2.8413	3.3374	4.0017	4.5132	5.0338
6	0.3430	0.3073	0.2656	0.2111	0.1599	0.1315	0.1091
6	2.1765	2.3530	2.5998	3.0199	3.5771	4.0027	4.4334
7	0.3823	0.3460	0.3032	0.2462	0.1917	0.1608	0.1361
7	2.0526	2.2083	2.4251	2.7918	3.2743	3.6404	4.0092
8	0.4154	0.3789	0.3355	0.2771	0.2203	0.1876	0.1611
8	1.9578	2.0979	2.2921	2.6190	3.0464	3.3690	3.6925
9	0.4437	0.4072	0.3636	0.3044	0.2460	0.2120	0.1842
9	1.8826	2.0104	2.1871	2.4832	2.8682	3.1575	3.4465
10	0.4683	0.4320	0.3884	0.3288	0.2694	0.2344	0.2055
10	1.8211	1.9391	2.1018	2.3732	2.7246	2.9875	3.2494
11	0.4898	0.4539	0.4104	0.3507	0.2906	0.2549	0.2252
11	1.7699	1.8798	2.0309	2.2822	2.6061	2.8476	3.0876
12	0.5090	0.4734	0.4302	0.3705	0.3100	0.2738	0.2436
12	1.7265	1.8294	1.9708	2.2053	2.5065	2.7303	2.9521
13	0.5262	0.4909	0.4481	0.3885	0.3279	0.2913	0.2606
13	1.6890	1.7862	1.9193	2.1395	2.4213	2.6302	2.8368
14	0.5416	0.5068	0.4644	0.4051	0.3443	0.3076	0.2765
14	1.6563	1.7484	1.8744	2.0823	2.3477	2.5438	2.7374
15	0.5557	0.5213	0.4792	0.4203	0.3596	0.3227	0.2914
15	1.6276	1.7153	1.8350	2.0322	2.2832	2.4683	2.6506
16	0.5685	0.5345	0.4929	0.4344	0.3738	0.3368	0.3053
16	1.6020	1.6858	1.8000	1.9878	2.2262	2.4016	2.5742
17	0.5803	0.5467	0.5055	0.4474	0.3871	0.3500	0.3184
17	1.5791	1.6593	1.7687	1.9481	2.1754	2.3423	2.5063
18	0.5912	0.5580	0.5172	0.4596	0.3995	0.3624	0.3308
18	1.5584	1.6355	1.7405	1.9125	2.1298	2.2892	2.4455

Table A.5 (Continued)
Factors for obtaining a confidence interval for the intensity at the time of the last failure for a time truncated power law process.

	Confidence Level						
n	**0.8**	**0.85**	**0.9**	**0.95**	**0.98**	**0.99**	**0.995**
19	0.6012	0.5685	0.5281	0.4710	0.4111	0.3742	0.3425
19	1.5396	1.6139	1.7150	1.8802	2.0886	2.2412	2.3907
20	0.6106	0.5782	0.5383	0.4816	0.4221	0.3852	0.3535
20	1.5225	1.5942	1.6917	1.8508	2.0512	2.1977	2.3410
21	0.6193	0.5873	0.5478	0.4916	0.4324	0.3956	0.3640
21	1.5067	1.5762	1.6704	1.8240	2.0171	2.1579	2.2957
22	0.6275	0.5959	0.5568	0.5010	0.4422	0.4056	0.3739
22	1.4923	1.5595	1.6508	1.7993	1.9857	2.1215	2.2542
23	0.6352	0.6039	0.5652	0.5099	0.4514	0.4150	0.3834
23	1.4789	1.5442	1.6326	1.7765	1.9568	2.0880	2.2160
24	0.6424	0.6115	0.5732	0.5184	0.4602	0.4239	0.3924
24	1.4664	1.5299	1.6158	1.7554	1.9301	2.0570	2.1808
25	0.6492	0.6186	0.5807	0.5264	0.4686	0.4324	0.4010
25	1.4549	1.5167	1.6002	1.7358	1.9053	2.0283	2.1481
26	0.6556	0.6254	0.5878	0.5339	0.4766	0.4406	0.4093
26	1.4441	1.5043	1.5856	1.7175	1.8822	2.0016	2.1177
27	0.6617	0.6318	0.5946	0.5412	0.4842	0.4483	0.4172
27	1.4339	1.4927	1.5720	1.7004	1.8606	1.9766	2.0894
28	0.6674	0.6378	0.6010	0.5481	0.4914	0.4558	0.4247
28	1.4244	1.4818	1.5592	1.6844	1.8404	1.9533	2.0629
29	0.6729	0.6436	0.6072	0.5546	0.4984	0.4629	0.4320
29	1.4155	1.4715	1.5471	1.6694	1.8214	1.9314	2.0381
30	0.6781	0.6491	0.6130	0.5609	0.5051	0.4698	0.4389
30	1.4071	1.4619	1.5358	1.6552	1.8036	1.9108	2.0148
31	0.6831	0.6544	0.6186	0.5669	0.5115	0.4763	0.4456
31	1.3991	1.4528	1.5251	1.6418	1.7868	1.8914	1.9928
32	0.6879	0.6594	0.6240	0.5727	0.5176	0.4827	0.4521
32	1.3915	1.4441	1.5149	1.6292	1.7709	1.8730	1.9721
33	0.6924	0.6643	0.6291	0.5783	0.5235	0.4888	0.4583
33	1.3844	1.4359	1.5053	1.6172	1.7558	1.8557	1.9524
34	0.6968	0.6689	0.6341	0.5836	0.5292	0.4946	0.4643
34	1.3776	1.4281	1.4962	1.6058	1.7415	1.8393	1.9339
35	0.7009	0.6733	0.6388	0.5887	0.5346	0.5003	0.4701
35	1.3711	1.4207	1.4875	1.5950	1.7280	1.8237	1.9162

Table A.5 (Continued)

Factors for obtaining a confidence interval for the intensity at the time of the last failure for a time truncated power law process.

	Confidence Level						
n	**0.8**	**0.85**	**0.9**	**0.95**	**0.98**	**0.99**	**0.995**
36	0.7049	0.6776	0.6433	0.5936	0.5399	0.5057	0.4757
36	1.3649	1.4137	1.4792	1.5847	1.7151	1.8088	1.8995
37	0.7088	0.6817	0.6477	0.5984	0.5450	0.5110	0.4811
37	1.3590	1.4069	1.4713	1.5749	1.7028	1.7947	1.8835
38	0.7125	0.6856	0.6519	0.6029	0.5499	0.5161	0.4864
38	1.3534	1.4005	1.4638	1.5655	1.6910	1.7812	1.8683
39	0.7160	0.6894	0.6560	0.6074	0.5547	0.5211	0.4915
39	1.3480	1.3943	1.4566	1.5565	1.6798	1.7683	1.8538
40	0.7195	0.6930	0.6599	0.6116	0.5593	0.5258	0.4964
40	1.3428	1.3884	1.4497	1.5480	1.6691	1.7560	1.8399
41	0.7228	0.6966	0.6637	0.6158	0.5638	0.5305	0.5011
41	1.3379	1.3828	1.4430	1.5397	1.6589	1.7443	1.8266
42	0.7260	0.7000	0.6674	0.6198	0.5681	0.5350	0.5058
42	1.3331	1.3774	1.4367	1.5319	1.6490	1.7330	1.8139
43	0.7291	0.7033	0.6709	0.6237	0.5723	0.5393	0.5103
43	1.3285	1.3721	1.4306	1.5243	1.6396	1.7221	1.8017
44	0.7321	0.7065	0.6743	0.6274	0.5763	0.5436	0.5146
44	1.3241	1.3671	1.4247	1.5170	1.6305	1.7117	1.7900
45	0.7350	0.7096	0.6777	0.6310	0.5803	0.5477	0.5189
45	1.3199	1.3623	1.4191	1.5100	1.6218	1.7017	1.7787
46	0.7378	0.7126	0.6809	0.6346	0.5841	0.5517	0.5230
46	1.3158	1.3576	1.4136	1.5033	1.6134	1.6921	1.7679
47	0.7405	0.7155	0.6840	0.6380	0.5878	0.5556	0.5270
47	1.3119	1.3531	1.4084	1.4968	1.6053	1.6828	1.7574
48	0.7431	0.7183	0.6870	0.6413	0.5914	0.5594	0.5310
48	1.3081	1.3488	1.4033	1.4905	1.5975	1.6739	1.7474
49	0.7456	0.7210	0.6900	0.6446	0.5950	0.5631	0.5348
49	1.3044	1.3446	1.3984	1.4844	1.5899	1.6653	1.7377
50	0.7481	0.7237	0.6929	0.6477	0.5984	0.5666	0.5385
50	1.3008	1.3406	1.3937	1.4786	1.5827	1.6570	1.7283
51	0.7505	0.7262	0.6956	0.6508	0.6017	0.5701	0.5421
51	1.2974	1.3366	1.3891	1.4729	1.5756	1.6489	1.7193
52	0.7529	0.7287	0.6983	0.6538	0.6050	0.5736	0.5457
52	1.2941	1.3329	1.3847	1.4675	1.5688	1.6411	1.7106

Table A.5 (Continued)
Factors for obtaining a confidence interval for the intensity at the time of the last failure for a time truncated power law process.

	Confidence Level						
n	**0.8**	**0.85**	**0.9**	**0.95**	**0.98**	**0.99**	**0.995**
53	0.7551	0.7312	0.7010	0.6567	0.6082	0.5769	0.5491
53	1.2909	1.3292	1.3804	1.4622	1.5623	1.6336	1.7021
54	0.7573	0.7335	0.7036	0.6595	0.6113	0.5801	0.5525
54	1.2877	1.3256	1.3762	1.4570	1.5559	1.6263	1.6940
55	0.7595	0.7359	0.7061	0.6623	0.6143	0.5833	0.5558
55	1.2847	1.3222	1.3722	1.4521	1.5497	1.6193	1.6860
56	0.7616	0.7381	0.7085	0.6650	0.6172	0.5864	0.5590
56	1.2818	1.3188	1.3683	1.4472	1.5437	1.6124	1.6784
57	0.7636	0.7403	0.7109	0.6676	0.6201	0.5894	0.5621
57	1.2789	1.3156	1.3645	1.4426	1.5379	1.6058	1.6709
58	0.7656	0.7424	0.7132	0.6702	0.6229	0.5924	0.5652
58	1.2761	1.3124	1.3608	1.4380	1.5323	1.5994	1.6637
59	0.7676	0.7445	0.7155	0.6727	0.6257	0.5953	0.5682
59	1.2734	1.3093	1.3572	1.4336	1.5268	1.5931	1.6567
60	0.7695	0.7466	0.7177	0.6751	0.6284	0.5981	0.5712
60	1.2708	1.3063	1.3538	1.4293	1.5215	1.5871	1.6499
61	0.7713	0.7486	0.7198	0.6775	0.6310	0.6009	0.5740
61	1.2683	1.3034	1.3504	1.4251	1.5163	1.5812	1.6433
62	0.7731	0.7505	0.7219	0.6799	0.6336	0.6036	0.5769
62	1.2658	1.3006	1.3471	1.4211	1.5113	1.5754	1.6368
63	0.7749	0.7524	0.7240	0.6822	0.6361	0.6062	0.5796
63	1.2633	1.2978	1.3439	1.4171	1.5064	1.5698	1.6306
64	0.7766	0.7542	0.7260	0.6844	0.6385	0.6088	0.5823
64	1.2610	1.2952	1.3407	1.4133	1.5016	1.5644	1.6245
65	0.7783	0.7561	0.7280	0.6866	0.6410	0.6114	0.5850
65	1.2587	1.2925	1.3377	1.4095	1.4970	1.5591	1.6186
66	0.7799	0.7578	0.7299	0.6887	0.6433	0.6139	0.5876
66	1.2564	1.2900	1.3347	1.4058	1.4925	1.5540	1.6128
67	0.7815	0.7596	0.7318	0.6908	0.6456	0.6163	0.5901
67	1.2543	1.2875	1.3318	1.4023	1.4881	1.5489	1.6072
68	0.7831	0.7613	0.7336	0.6929	0.6479	0.6187	0.5926
68	1.2521	1.2851	1.3290	1.3988	1.4838	1.5440	1.6017
69	0.7846	0.7629	0.7355	0.6949	0.6501	0.6210	0.5951
69	1.2500	1.2827	1.3262	1.3954	1.4796	1.5392	1.5963

Table A.5 (Continued)

Factors for obtaining a confidence interval for the intensity at the time of the last failure for a time truncated power law process.

	Confidence Level						
n	**0.8**	**0.85**	**0.9**	**0.95**	**0.98**	**0.99**	**0.995**
70	0.7862	0.7646	0.7372	0.6969	0.6523	0.6233	0.5975
70	1.2480	1.2804	1.3235	1.3921	1.4755	1.5346	1.5911
71	0.7876	0.7661	0.7390	0.6988	0.6544	0.6256	0.5998
71	1.2460	1.2781	1.3209	1.3888	1.4715	1.5300	1.5860
72	0.7891	0.7677	0.7407	0.7007	0.6565	0.6278	0.6022
72	1.2441	1.2759	1.3183	1.3857	1.4676	1.5256	1.5810
73	0.7905	0.7692	0.7423	0.7026	0.6586	0.6300	0.6044
73	1.2422	1.2738	1.3158	1.3826	1.4638	1.5212	1.5762
74	0.7919	0.7707	0.7440	0.7044	0.6606	0.6321	0.6067
74	1.2403	1.2716	1.3134	1.3796	1.4600	1.5170	1.5714
75	0.7932	0.7722	0.7456	0.7062	0.6626	0.6342	0.6089
75	1.2385	1.2696	1.3109	1.3766	1.4564	1.5129	1.5668
76	0.7946	0.7737	0.7472	0.7080	0.6646	0.6363	0.6110
76	1.2367	1.2676	1.3086	1.3737	1.4528	1.5088	1.5623
77	0.7959	0.7751	0.7487	0.7097	0.6665	0.6383	0.6132
77	1.2350	1.2656	1.3063	1.3709	1.4493	1.5048	1.5578
78	0.7992	0.7765	0.7502	0.7114	0.6684	0.6403	0.6152
78	1.2333	1.2636	1.3040	1.3681	1.4459	1.5009	1.5535
79	0.7984	0.7778	0.7517	0.7130	0.6702	0.6423	0.6173
79	1.2316	1.2617	1.3018	1.3654	1.4426	1.4971	1.5492
80	0.7997	0.7792	0.7532	0.7147	0.6720	0.6442	0.6193
80	1.2300	1.2599	1.2997	1.3628	1.4393	1.4934	1.5451
81	0.8009	0.7805	0.7546	0.7163	0.6738	0.6461	0.6213
81	1.2283	1.2581	1.2975	1.3602	1.4361	1.4898	1.5410
82	0.8021	0.7818	0.7560	0.7179	0.6756	0.6480	0.6232
82	1.2268	1.2563	1.2955	1.3576	1.4330	1.4862	1.5370
83	0.8033	0.7830	0.7574	0.7194	0.6773	0.6498	0.6252
83	1.2252	1.2545	1.2934	1.3551	1.4299	1.4827	1.5331
84	0.8044	0.7843	0.7588	0.7210	0.6790	0.6516	0.6271
84	1.2237	1.2528	1.2914	1.3527	1.4269	1.4793	1.5293
85	0.8055	0.7855	0.7601	0.7225	0.6807	0.6534	0.6289
85	1.2222	1.2511	1.2895	1.3503	1.4239	1.4759	1.5256
86	0.8067	0.7867	0.7614	0.7239	0.6823	0.6551	0.6308
86	1.2208	1.2495	1.2876	1.3479	1.4210	1.4727	1.5219

Table A.5 (Continued)
Factors for obtaining a confidence interval for the intensity at the time of the last failure for a time truncated power law process.

	Confidence Level						
n	0.8	0.85	0.9	0.95	0.98	0.99	0.995
87	0.8077	0.7879	0.7627	0.7254	0.6839	0.6568	0.6326
87	1.2194	1.2478	1.2857	1.3456	1.4182	1.4694	1.5183
88	0.8088	0.7891	0.7640	0.7268	0.6855	0.6585	0.6343
88	1.2180	1.2462	1.2838	1.3433	1.4154	1.4663	1.5147
89	0.8099	0.7902	0.7653	0.7282	0.6871	0.6602	0.6361
89	1.2166	1.2447	1.2820	1.3411	1.4127	1.4632	1.5113
90	0.8109	0.7913	0.7665	0.7296	0.6886	0.6618	0.6378
90	1.2152	1.2431	1.2802	1.3389	1.4100	1.4601	1.5079
91	0.8119	0.7925	0.7677	0.7310	0.6902	0.6635	0.6395
91	1.2139	1.2416	1.2785	1.3368	1.4073	1.4571	1.5045
92	0.8130	0.7935	0.7689	0.7323	0.6917	0.6651	0.6412
92	1.2126	1.2402	1.2768	1.3347	1.4047	1.4542	1.5013
93	0.8139	0.7946	0.7701	0.7337	0.6931	0.6666	0.6428
93	1.2113	1.2387	1.2751	1.3326	1.4022	1.4513	1.4980
94	0.8149	0.7957	0.7712	0.7350	0.6946	0.6682	0.6445
94	1.2101	1.2373	1.2734	1.3306	1.3997	1.4484	1.4949
95	0.8159	0.7967	0.7724	0.7362	0.6960	0.6697	0.6461
95	1.2088	1.2359	1.2718	1.3286	1.3972	1.4457	1.4918
96	0.8168	0.7977	0.7735	0.7375	0.6974	0.6712	0.6476
96	1.2076	1.2345	1.2702	1.3266	1.3948	1.4429	1.4887
97	0.8178	0.7988	0.7746	0.7388	0.6988	0.6727	0.6492
97	1.2064	1.2331	1.2686	1.3247	1.3925	1.4402	1.4857
98	0.8187	0.7997	0.7757	0.7400	0.7002	0.6742	0.6507
98	1.2052	1.2318	1.2671	1.3228	1.3901	1.4376	1.4828
99	0.8196	0.8007	0.7768	0.7412	0.7016	0.6756	0.6523
99	1.2041	1.2305	1.2655	1.3209	1.3878	1.4350	1.4799
100	0.8205	0.8017	0.7778	0.7424	0.7029	0.6770	0.6538
100	1.2029	1.2292	1.2640	1.3191	1.3856	1.4324	1.4770

Table A.6
Critical values for the Cramér-von Mises goodness-of-fit test.

m	0.2	0.15	0.1	0.08	0.05	0.03
2	0.137	0.149	0.161	0.166	0.175	0.180
3	0.120	0.135	0.154	0.164	0.183	0.200
4	0.120	0.134	0.154	0.166	0.191	0.219
5	0.121	0.137	0.159	0.172	0.199	0.229
6	0.122	0.139	0.162	0.175	0.203	0.235
7	0.123	0.140	0.164	0.178	0.207	0.240
8	0.124	0.141	0.165	0.179	0.209	0.242
9	0.125	0.142	0.167	0.181	0.211	0.244
10	0.125	0.143	0.168	0.182	0.212	0.246
11	0.126	0.143	0.168	0.182	0.213	0.247
12	0.126	0.144	0.169	0.183	0.214	0.248
13	0.126	0.144	0.169	0.184	0.215	0.249
14	0.126	0.144	0.170	0.184	0.215	0.249
15	0.127	0.145	0.170	0.184	0.216	0.250
16	0.127	0.145	0.170	0.185	0.216	0.251
17	0.127	0.145	0.170	0.185	0.216	0.251
18	0.127	0.145	0.171	0.185	0.217	0.252
19	0.127	0.145	0.171	0.185	0.217	0.252
20	0.127	0.145	0.171	0.186	0.217	0.252
21	0.128	0.145	0.171	0.186	0.217	0.252
22	0.128	0.146	0.171	0.186	0.217	0.252
23	0.128	0.146	0.171	0.186	0.218	0.253
24	0.128	0.146	0.172	0.186	0.218	0.253
25	0.128	0.146	0.172	0.186	0.218	0.253
26	0.128	0.146	0.172	0.187	0.218	0.253
27	0.128	0.146	0.172	0.186	0.217	0.252
28	0.128	0.146	0.172	0.186	0.217	0.252
29	0.128	0.146	0.172	0.186	0.217	0.252
30	0.128	0.146	0.172	0.186	0.217	0.252
31	0.128	0.146	0.172	0.186	0.217	0.252
32	0.128	0.146	0.172	0.186	0.218	0.253
33	0.128	0.146	0.172	0.186	0.218	0.253
34	0.128	0.146	0.172	0.187	0.218	0.254
35	0.128	0.146	0.172	0.187	0.219	0.255

m	0.02	0.01	0.005	0.002	0.001
2	0.183	0.186	0.188	0.189	0.189
3	0.212	0.230	0.246	0.266	0.277
4	0.242	0.280	0.321	0.374	0.408
5	0.254	0.296	0.342	0.410	0.462
6	0.261	0.308	0.356	0.423	0.476
7	0.267	0.314	0.364	0.434	0.487
8	0.270	0.318	0.369	0.438	0.492
9	0.272	0.321	0.371	0.441	0.497
10	0.274	0.323	0.375	0.447	0.502
11	0.275	0.325	0.377	0.448	0.503
12	0.276	0.325	0.377	0.449	0.505
13	0.278	0.328	0.380	0.449	0.504
14	0.278	0.328	0.380	0.451	0.505
15	0.279	0.329	0.380	0.451	0.508
16	0.280	0.329	0.380	0.450	0.504
17	0.280	0.330	0.381	0.453	0.507
18	0.280	0.330	0.381	0.453	0.507
19	0.280	0.330	0.382	0.453	0.506
20	0.281	0.332	0.384	0.453	0.508
21	0.281	0.332	0.384	0.453	0.506
22	0.281	0.331	0.383	0.453	0.507
23	0.281	0.331	0.384	0.454	0.509
24	0.282	0.333	0.384	0.455	0.507
25	0.282	0.332	0.384	0.453	0.510
26	0.282	0.333	0.384	0.453	0.506
27	0.280	0.330	0.381	0.449	0.503
28	0.279	0.328	0.379	0.449	0.502
29	0.279	0.329	0.380	0.449	0.503
30	0.280	0.329	0.381	0.453	0.508
31	0.281	0.331	0.382	0.453	0.509
32	0.282	0.332	0.384	0.453	0.507
33	0.281	0.332	0.384	0.455	0.509
34	0.283	0.334	0.386	0.458	0.514
35	0.284	0.336	0.388	0.460	0.516

Table A.6 (Continued)
Critical values for the Cramér-von Mises goodness-of-fit test.

m	0.2	0.15	0.1	0.08	0.05	0.03
36	0.128	0.146	0.173	0.187	0.219	0.255
38	0.128	0.146	0.173	0.188	0.220	0.256
40	0.129	0.147	0.173	0.188	0.220	0.257
42	0.129	0.147	0.173	0.188	0.221	0.257
44	0.129	0.147	0.173	0.188	0.221	0.258
46	0.129	0.147	0.174	0.189	0.222	0.259
48	0.129	0.147	0.174	0.189	0.222	0.259
50	0.129	0.148	0.174	0.190	0.222	0.259
52	0.129	0.148	0.174	0.189	0.222	0.259
54	0.129	0.147	0.174	0.189	0.222	0.259
56	0.129	0.148	0.174	0.189	0.222	0.258
58	0.129	0.148	0.175	0.190	0.222	0.258
60	0.129	0.148	0.174	0.190	0.222	0.258
62	0.129	0.148	0.175	0.190	0.222	0.258
64	0.129	0.148	0.174	0.189	0.222	0.258
66	0.129	0.148	0.174	0.189	0.222	0.258
68	0.129	0.148	0.174	0.189	0.222	0.257
70	0.130	0.148	0.174	0.189	0.221	0.257
72	0.129	0.148	0.174	0.189	0.222	0.257
74	0.129	0.148	0.174	0.189	0.221	0.257
76	0.129	0.148	0.174	0.189	0.221	0.257
78	0.129	0.148	0.174	0.189	0.221	0.257
80	0.129	0.148	0.174	0.189	0.221	0.257
82	0.130	0.148	0.174	0.189	0.221	0.257
84	0.129	0.148	0.174	0.189	0.221	0.257
86	0.129	0.148	0.174	0.189	0.221	0.257
88	0.129	0.148	0.174	0.189	0.221	0.256
90	0.129	0.148	0.174	0.189	0.221	0.257
92	0.129	0.147	0.174	0.189	0.221	0.256
94	0.129	0.147	0.174	0.189	0.220	0.256
96	0.129	0.147	0.174	0.188	0.220	0.256
98	0.129	0.147	0.174	0.188	0.220	0.256
100	0.129	0.147	0.174	0.189	0.220	0.256

m	0.02	0.01	0.005	0.002	0.001
36	0.284	0.336	0.389	0.461	0.515
38	0.285	0.338	0.391	0.462	0.515
40	0.287	0.339	0.392	0.463	0.517
42	0.287	0.340	0.393	0.465	0.517
44	0.288	0.340	0.393	0.465	0.519
46	0.289	0.340	0.393	0.464	0.517
48	0.288	0.340	0.393	0.465	0.518
50	0.289	0.341	0.393	0.465	0.521
52	0.288	0.341	0.394	0.465	0.520
54	0.288	0.339	0.392	0.464	0.518
56	0.288	0.340	0.392	0.462	0.515
58	0.288	0.339	0.392	0.462	0.516
60	0.288	0.340	0.393	0.465	0.517
62	0.287	0.338	0.391	0.462	0.514
64	0.288	0.339	0.391	0.462	0.516
66	0.287	0.338	0.390	0.461	0.514
68	0.286	0.338	0.390	0.461	0.513
70	0.287	0.338	0.390	0.460	0.512
72	0.286	0.336	0.389	0.459	0.512
74	0.286	0.336	0.388	0.457	0.511
76	0.286	0.337	0.388	0.458	0.512
78	0.286	0.337	0.388	0.457	0.510
80	0.286	0.337	0.389	0.458	0.512
82	0.286	0.336	0.389	0.457	0.510
84	0.286	0.336	0.388	0.457	0.508
86	0.285	0.335	0.386	0.454	0.506
88	0.285	0.336	0.388	0.457	0.508
90	0.285	0.335	0.387	0.457	0.510
92	0.285	0.336	0.387	0.455	0.506
94	0.285	0.335	0.386	0.455	0.506
96	0.284	0.334	0.385	0.453	0.506
98	0.285	0.335	0.386	0.454	0.503
100	0.284	0.335	0.386	0.452	0.505

Table A.7
Critical values for Lilliefors' goodness-of-fit test.

m	0.2	0.15	0.1	0.08	0.05	0.03
2	0.551	0.573	0.593	0.602	0.613	0.621
3	0.451	0.479	0.511	0.526	0.551	0.571
4	0.401	0.420	0.444	0.457	0.484	0.512
5	0.360	0.379	0.404	0.417	0.442	0.466
6	0.332	0.350	0.373	0.385	0.408	0.432
7	0.310	0.326	0.348	0.359	0.381	0.403
8	0.291	0.307	0.327	0.338	0.359	0.380
9	0.276	0.291	0.310	0.320	0.340	0.360
10	0.263	0.277	0.295	0.305	0.324	0.344
11	0.251	0.265	0.283	0.292	0.310	0.329
12	0.241	0.254	0.271	0.280	0.298	0.316
13	0.232	0.245	0.261	0.270	0.287	0.304
14	0.224	0.236	0.252	0.261	0.277	0.294
15	0.217	0.229	0.244	0.252	0.269	0.285
16	0.210	0.222	0.237	0.245	0.261	0.277
17	0.205	0.216	0.230	0.238	0.253	0.269
18	0.199	0.210	0.224	0.232	0.246	0.262
19	0.194	0.204	0.218	0.226	0.240	0.255
20	0.189	0.200	0.213	0.220	0.234	0.249
21	0.185	0.195	0.208	0.215	0.229	0.243
22	0.181	0.191	0.204	0.210	0.224	0.238
23	0.177	0.187	0.199	0.206	0.219	0.233
24	0.173	0.183	0.195	0.202	0.215	0.228
25	0.170	0.179	0.192	0.198	0.211	0.224
26	0.167	0.176	0.188	0.194	0.207	0.220
27	0.164	0.173	0.185	0.191	0.203	0.215
28	0.161	0.170	0.181	0.187	0.199	0.211
29	0.158	0.167	0.178	0.184	0.196	0.208
30	0.156	0.164	0.175	0.181	0.193	0.204
31	0.153	0.162	0.173	0.178	0.190	0.201
32	0.151	0.159	0.170	0.176	0.187	0.198
33	0.149	0.157	0.168	0.173	0.184	0.195
34	0.147	0.155	0.165	0.171	0.182	0.193
35	0.145	0.153	0.163	0.169	0.179	0.190

m	**0.02**	**0.01**	**0.005**	**0.002**	**0.001**
2	0.625	0.628	0.630	0.631	0.632
3	0.584	0.600	0.612	0.622	0.629
4	0.531	0.557	0.578	0.602	0.621
5	0.484	0.513	0.536	0.562	0.580
6	0.448	0.475	0.500	0.530	0.549
7	0.420	0.446	0.469	0.497	0.517
8	0.396	0.420	0.443	0.471	0.490
9	0.375	0.399	0.421	0.447	0.466
10	0.358	0.381	0.402	0.428	0.446
11	0.343	0.365	0.386	0.411	0.428
12	0.330	0.351	0.371	0.396	0.412
13	0.317	0.338	0.357	0.381	0.397
14	0.307	0.327	0.346	0.369	0.385
15	0.297	0.317	0.335	0.358	0.373
16	0.288	0.307	0.325	0.347	0.362
17	0.281	0.299	0.316	0.337	0.352
18	0.273	0.291	0.308	0.328	0.343
19	0.266	0.283	0.300	0.321	0.335
20	0.259	0.277	0.293	0.313	0.326
21	0.254	0.271	0.286	0.306	0.319
22	0.248	0.265	0.280	0.299	0.312
23	0.243	0.259	0.275	0.294	0.307
24	0.238	0.254	0.269	0.288	0.300
25	0.233	0.249	0.264	0.282	0.294
26	0.229	0.244	0.259	0.276	0.289
27	0.224	0.239	0.253	0.270	0.283
28	0.220	0.235	0.248	0.265	0.276
29	0.216	0.231	0.244	0.260	0.272
30	0.213	0.227	0.240	0.256	0.268
31	0.210	0.224	0.237	0.253	0.264
32	0.207	0.221	0.234	0.250	0.261
33	0.204	0.217	0.230	0.246	0.257
34	0.201	0.215	0.228	0.243	0.254
35	0.199	0.212	0.225	0.240	0.251

Table A.7 (Continued)
Critical values for Lilliefors' goodness-of-fit test.

m	0.2	0.15	0.1	0.08	0.05	0.03
36	0.143	0.151	0.161	0.166	0.177	0.188
37	0.141	0.149	0.159	0.164	0.175	0.186
38	0.139	0.147	0.157	0.162	0.173	0.183
39	0.138	0.145	0.155	0.160	0.170	0.181
40	0.136	0.143	0.153	0.158	0.169	0.179
41	0.134	0.142	0.151	0.156	0.167	0.177
42	0.133	0.140	0.150	0.155	0.165	0.175
43	0.131	0.138	0.148	0.153	0.163	0.173
44	0.130	0.137	0.146	0.151	0.161	0.171
45	0.128	0.135	0.145	0.150	0.159	0.169
46	0.127	0.134	0.143	0.148	0.158	0.168
47	0.126	0.133	0.142	0.147	0.156	0.166
48	0.124	0.131	0.140	0.145	0.155	0.165
49	0.123	0.130	0.139	0.144	0.153	0.163
50	0.122	0.129	0.138	0.142	0.152	0.161
51	0.121	0.127	0.136	0.141	0.150	0.160
52	0.120	0.126	0.135	0.140	0.149	0.158
53	0.119	0.125	0.134	0.138	0.148	0.157
54	0.117	0.124	0.133	0.137	0.146	0.156
55	0.116	0.123	0.132	0.136	0.145	0.154
56	0.115	0.122	0.130	0.135	0.144	0.153
57	0.115	0.121	0.129	0.134	0.142	0.151
58	0.114	0.120	0.128	0.133	0.141	0.150
59	0.113	0.119	0.127	0.131	0.140	0.149
60	0.112	0.118	0.126	0.130	0.139	0.148
61	0.111	0.117	0.125	0.129	0.138	0.147
62	0.110	0.116	0.124	0.128	0.137	0.146
63	0.109	0.115	0.123	0.127	0.136	0.144
64	0.108	0.114	0.122	0.126	0.135	0.143
65	0.107	0.113	0.121	0.125	0.134	0.142
66	0.107	0.113	0.120	0.124	0.133	0.141
67	0.106	0.112	0.120	0.124	0.132	0.140
68	0.105	0.111	0.119	0.123	0.131	0.139
69	0.104	0.110	0.118	0.122	0.130	0.138
70	0.104	0.109	0.117	0.121	0.129	0.137

m	**0.02**	**0.01**	**0.005**	**0.002**	**0.001**
36	0.196	0.210	0.222	0.238	0.248
37	0.194	0.207	0.220	0.235	0.246
38	0.192	0.205	0.217	0.233	0.244
39	0.189	0.202	0.214	0.229	0.240
40	0.187	0.200	0.213	0.227	0.238
41	0.185	0.198	0.210	0.225	0.235
42	0.183	0.196	0.208	0.223	0.233
43	0.181	0.194	0.205	0.220	0.230
44	0.179	0.192	0.203	0.218	0.228
45	0.177	0.189	0.201	0.215	0.225
46	0.175	0.188	0.199	0.213	0.223
47	0.174	0.186	0.197	0.211	0.221
48	0.172	0.184	0.195	0.209	0.219
49	0.170	0.182	0.193	0.207	0.217
50	0.169	0.180	0.191	0.205	0.215
51	0.167	0.179	0.190	0.203	0.213
52	0.166	0.177	0.188	0.202	0.211
53	0.164	0.176	0.186	0.200	0.209
54	0.163	0.174	0.184	0.198	0.207
55	0.161	0.172	0.183	0.196	0.205
56	0.160	0.171	0.181	0.194	0.203
57	0.158	0.169	0.180	0.192	0.201
58	0.157	0.168	0.178	0.191	0.200
59	0.156	0.167	0.177	0.189	0.198
60	0.155	0.165	0.176	0.188	0.196
61	0.153	0.164	0.174	0.186	0.195
62	0.152	0.163	0.172	0.185	0.193
63	0.151	0.161	0.171	0.183	0.192
64	0.150	0.160	0.170	0.182	0.190
65	0.149	0.159	0.169	0.181	0.189
66	0.147	0.158	0.167	0.179	0.187
67	0.146	0.157	0.166	0.178	0.186
68	0.145	0.155	0.165	0.176	0.184
69	0.144	0.154	0.164	0.175	0.183
70	0.143	0.153	0.163	0.174	0.183

Table A.7 (Continued)
Critical values for Lilliefors' goodness-of-fit test.

m	**0.2**	**0.15**	**0.1**	**0.08**	**0.05**	**0.03**
71	0.103	0.109	0.116	0.120	0.128	0.136
72	0.102	0.108	0.115	0.119	0.127	0.135
73	0.102	0.107	0.115	0.118	0.126	0.134
74	0.101	0.106	0.114	0.118	0.125	0.133
75	0.100	0.106	0.113	0.117	0.125	0.132
76	0.0996	0.105	0.112	0.116	0.124	0.132
77	0.0990	0.104	0.112	0.115	0.123	0.131
78	0.0984	0.104	0.111	0.115	0.122	0.130
79	0.0978	0.103	0.110	0.114	0.122	0.129
80	0.0972	0.103	0.110	0.113	0.121	0.128
81	0.0965	0.102	0.109	0.113	0.120	0.127
82	0.0960	0.101	0.108	0.112	0.119	0.127
83	0.0954	0.101	0.108	0.111	0.119	0.126
84	0.0948	0.100	0.107	0.111	0.118	0.125
85	0.0943	0.0995	0.106	0.110	0.117	0.125
86	0.0938	0.0990	0.106	0.109	0.116	0.124
87	0.0932	0.0984	0.105	0.109	0.116	0.123
88	0.0927	0.0979	0.105	0.108	0.115	0.122
89	0.0922	0.0973	0.104	0.108	0.115	0.122
90	0.0917	0.0968	0.103	0.107	0.114	0.121
91	0.0912	0.0962	0.103	0.106	0.113	0.120
92	0.0907	0.0957	0.102	0.106	0.113	0.120
93	0.0903	0.0953	0.102	0.105	0.112	0.119
94	0.0898	0.0947	0.101	0.105	0.112	0.118
95	0.0893	0.0942	0.101	0.104	0.111	0.118
96	0.0889	0.0937	0.100	0.104	0.110	0.117
97	0.0884	0.0933	0.0997	0.103	0.110	0.117
98	0.0880	0.0929	0.0992	0.103	0.109	0.116
99	0.0875	0.0923	0.0986	0.102	0.109	0.115
100	0.0871	0.0919	0.0982	0.102	0.108	0.115

m	0.02	0.01	0.005	0.002	0.001
71	0.142	0.152	0.162	0.173	0.181
72	0.141	0.151	0.161	0.172	0.180
73	0.140	0.150	0.159	0.170	0.178
74	0.139	0.149	0.158	0.169	0.177
75	0.138	0.148	0.157	0.168	0.176
76	0.138	0.147	0.156	0.167	0.175
77	0.137	0.146	0.155	0.166	0.174
78	0.136	0.145	0.154	0.165	0.172
79	0.135	0.144	0.153	0.164	0.171
80	0.134	0.143	0.152	0.163	0.170
81	0.133	0.142	0.151	0.162	0.169
82	0.133	0.142	0.150	0.161	0.168
83	0.132	0.141	0.149	0.160	0.167
84	0.131	0.140	0.148	0.159	0.166
85	0.130	0.139	0.148	0.158	0.165
86	0.129	0.138	0.147	0.157	0.164
87	0.129	0.137	0.146	0.156	0.163
88	0.128	0.137	0.145	0.155	0.162
89	0.127	0.136	0.144	0.154	0.161
90	0.126	0.135	0.144	0.154	0.161
91	0.126	0.134	0.143	0.152	0.160
92	0.125	0.134	0.142	0.152	0.159
93	0.124	0.133	0.141	0.151	0.158
94	0.124	0.132	0.140	0.150	0.157
95	0.123	0.132	0.140	0.149	0.156
96	0.122	0.131	0.139	0.148	0.155
97	0.122	0.130	0.138	0.148	0.155
98	0.121	0.130	0.137	0.147	0.153
99	0.121	0.129	0.137	0.146	0.153
100	0.120	0.128	0.136	0.146	0.152

References

1. Abramowitz, M. and Stegun, I. A., Eds. (1972) *Handbook of Mathematical Functions with Formulas, Graphs, and Mathematical Tables*, Ninth Printing, Dover: New York.

2. Agustin, Z. and Peña, E. A. (1999) "Order Statistic Properties, Random Generation, and Goodness-of-Fit Testing for a Minimal Repair Model," *Journal of the American Statistical Association*, **94**, pp. 266–272.

3. ANSI/IEC/ASQ D61164-1997 (1997) *Reliability Growth—Statistical Test and Estimation Methods*, American Society for Quality: Milwaukee.

4. Ascher, H. (1987) "MIL-STD-781C: A Vicious Circle," *IEEE Transactions on Reliability*, **R-36**, pp. 397–402.

5. Ascher, H. and Feingold, H. (1984) *Repairable Systems Reliability: Modeling, Inference, Misconceptions and Their Causes*, Marcel Dekker: New York.

6. Bain, L. J. and Engelhardt, M. (1980). "Inferences on the Parameters and Current System Reliability for a Time Truncated Weibull Process." *Technometrics* **22**, pp. 421–426.

7. Bain, L. J. and Engelhardt, M. (1982) "Sequential Probability Ratio Tests for the Shape Parameter of a Nonhomogeneous Poisson Process," *IEEE Transactions on Reliability*, **31**, pp. 79–83.

8. Bain, L. J. and Engelhardt, M. (1991) *Statistical Analysis of Reliability and Life Testing Models: Theory and Methods*, Second Edition, Marcel-Dekker: New York.

9. Bain, L. J. and Engelhardt, M. (1992) *Introduction to Probability and Mathematical Statistics*, Second Edition, PWS-Kent: Boston.

10. Bain, L. J., Engelhardt, M. and Wright, F. T. (1985) "Tests for an Increasing Trend in the Intensity of a Poisson Process: A Power Study," *Journal of the American Statistical Association*, **80**, pp. 419–422.

11. Baker, R. D. (1996) "Some New Tests of the Power Law Process," *Technometrics*, **38**, pp. 256–265.

12. Balaban, H. S. (1978) "Reliability Growth Models," *Journal of Environmental Sciences*, **21**, pp. 11–18.

13. Bar-Lev, S. Lavi, I. and Reiser, B. (1992) "Bayesian Inference for the Power Law Process," *Annals of the Institute of Statistical Mathematics*, **44**, pp. 632–639.

14. Barlow, R. E., Bartholomew, D. J., Bremner, J. M., and Brunk, H. D. (1972) *Statistical Inference under Order Restrictions*, Wiley: New York.

15. Barlow, R. E. and Campo, R. (1975) "Total Time on Test Processes and Applications to Failure Data Analysis," *Reliability and Fault Tree Analysis*, Barlow, Fussel and Singpurwalla, Editors, SIAM: Philadelphia, pp. 451–481.

16. Bartlett, M. S. (1937) "Properties of Sufficiency and Statistical Tests," *Proceedings of the Royal Society of London A*, **160**, pp. 268–282.

17. Bartlett, M. S. (1963) "The Spectral Analysis of Point Processes," *Journal of the Royal Statistical Society, Series B*, **25**, pp. 264–296.

18. Basu, A. P. and Rigdon, S. E. (1984) "Examples of Parametric Empirical Bayes Methods for the Estimation of Failure Processes for Repairable Systems," in *Reliability and Quality Control*, Ed. A. P. Basu, Elsevier: North Holland, pp. 47–55.

19. Beiser, J. A. and Rigdon, S. E. (1992) "Bayesian Prediction for the Number of Failures of a Repairable System Modeled by a Homogeneous Poisson Process," In *Modeling and Simulation Vol 23, Part 4: Control, Signal Processing, Robotics, Systems, Power*, Eds. W. Vogt and M. H. Mickle, pp. 1783–1789.

20. Beiser, J. A. and Rigdon, S. E. (1997) "Bayes Prediction for the Number of Failures of a Repairable System," *IEEE Transactions on Reliability*, **46**, pp. 291–295.

21. Berman, M. (1981) "Inhomogeneous and Modulated Gamma Processes," *Biometrika*, **68**, pp. 143–152.

22. Bernardo, J. M. (1976), "Algorithm 103, Psi (Digamma Function)," *Journal of the Royal Statistics Society, Series C*, **25**, pp. 315–317.

23. Black, S. E. and Rigdon, S. E. (1996) "Statistical Inference for a Modulated Power Law Process," *Journal of Quality Technology*," **28**, pp. 81–90.

24. Block, H. W., Borges, W. S. and Savits, T. H. (1985) "Age-Dependent Minimal Repair," *Journal of Applied Probability*, **22**, pp. 390–385.

25. Boswell, M. T. (1966) "Estimating and Testing Trend in a Stochastic Process of Poisson Type," *Annals of Mathematical Statistics*, **37**, pp. 1564–1573.

26. Box, G. E. P. and Jenkins, G. M. (1976) *Time Series Analysis: Forecasting and Control*, Revised Edition, Holden Day: Oakland.

27. Brown, M. and Proschan, F. (1983) "Imperfect Repair," *Journal of Applied Probability*, **20**, pp. 851–859.

28. Buchanan, J. T. and Turner, P. R. (1992) *Numerical Methods and Analysis*, McGraw-Hill: New York.

29. Butler, J. P. and Rees, D. A. (1974) "Exploration of Statistical Fatigue Failure Characteristics of 0.063-Inch Mill Annealed Ti6A1-4V Sheet and 0.050-Inch Heat-Treated 17-7 Ph Steel Sheet Under Simulated Flight-by-Flight Loading." Technical Report AFML-TR-74-269, Air Force Material Command, Wright-Patterson Air Force Base, Ohio.

30. Buynoski, M. S. (1990) "Comment on Bayes Is for the Birds," *IEEE Transactions on Reliability*, **R39**, p. 3.

31. Calabria, R., Guida, M. and Pulcini, G. (1990) "Bayes Estimation of Prediction Intervals for a Power Law Process," *Communications in Statistics: Theory and Methods*, **19**, pp. 3023–3035.

32. Calabria, R., Guida, M., and Pulcini, G. (1992) "A Bayes Procedure for Estimation of Current System Reliability," *IEEE Transactions on Reliability*, **41**, pp. 616–619. [Correction: Calabria, R. et al. (1993) *IEEE Transactions on Reliability*, **42**, p. 326.]

33. Carlin, B. P. and Gelfand, A. E. (1990) "Approaches for Empirical Bayes Confidence Intervals," *Journal of the American Statistical Association*, **85**, pp. 105–114.

34. Cox, D. R. (1972) "The Statistical Analysis of Dependencies in Point Processes," in *Stochastic Point Processes: Statistical Analysis, Theory, and Applications*, P. A. W. Lewis, Editor. Wiley: New York.

35. Cox, D. R. and Lewis, P. A. W. (1966) *The Statistical Analysis of Series of Events*, Chapman and Hall: London

36. Cox, D. R. and Miller, H. D. (1965) *The Theory of Stochastic Processes*, Methuen: London.

37. Crow, L. R. (1974) "Reliability Analysis for Complex Systems," *Reliability and Biometry*, Eds. F. Proschan and R. J. Serfling, pp. 379–410.

38. Crow, L. R. (1982) "Confidence Interval Procedures for the Weibull Process with Applications to Reliability Growth." *Technometrics* **24**, pp. 67–72.

39. Crow, L. R. (1990) "Evaluating the Reliability of Repairable Systems," *1990 Annual Reliability and Maintainability Symposium*, pp. 275–279.

40. Davis, D. (1952) "An Analysis of Some Failure Data," *Journal of the American Statistical Association*, **47**, 113–134.

41. Dempster, A. P., Laird, N. and Rubin, D. B. (1977) "Maximum Likelihood Estimation from Incomplete Data Via the EM Algorithm, with Discussion," *Journal of the Royal Statistical Society, Series B*, **39**, pp. 1–38.

42. Doganaksoy, N. and Nelson, W. (1998) "A Method to Compare Two Samples of Recurrence Data," *Lifetime Data Analysis*, **4**, pp. 51–63.

43. Dougherty, E. (1991) "Bayesianism Is for Turkeys," *IEEE Transactions on Reliability*, **40**, pp. 3–6.

44. Duane, J. T. (1964) "Learning Curve Approach to Reliability Monitoring," *IEEE Transactions on Aerospace*, **2**, pp. 563–566.

45. Easterling, R. G. (1972) "A Personal View of the Bayesian Controversy in Reliability and Statistics," *IEEE Transactions on Reliability*, **R-21**, pp. 186–194.

46. Ebrahimi, N. (1989) "How to Define System Improvement and Deterioration for a Repairable System," *IEEE Transactions on Reliability*, **R-38**, pp. 214–217.

47. Ebrahimi, N. (1996) "How to Model Reliability-Growth When Times of Design Modifications Are Known," *IEEE Transactions on Reliability*, **45**, pp. 54–58.

48. Engelhardt, M. (1995) "Models and Analyses for the Reliability of a Single Repairable System," in *Recent Advances in Life-Testing and Reliability*, Ed. N. Balakrishnan, CRC Press: Boca Raton, FL, pp. 79–106.

49. Engelhardt, M. and Bain, L. J. (1978) "Prediction Intervals for the Weibull Process," *Technometrics*, **20**, pp. 167–169.

50. Engelhardt, M. and Bain, L. J. (1987) "Statistical Analysis of a Compound Power-Law Model for Repairable Systems," *IEEE Transactions on Reliability*, **R-36**, pp. 392–396.

51. Engelhardt, M. and Bain, L. J. (1992) "Statistical Analysis of a Weibull Process with Left-Censored Data," *Proceedings of the NATO Advanced Studies Workshop on Survival Analysis and Related Topics,* Eds. J. P. Klein and P. Goel, NATO ASI, Series E: Applied Sciences, **221**, Kluwer: Dordrecht, The Netherlands.

52. Engelhardt, M., Williams, D. H., and Bain, L. J. (1993) "Statistical Analysis of a Power-Law Process with Left-Truncated Data," In *Advances in Reliability*, Ed. A. P. Basu, Elsevier: Dordrecht, The Netherlands, pp. 105–121.

53. Evans, R. A. (1989) "Bayes Is for the Birds," *IEEE Transactions on Reliability*, **38**, p. 401.

54. Feller, W. (1971) *An Introduction to Probability Theory and Its Applications, Volume II*, Second Edition. Wiley: New York.

55. Finkelstein, J. M. (1976) "Confidence Bounds on the Parameters of the Weibull Process," *Technometrics*, **18**, pp. 115–117.

56. Fries, A. and Sen, A. (1996) "A Survey of Discrete Reliability-Growth Models," *IEEE Transactions on Reliability*, **45**, pp. 582–604.

57. Gaver, D. P. and O'Muircheartaigh, I. G. (1987) "Robust Empirical Bayes Analyses of Event Rates," *Technometrics*, **29**, pp. 1–15.

58. Goel, A. L. and Okumoto, K. (1970) "Time-Dependent Error-Detection Rate Model for Software Reliability and Other Performance Models," *IEEE Transactions on Reliability*, **28**, pp. 206–211.

59. Greenwood, J. A. and Durand, D. (1960) "Aids for Fitting the Gamma Distribution by Maximum Likelihood," *Technometrics*, **2**, pp. 55–65.

60. Guida, M. Calabria, R. and Pulcini, G. (1989) "Bayes Inference for a Nonhomogeneous Poisson Process with Power Intensity Law," *IEEE Transactions on Reliability*, **38**, pp. 603–609.

61. Hand, D. J. et al., (1994) *A Handbook of Small Data Sets*, Chapman & Hall: London.

62. Hossain, S. A. and Dahiya, R. C. (1993) "Estimating the Parameters of a Non-Homogeneous Poisson-Process Model for Software Reliability," *IEEE Transactions on Reliability*, **42**, pp. 604–612.

63. Jang, J. S. and Bai, D. S. (1987) "Efficient Sequential Estimation in a Nonhomogeneous Poisson Process," *IEEE Transactions on Reliability*, **36**, pp. 255–258.

64. Kaplan, S. (1990) "Bayes Is for Eagles," *IEEE Transactions on Reliability*, **39**, pp. 130–131.

65. Kreider, D. L., Kuller, R. G., Ostberg, D. R., and Perkins, F. W. (1966) *Introduction to Linear Analysis*, Addison-Wesley: Reading, MA.

66. Lakey, M. J. and Rigdon, S. E. (1992) "The Modulated Power Law Process," *Transactions of the 46th Annual Quality Congress*, pp. 559–563.

67. Lawless, J. F. and Nadeau, C. (1995) "Some Simple Robust Methods for the Analysis of Recurrent Events," *Technometrics*, **37**, pp. 158–168.

68. Lee, L. (1980a) "Testing Adequacy of the Weibull and Log Linear Rate Models for a Poisson Process," *Technometrics*, **22**, pp. 195–199.

69. Lee, L. (1980b) "Comparing Rates of Several Independent Weibull Processes," *Technometrics*, **22**, pp. 427–430.

70. Lee, L. and Lee, S. K. (1978) "Some Results on Inference for the Weibull Process," *Technometrics*, **20**, pp. 41–45.

71. Lewis, P. A. W. (1964) "A Branching Poisson Process Model for the Analysis of Computer Failure Patterns, with Discussion," *Journal of the Royal Statistical Society, Series B*, **26**, pp. 398–456.

72. Lewis, P. A. W. (1967) "A Non-Homogeneous Branching Poisson Process," *Journal of the Royal Statistical Society, Series B*, **29**, pp. 343–354.

73. Lewis, P. A. W., Editor (1972) *Stochastic Point Processes: Statistical Analysis, Theory, and Applications*, Wiley: New York.

74. Lewis, P. A. W. and Shedler, G. S. (1976) "Statistical Analysis of Non-stationary Series of Events in a Data Base System," *IBM Journal of Research and Development,* **20**, pp. 465–482.

75. Lilliefors, H. W. (1969) "On the Kolmogorov-Smirnov Test for the Exponential Distribution with Mean Unknown," *Journal of the American Statistical Association,* **64**, pp. 387–389.

76. Lin, H.-H. and Chen, K.-H. (1993) "Nonhomogeneous Poisson Process Software-Debugging Models with Linear Dependence," *IEEE Transactions on Reliability*, **42**, pp. 613–617.

77. Lloyd, D. K. and Lipow, M. (1962) *Reliability: Management, Methods, and Mathematics*, Prentice-Hall: Englewood Cliffs, NJ.

78. Lucas, J. M. (1985) "Counted Data CUSUM's," *Technometrics*, **27**, pp. 129–144.

79. Martz, H. F. (1975) "Pooling Life Test Data by Means of the Empirical Bayes Method," *IEEE Transactions on Reliability*, **R-24**, pp. 27–30.

80. Martz, H. F. and Waller, R. A. (1982) *Bayesian Reliability Analysis*, Wiley: New York.

81. MIL-HDBK-781 (1997) *Handbook for Reliability Test Methods, Plans, and Environments for Engineering, Development Qualification, and Production*, United States Government Printing Office: Washington, DC.

82. MIL-STD-781C (1977) *Reliability Design Qualification and Production Acceptance Tests: Exponential Distribution*, United States Government Printing Office: Washington, DC.

83. Molitor, J. T. and Rigdon, S. E. (1993) "A Comparison of Estimators of the Shape Parameter of the Power Law Process," *Proceedings of the Section on Physical and Engineering Sciences*, American Statistical Association: Alexandria, VA, pp. 143–147.

84. Møller, S. K. (1976) "The Rasch-Weibull Process," *Scandinavian Journal of Statistics*, **3**, pp. 107–115.

85. Morris, C. N. (1983a) "Parametric Empirical Bayes Inference: Theory and Applications, with Discussion," *Journal of the American Statistical Association*, **78**, pp. 47–65.

86. Morris, C. N. (1983b) "Parametric Empirical Bayes Confidence Intervals," in *Scientific Inference, Data Analysis, and Robustness*, Academic Press: New York, pp. 25–50.

87. Musa, J. D. (1979) *Software Reliability Data*, Data Analysis Center for Software, Rome Air Development Center, Rome, NY.

88. Nelder, J. A. and Mead, R. (1965) "A Simplex Method for Function Minimization," *Computer Journal*, **7**, pp. 308–313.

89. Nelson, W. (1995) "Confidence Limits for the Recurrence Data – Applied to Cost or Number of Product Repairs," *Technometrics*, **37**, pp. 147–157.

90. Newby, M. J. (1991) "Repairable Systems and Growth Models," in *Bayesian Methods in Reliability*, Eds. P. Sander and R. Badoux, Kluwer: Dordrecht, pp. 101–117.

91. Pang, C.-F., Thompson, W. A. Jr., and Fallahi, H. (1989) "Maximum Likelihood Estimation of Convex Arrival Rate," *Biometrika*, **76**, pp. 626–628.

92. Park, W. J. and Kim, Y. G. (1992) "Goodness-of-Fit Tests for the Power-Law Process," *IEEE Transactions on Reliability*, **41**, pp. 107–111.

93. Park, W. J. and Seoh, M. (1994) "More Goodness-of-Fit Tests for the Power-Law Process," *IEEE Transactions on Reliability*, **43**, pp. 275–278.

94. Pearson, E. S. and Hartley, H. O. (1966) *Biometrika Tables for Statisticians*, Vol. 1, 3rd Edition, Cambridge University Press: Cambridge.

95. Presnell, B., Hollander, M., and Sethurman, J. (1994) "Testing the Minimal Repair Assumption in an Imperfect Repair Model," *Journal of the American Statistical Association*, **89**, pp. 289–297.

96. Proschan, F. (1963) "Theoretical Explanation of Observed Decreasing Failure Rate," *Technometrics*, **18**, pp. 375–383.

97. Rigdon, S. E. (1989) "Testing Goodness-of-Fit for the Power Law Process," *Communications in Statistics*, **A-18**, pp. 4665–4676.

98. Rigdon, S. E. (1997) "Tables for Obtaining Confidence Intervals for the Intensity Function of the Power Law Process," Unpublished manuscript.

99. Rigdon, S. E. and Basu, A. P. (1988) " Estimating the Intensity Function of a Weibull Poisson Process at the Current Time: Failure Truncated Case," *Journal of Statistical Computation and Simulation*, **30**, pp. 17–38.

100. Rigdon, S. E. and Basu, A. P. (1989a) "The Power Law Process: A Model for the Reliability of Repairable Systems," *Journal of Quality Technology*, **10**, pp. 251–260.

101. Rigdon, S. E. and Basu, A. P. (1989b) "Mean Squared Errors of Estimators of the Intensity Function of a Nonhomogeneous Poisson Process," *Statistics and Probability Letters*, **8**, pp. 445–449.

102. Rigdon, S. E. and Basu, S. E. (1990a) "The Effect of Assuming a Homogeneous Poisson Process when the True Process Is a Power Law Process," *Journal of Quality Technology*, **11**, pp. 111–117.

103. Rigdon, S. E. and Basu, A. P. (1990b) "Estimating the Intensity Function of a Power Law Process at the Current Time: Time Truncated Case," *Communications in Statistics: Computation and Simulation*, **B-19**, pp. 1079–1104.

104. Robbins, H. (1955) "An Empirical Bayes Approach to Statistics," *Proceedings of the Third Berkeley Symposium on Probability and Mathematical Statistics*, **1**, pp. 157–163.

105. Rosenblatt, M. (1956) "Remarks on Some Non-Parametric Estimates of a Density Function," *Annals of Mathematical Statistics*, **27**, pp. 832–837.

106. Sen, A. (1998) "Estimation of Current Reliability in a Duane-Based Reliability Growth Model, *Technometrics*, **40**, pp. 334–344.

107. Sen, A. and Bhattacharyya, G. K. (1993) "A Piecewise Exponential Model for Reliability Growth and Associated Inferences," in *Advances in Reliability*, Ed. A. P. Basu, Elsevier: North Holland, pp. 331–355.

108. Schneider, B. E. (1978) "Algorithm AS 121, Tri-Gamma Function," *Journal of the Royal Statistical, Series C*, **27**, pp. 97–99.

109. Shapiro, S. S. and Wilk, M. B. (1972) "An Analysis of Variance Test for the Exponential Distribution," *Technometrics*, **14**, pp. 355–370.

110. Singpurwalla, N. (1978) "Estimating Reliability Growth (or Deterioration) Using Time Series Analysis." *Naval Research Logistics Quarterly*, **25**, pp. 1–14.

111. Stephens, M. A. (1974) "EDF Statistics for Goodness of Fit and Some Comparisons." *Journal of the American Statistical Association* **69**, pp. 730–737.

112. Stephens, M. A. (1978) "On the W Test for Exponentiality with Origin Known," *Technometrics*, **20**, pp. 33–35.

113. Thompson, W. A. Jr.(1981) "On the Foundations of Reliability," *Technometrics*, **23**, pp. 1–14.

114. Thompson, W. A., Jr. (1988) *Point Processes Models with Applications to Safety and Reliability*, Chapman and Hall: London.

115. van Pul, M. C. J. (1993) *Statistical Analysis of Software Reliability Models*, Stichting Mathematisch Centrum: Amsterdam.

116. Whitaker, L. R. and Samaniego, F. J. (1989) "Estimating the Reliability of Systems Subject to Imperfect Repair," *Journal of the American Statistical Association*, pp. 301–309.

117. Wolfram, S. (1996) *The Mathematica Book, Third Edition,* Wolfram: Cambridge.

Index

F

G

H

I

K

L

M

N

P

R

S

T

U

W

WILEY SERIES IN PROBABILITY AND STATISTICS
ESTABLISHED BY WALTER A. SHEWHART AND SAMUEL S. WILKS

Probability and Statistics Section

*ANDERSON · The Statistical Analysis of Time Series
ARNOLD, BALAKRISHNAN, and NAGARAJA · A First Course in Order Statistics
ARNOLD, BALAKRISHNAN, and NAGARAJA · Records
BACCELLI, COHEN, OLSDER, and QUADRAT · Synchronization and Linearity: An Algebra for Discrete Event Systems
BARNETT · Comparative Statistical Inference, *Third Edition*
BASILEVSKY · Statistical Factor Analysis and Related Methods: Theory and Applications
BERNARDO and SMITH · Bayesian Statistical Concepts and Theory
BILLINGSLEY · Convergence of Probability Measures, *Second Edition*
BOROVKOV · Asymptotic Methods in Queuing Theory
BOROVKOV · Ergodicity and Stability of Stochastic Processes
BRANDT, FRANKEN, and LISEK · Stationary Stochastic Models
CAINES · Linear Stochastic Systems
CAIROLI and DALANG · Sequential Stochastic Optimization
CONSTANTINE · Combinatorial Theory and Statistical Design
COOK · Regression Graphics
COVER and THOMAS · Elements of Information Theory
CSÖRGŐ and HORVÁTH · Weighted Approximations in Probability Statistics
CSÖRGŐ and HORVÁTH · Limit Theorems in Change Point Analysis
*DANIEL · Fitting Equations to Data: Computer Analysis of Multifactor Data, *Second Edition*
DETTE and STUDDEN · The Theory of Canonical Moments with Applications in Statistics, Probability, and Analysis
DEY and MUKERJEE · Fractional Factorial Plans
*DOOB · Stochastic Processes
DRYDEN and MARDIA · Statistical Shape Analysis
DUPUIS and ELLIS · A Weak Convergence Approach to the Theory of Large Deviations
ETHIER and KURTZ · Markov Processes: Characterization and Convergence
FELLER · An Introduction to Probability Theory and Its Applications, Volume 1, *Third Edition,* Revised; Volume II, *Second Edition*
FULLER · Introduction to Statistical Time Series, *Second Edition*
FULLER · Measurement Error Models
GHOSH, MUKHOPADHYAY, and SEN · Sequential Estimation
GIFI · Nonlinear Multivariate Analysis
GUTTORP · Statistical Inference for Branching Processes
HALL · Introduction to the Theory of Coverage Processes
HAMPEL · Robust Statistics: The Approach Based on Influence Functions
HANNAN and DEISTLER · The Statistical Theory of Linear Systems
HUBER · Robust Statistics

*Now available in a lower priced paperback edition in the Wiley Classics Library.

Probability and Statistics (Continued)

HUSKOVA, BERAN, and DUPAC · Collected Works of Jaroslav Hajek—with Commentary
IMAN and CONOVER · A Modern Approach to Statistics
JUREK and MASON · Operator-Limit Distributions in Probability Theory
KASS and VOS · Geometrical Foundations of Asymptotic Inference
KAUFMAN and ROUSSEEUW · Finding Groups in Data: An Introduction to Cluster Analysis
KELLY · Probability, Statistics, and Optimization
KENDALL, BARDEN, CARNE, and LE · Shape and Shape Theory
LINDVALL · Lectures on the Coupling Method
McFADDEN · Management of Data in Clinical Trials
MANTON, WOODBURY, and TOLLEY · Statistical Applications Using Fuzzy Sets
MORGENTHALER and TUKEY · Configural Polysampling: A Route to Practical Robustness
MUIRHEAD · Aspects of Multivariate Statistical Theory
OLIVER and SMITH · Influence Diagrams, Belief Nets and Decision Analysis
*PARZEN · Modern Probability Theory and Its Applications
PRESS · Bayesian Statistics: Principles, Models, and Applications
PUKELSHEIM · Optimal Experimental Design
RAO · Asymptotic Theory of Statistical Inference
RAO · Linear Statistical Inference and Its Applications, *Second Edition*
RAO and SHANBHAG · Choquet-Deny Type Functional Equations with Applications to Stochastic Models
ROBERTSON, WRIGHT, and DYKSTRA · Order Restricted Statistical Inference
ROGERS and WILLIAMS · Diffusions, Markov Processes, and Martingales, Volume I: Foundations, *Second Edition:* Volume II: Îto Calculus
RUBINSTEIN and SHAPIRO · Discrete Event Systems: Sensitivity Analysis and Stochastic Optimization by the Score Function Method
RUZSA and SZEKELY · Algebraic Probability Theory
SCHEFFE · The Analysis of Variance
SEBER · Linear Regression Analysis
SEBER · Multivariate Observations
SEBER and WILD · Nonlinear Regression
SERFLING · Approximation Theorems of Mathematical Statistics
SHORACK and WELLNER · Empirical Processes with Applications to Statistics
SMALL and McLEISH · Hilbert Space Methods in Probability and Statistical Inference
STAPLETON · Linear Statistical Models
STAUDTE and SHEATHER · Robust Estimation and Testing
STOYANOV · Counterexamples in Probability
TANAKA · Time Series Analysis: Nonstationary and Noninvertible Distribution Theory
THOMPSON and SEBER · Adaptive Sampling
WELSH · Aspects of Statistical Inference
WHITTAKER · Graphical Models in Applied Multivariate Statistics
YANG · The Construction Theory of Denumerable Markov Processes

Applied Probability and Statistics Section

ABRAHAM and LEDOLTER · Statistical Methods for Forecasting
AGRESTI · Analysis of Ordinal Categorical Data
AGRESTI · Categorical Data Analysis

*Now available in a lower priced paperback edition in the Wiley Classics Library.

Applied Probability and Statistics (Continued)

ANDERSON, AUQUIER, HAUCK, OAKES, VANDAELE, and WEISBERG · Statistical Methods for Comparative Studies
ARMITAGE and DAVID (editors) · Advances in Biometry
*ARTHANARI and DODGE · Mathematical Programming in Statistics
ASMUSSEN · Applied Probability and Queues
*BAILEY · The Elements of Stochastic Processes with Applications to the Natural Sciences
BARNETT and LEWIS · Outliers in Statistical Data, *Third Edition*
BARTHOLOMEW, FORBES, and McLEAN · Statistical Techniques for Manpower Planning, *Second Edition*
BASU and RIGDON · Statistical Methods for the Reliability of Repairable Systems
BATES and WATTS · Nonlinear Regression Analysis and Its Applications
BECHHOFER, SANTNER, and GOLDSMAN · Design and Analysis of Experiments for Statistical Selection, Screening, and Multiple Comparisons
BELSLEY · Conditioning Diagnostics: Collinearity and Weak Data in Regression
BELSLEY, KUH, and WELSCH · Regression Diagnostics: Identifying Influential Data and Sources of Collinearity
BHAT · Elements of Applied Stochastic Processes, *Second Edition*
BHATTACHARYA and WAYMIRE · Stochastic Processes with Applications
BIRKES and DODGE · Alternative Methods of Regression
BLISCHKE AND MURTHY · Reliability: Modeling, Prediction, and Optimization
BLOOMFIELD · Fourier Analysis of Time Series: An Introduction, *Second Edition*
BOLLEN · Structural Equations with Latent Variables
BOULEAU · Numerical Methods for Stochastic Processes
BOX · Bayesian Inference in Statistical Analysis
BOX and DRAPER · Empirical Model-Building and Response Surfaces
*BOX and DRAPER · Evolutionary Operation: A Statistical Method for Process Improvement
BUCKLEW · Large Deviation Techniques in Decision, Simulation, and Estimation
BUNKE and BUNKE · Nonlinear Regression, Functional Relations and Robust Methods: Statistical Methods of Model Building
CHATTERJEE and HADI · Sensitivity Analysis in Linear Regression
CHERNICK · Bootstrap Methods: A Practitioner's Guide
CHILÈS and DELFINER · Geostatistics: Modeling Spatial Uncertainty
CHOW and LIU · Design and Analysis of Clinical Trials: Concepts and Methodologies
CLARKE and DISNEY · Probability and Random Processes: A First Course with Applications, *Second Edition*
*COCHRAN and COX · Experimental Designs, *Second Edition*
CONOVER · Practical Nonparametric Statistics, *Second Edition*
CORNELL · Experiments with Mixtures, Designs, Models, and the Analysis of Mixture Data, *Second Edition*
*COX · Planning of Experiments
CRESSIE · Statistics for Spatial Data, *Revised Edition*
DANIEL · Applications of Statistics to Industrial Experimentation
DANIEL · Biostatistics: A Foundation for Analysis in the Health Sciences, *Sixth Edition*
DAVID · Order Statistics, *Second Edition*
*DEGROOT, FIENBERG, and KADANE · Statistics and the Law
DODGE · Alternative Methods of Regression
DOWDY and WEARDEN · Statistics for Research, *Second Edition*
DUNN and CLARK · Applied Statistics: Analysis of Variance and Regression, *Second Edition*
*ELANDT-JOHNSON and JOHNSON · Survival Models and Data Analysis
EVANS, PEACOCK, and HASTINGS · Statistical Distributions, *Second Edition*

*Now available in a lower priced paperback edition in the Wiley Classics Library.

Applied Probability and Statistics (Continued)

*FLEISS · The Design and Analysis of Clinical Experiments
FLEISS · Statistical Methods for Rates and Proportions, *Second Edition*
FLEMING and HARRINGTON · Counting Processes and Survival Analysis
GALLANT · Nonlinear Statistical Models
GLASSERMAN and YAO · Monotone Structure in Discrete-Event Systems
GNANADESIKAN · Methods for Statistical Data Analysis of Multivariate Observations, *Second Edition*
GOLDSTEIN and LEWIS · Assessment: Problems, Development, and Statistical Issues
GREENWOOD and NIKULIN · A Guide to Chi-Squared Testing
*HAHN · Statistical Models in Engineering
HAHN and MEEKER · Statistical Intervals: A Guide for Practitioners
HAND · Construction and Assessment of Classification Rules
HAND · Discrimination and Classification
HEIBERGER · Computation for the Analysis of Designed Experiments
HEDAYAT and SINHA · Design and Inference in Finite Population Sampling
HINKELMAN and KEMPTHORNE: · Design and Analysis of Experiments, Volume 1: Introduction to Experimental Design
HOAGLIN, MOSTELLER, and TUKEY · Exploratory Approach to Analysis of Variance
HOAGLIN, MOSTELLER, and TUKEY · Exploring Data Tables, Trends and Shapes
HOAGLIN, MOSTELLER, and TUKEY · Understanding Robust and Exploratory Data Analysis
HOCHBERG and TAMHANE · Multiple Comparison Procedures
HOCKING · Methods and Applications of Linear Models: Regression and the Analysis of Variables
HOGG and KLUGMAN · Loss Distributions
HOSMER and LEMESHOW · Applied Logistic Regression
HØYLAND and RAUSAND · System Reliability Theory: Models and Statistical Methods
HUBERTY · Applied Discriminant Analysis
JACKSON · A User's Guide to Principle Components
JOHN · Statistical Methods in Engineering and Quality Assurance
JOHNSON · Multivariate Statistical Simulation
JOHNSON and KOTZ · Distributions in Statistics
Continuous Multivariate Distributions
JOHNSON, KOTZ, and BALAKRISHNAN · Continuous Univariate Distributions, Volume 1, *Second Edition*
JOHNSON, KOTZ, and BALAKRISHNAN · Continuous Univariate Distributions, Volume 2, *Second Edition*
JOHNSON, KOTZ, and BALAKRISHNAN · Discrete Multivariate Distributions
JOHNSON, KOTZ, and KEMP · Univariate Discrete Distributions, *Second Edition*
JUREČKOVÁ and SEN · Robust Statistical Procedures: Aymptotics and Interrelations
KADANE · Bayesian Methods and Ethics in a Clinical Trial Design
KADANE AND SCHUM · A Probabilistic Analysis of the Sacco and Vanzetti Evidence
KALBFLEISCH and PRENTICE · The Statistical Analysis of Failure Time Data
KELLY · Reversability and Stochastic Networks
KHURI, MATHEW, and SINHA · Statistical Tests for Mixed Linear Models
KLUGMAN, PANJER, and WILLMOT · Loss Models: From Data to Decisions
KLUGMAN, PANJER, and WILLMOT · Solutions Manual to Accompany Loss Models: From Data to Decisions
KOVALENKO, KUZNETZOV, and PEGG · Mathematical Theory of Reliability of Time-Dependent Systems with Practical Applications
LACHIN · Biostatistical Methods: The Assessment of Relative Risks
LAD · Operational Subjective Statistical Methods: A Mathematical, Philosophical, and Historical Introduction

*Now available in a lower priced paperback edition in the Wiley Classics Library.

Applied Probability and Statistics (Continued)

LANGE, RYAN, BILLARD, BRILLINGER, CONQUEST, and GREENHOUSE
Case Studies in Biometry

LAWLESS · Statistical Models and Methods for Lifetime Data

LEE · Statistical Methods for Survival Data Analysis, *Second Edition*

LePAGE and BILLARD · Exploring the Limits of Bootstrap

LINHART and ZUCCHINI · Model Selection

LITTLE and RUBIN · Statistical Analysis with Missing Data

LLOYD · The Statistical Analysis of Categorical Data

MAGNUS and NEUDECKER · Matrix Differential Calculus with Applications in Statistics and Econometrics, *Revised Edition*

MALLER and ZHOU · Survival Analysis with Long Term Survivors

MANN, SCHAFER, and SINGPURWALLA · Methods for Statistical Analysis of Reliability and Life Data

McLACHLAN and KRISHNAN · The EM Algorithm and Extensions

McLACHLAN · Discriminant Analysis and Statistical Pattern Recognition

McNEIL · Epidemiological Research Methods

MEEKER and ESCOBAR · Statistical Methods for Reliability Data

*MILLER · Survival Analysis, *Second Edition*

MONTGOMERY and PECK · Introduction to Linear Regression Analysis, *Second Edition*

MYERS and MONTGOMERY · Response Surface Methodology: Process and Product in Optimization Using Designed Experiments

NELSON · Accelerated Testing, Statistical Models, Test Plans, and Data Analyses

NELSON · Applied Life Data Analysis

OCHI · Applied Probability and Stochastic Processes in Engineering and Physical Sciences

OKABE, BOOTS, and SUGIHARA · Spatial Tesselations: Concepts and Applications of Voronoi Diagrams

PANKRATZ · Forecasting with Dynamic Regression Models

PANKRATZ · Forecasting with Univariate Box-Jenkins Models: Concepts and Cases

PIANTADOSI · Clinical Trials: A Methodologic Perspective

PORT · Theoretical Probability for Applications

PUTERMAN · Markov Decision Processes: Discrete Stochastic Dynamic Programming

RACHEV · Probability Metrics and the Stability of Stochastic Models

RÉNYI · A Diary on Information Theory

RIPLEY · Spatial Statistics

RIPLEY · Stochastic Simulation

ROLSKI, SCHMIDLI, SCHMIDT, and TEUGELS · Stochastic Processes for Insurance and Finance

ROUSSEEUW and LEROY · Robust Regression and Outlier Detection

RUBIN · Multiple Imputation for Nonresponse in Surveys

RUBINSTEIN · Simulation and the Monte Carlo Method

RUBINSTEIN and MELAMED · Modern Simulation and Modeling

RYAN · Statistical Methods for Quality Improvement, *Second Edition*

SCHUSS · Theory and Applications of Stochastic Differential Equations

SCOTT · Multivariate Density Estimation: Theory, Practice, and Visualization

*SEARLE · Linear Models

SEARLE · Linear Models for Unbalanced Data

SEARLE, CASELLA, and McCULLOCH · Variance Components

SENNOTT · Stochastic Dynamic Programming and the Control of Queueing Systems

STOYAN, KENDALL, and MECKE · Stochastic Geometry and Its Applications, *Second Edition*

STOYAN and STOYAN · Fractals, Random Shapes and Point Fields: Methods of Geometrical Statistics

*Now available in a lower priced paperback edition in the Wiley Classics Library.

Applied Probability and Statistics (Continued)

THOMPSON · Empirical Model Building
THOMPSON · Sampling
THOMPSON · Simulation: A Modeler's Approach
TIJMS · Stochastic Modeling and Analysis: A Computational Approach
TIJMS · Stochastic Models: An Algorithmic Approach
TITTERINGTON, SMITH, and MAKOV · Statistical Analysis of Finite Mixture Distributions
UPTON and FINGLETON · Spatial Data Analysis by Example, Volume 1: Point Pattern and Quantitative Data
UPTON and FINGLETON · Spatial Data Analysis by Example, Volume II: Categorical and Directional Data
VAN RIJCKEVORSEL and DE LEEUW · Component and Correspondence Analysis
VIDAKOVIC · Statistical Modeling by Wavelets
WEISBERG · Applied Linear Regression, *Second Edition*
WESTFALL and YOUNG · Resampling-Based Multiple Testing: Examples and Methods for *p*-Value Adjustment
WHITTLE · Systems in Stochastic Equilibrium
WOODING · Planning Pharmaceutical Clinical Trials: Basic Statistical Principles
WOOLSON · Statistical Methods for the Analysis of Biomedical Data
*ZELLNER · An Introduction to Bayesian Inference in Econometrics

Texts and References Section

AGRESTI · An Introduction to Categorical Data Analysis
ANDERSON · An Introduction to Multivariate Statistical Analysis, *Second Edition*
ANDERSON and LOYNES · The Teaching of Practical Statistics
ARMITAGE and COLTON · Encyclopedia of Biostatistics: Volumes 1 to 6 with Index
BARTOSZYNSKI and NIEWIADOMSKA-BUGAJ · Probability and Statistical Inference
BENDAT and PIERSOL · Random Data: Analysis and Measurement Procedures, *Third Edition*
BERRY, CHALONER, and GEWEKE · Bayesian Analysis in Statistics and Econometrics: Essays in Honor of Arnold Zellner
BHATTACHARYA and JOHNSON · Statistical Concepts and Methods
BILLINGSLEY · Probability and Measure, *Second Edition*
BOX · R. A. Fisher, the Life of a Scientist
BOX, HUNTER, and HUNTER · Statistics for Experimenters: An Introduction to Design, Data Analysis, and Model Building
BOX and LUCEÑO · Statistical Control by Monitoring and Feedback Adjustment
BROWN and HOLLANDER · Statistics: A Biomedical Introduction
CHATTERJEE and PRICE · Regression Analysis by Example, *Third Edition*
COOK and WEISBERG · Applied Regression Including Computing and Graphics
COOK and WEISBERG · An Introduction to Regression Graphics
COX · A Handbook of Introductory Statistical Methods
DILLON and GOLDSTEIN · Multivariate Analysis: Methods and Applications
*DODGE and ROMIG · Sampling Inspection Tables, *Second Edition*
DRAPER and SMITH · Applied Regression Analysis, *Third Edition*
DUDEWICZ and MISHRA · Modern Mathematical Statistics
DUNN · Basic Statistics: A Primer for the Biomedical Sciences, *Second Edition*
FISHER and VAN BELLE · Biostatistics: A Methodology for the Health Sciences
FREEMAN and SMITH · Aspects of Uncertainty: A Tribute to D. V. Lindley
GROSS and HARRIS · Fundamentals of Queueing Theory, *Third Edition*

*Now available in a lower priced paperback edition in the Wiley Classics Library.

Texts and References (Continued)

HALD · A History of Probability and Statistics and their Applications Before 1750
HALD · A History of Mathematical Statistics from 1750 to 1930
HELLER · MACSYMA for Statisticians
HOEL · Introduction to Mathematical Statistics, *Fifth Edition*
HOLLANDER and WOLFE · Nonparametric Statistical Methods, *Second Edition*
HOSMER and LEMESHOW · Applied Survival Analysis: Regression Modeling of Time to Event Data
JOHNSON and BALAKRISHNAN · Advances in the Theory and Practice of Statistics: A Volume in Honor of Samuel Kotz
JOHNSON and KOTZ (editors) · Leading Personalities in Statistical Sciences: From the Seventeenth Century to the Present
JUDGE, GRIFFITHS, HILL, LÜTKEPOHL, and LEE · The Theory and Practice of Econometrics, *Second Edition*
KHURI · Advanced Calculus with Applications in Statistics
KOTZ and JOHNSON (editors) · Encyclopedia of Statistical Sciences: Volumes 1 to 9 wtih Index
KOTZ and JOHNSON (editors) · Encyclopedia of Statistical Sciences: Supplement Volume
KOTZ, REED, and BANKS (editors) · Encyclopedia of Statistical Sciences: Update Volume 1
KOTZ, REED, and BANKS (editors) · Encyclopedia of Statistical Sciences: Update Volume 2
LAMPERTI · Probability: A Survey of the Mathematical Theory, *Second Edition*
LARSON · Introduction to Probability Theory and Statistical Inference, *Third Edition*
LE · Applied Categorical Data Analysis
LE · Applied Survival Analysis
MALLOWS · Design, Data, and Analysis by Some Friends of Cuthbert Daniel
MARDIA · The Art of Statistical Science: A Tribute to G. S. Watson
MASON, GUNST, and HESS · Statistical Design and Analysis of Experiments with Applications to Engineering and Science
MURRAY · X-STAT 2.0 Statistical Experimentation, Design Data Analysis, and Nonlinear Optimization
PURI, VILAPLANA, and WERTZ · New Perspectives in Theoretical and Applied Statistics
RENCHER · Linear Models in Statistics
RENCHER · Methods of Multivariate Analysis
RENCHER · Multivariate Statistical Inference with Applications
ROSS · Introduction to Probability and Statistics for Engineers and Scientists
ROHATGI · An Introduction to Probability Theory and Mathematical Statistics
RYAN · Modern Regression Methods
SCHOTT · Matrix Analysis for Statistics
SEARLE · Matrix Algebra Useful for Statistics
STYAN · The Collected Papers of T. W. Anderson: 1943–1985
TIERNEY · LISP-STAT: An Object-Oriented Environment for Statistical Computing and Dynamic Graphics
WONNACOTT and WONNACOTT · Econometrics, *Second Edition*

*Now available in a lower priced paperback edition in the Wiley Classics Library.

WILEY SERIES IN PROBABILITY AND STATISTICS
ESTABLISHED BY WALTER A. SHEWHART AND SAMUEL S. WILKS

Survey Methodology Section

BIEMER, GROVES, LYBERG, MATHIOWETZ, and SUDMAN · Measurement Errors in Surveys
COCHRAN · Sampling Techniques, *Third Edition*
COUPER, BAKER, BETHLEHEM, CLARK, MARTIN, NICHOLLS, and O'REILLY (editors) · Computer Assisted Survey Information Collection
COX, BINDER, CHINNAPPA, CHRISTIANSON, COLLEDGE, and KOTT (editors) · Business Survey Methods
*DEMING · Sample Design in Business Research
DILLMAN · Mail and Telephone Surveys: The Total Design Method
GROVES and COUPER · Nonresponse in Household Interview Surveys
GROVES · Survey Errors and Survey Costs
GROVES, BIEMER, LYBERG, MASSEY, NICHOLLS, and WAKSBERG · Telephone Survey Methodology
*HANSEN, HURWITZ, and MADOW · Sample Survey Methods and Theory, Volume 1: Methods and Applications
*HANSEN, HURWITZ, and MADOW · Sample Survey Methods and Theory, Volume II: Theory
KISH · Statistical Design for Research
*KISH · Survey Sampling
KORN and GRAUBARD · Analysis of Health Surveys
LESSLER and KALSBEEK · Nonsampling Error in Surveys
LEVY and LEMESHOW · Sampling of Populations: Methods and Applications, *Third Edition*
LYBERG, BIEMER, COLLINS, de LEEUW, DIPPO, SCHWARZ, TREWIN (editors) · Survey Measurement and Process Quality
SIRKEN, HERRMANN, SCHECHTER, SCHWARZ, TANUR, and TOURANGEAU (editors) · Cognition and Survey Research

*Now available in a lower priced paperback edition in the Wiley Classics Library.

DEFINITIONS

Theorem

29 pg 73

1 pg 9
2 pg 10
3 p12
4 p13
5 p15
6 p18
7 p20
8 p22
9 p23
10 p25
11 p25
12
13 p28
14 p33
15 p36
16
17 pg 43
18 pg 45
19
20 ~~p 50~~
21 p50
22 p51
23 pg 54
24
25 pg 58
26 pg 59
27
28 pg 69